"This book offers insights into all the wonderful sensory eleme
ories of a dining experience. For the last two decades the culina
with modernist cooking and the role science can play in the kit
more chefs are realising that science also has a place at the table; that a dining experience
is made up of far more than just good food and that by understanding how we use our
senses to interact with and appreciate food we may be able to further enhance our guest's
dining experience. This book looks at many topics which will become increasingly relevant
to both chefs and society as a whole in the coming years."

**Jozef Youseff, chef and author of *Molecular Gastronomy at Home*
(www.kitchen-theory.com)**

"Whether your idea of a good night is the local café or the latest Michelin-starred restaurant, it is unlikely that you'll be unaware of the cultural phenomenon that restaurant food
has become in recent years. The secrets of chefs – amateur and celebrity – have been laid
bare in myriad weighty books and glossy TV programmes. *The Perfect Meal* goes beyond
the exotic ingredients and creative insights of the chef and into the realm of the diner's psychology. Using an accessible writing style that neither talks down to the reader nor dumbs
down the science, the authors take us into the relatively unexplored world of the dining
context: the gastrophysics of the visual, acoustic, tactile – not to mention taste and smell –
influences that we are exposed to in every dining experience. This is a new approach to
thinking about dining that will appeal to and inform anyone who has ever been convinced
to buy a cookbook by its illustrations or who persists, despite failure, in trying to make a
reservation at The Fat Duck."

John Prescott, Ph.D, author of *Taste Matters: Why We Eat the Foods We Do*

"In striving for a hypothetical level of delicious perfection we are forced to look beyond
culinary technique and ingredients. Focussing on the brain's interpretation of our eating
experience, pioneers Spence and Piqueras-Fiszman acknowledge the interdisciplinary
nature of gastronomy, rendering a complex area of study both digestible and applicable.

This valuable input furthers the development of co-evolving disciplines; the scientific
study of our brains, and the relentless creativity, experimentation and intuition so important
in producing a truly great meal."

**Ben Reade, Gastronome and Head of Culinary Research and Development at Nordic
Food Lab**

"Not many people are as ready to realise the importance of the senses, not only in cooking
but in eating, as Charles Spence and Betina Piqueras-Fiszman. 'The senses as the starting
point for creating' was one of the most important creative methods of elBulli and one of
the pillars of our cooking.

This book demonstrates beyond doubt that gastronomy is the most complex creative
discipline that exists. Therefore, I invite you to enjoy the secrets of the fascinating world
of the senses from Charles' and Betina's hand, something which is fundamental to enjoy
gastronomy."

Ferran Adrià, elBullifoundation

The Perfect Meal

The Perfect Meal

The Multisensory Science of Food and Dining

Charles Spence

Oxford University, UK

Betina Piqueras-Fiszman

Wageningen UR, Netherlands

Library of Congress Cataloging-in-Publication Data

Spence, Charles.
  The perfect meal : the multisensory science of food and dining / Charles Spence, Betina Piqueras-Fiszman.
       pages cm
  Includes index.
    ISBN 978-1-118-49082-2 (paperback)
  1. Gastronomy. 2. Dinners and dining. 3. Food–Sensory evaluation. 4. Senses and sensation.
  5. Intersensory effects. I. Piqueras-Fiszman, Betina. II. Title.
    TX631.S68 2014
    641.01′3–dc23
                                                              2014013901

A catalogue record for this book is available from the British Library.

Contents

Foreword   xiii

Preface   xvii

1   Introducing the Perfect Meal   **1**

1.1   Introduction   1
1.2   A brief history of culinary movements   2
    1.2.1   Nouvelle Cuisine   2
    1.2.2   The rise of molecular gastronomy   3
    1.2.3   Molecular gastronomy or modernist cuisine?   5
    1.2.4   On the rise of the celebrity chef   7
1.3   The search for novelty and surprise   8
    1.3.1   The taste of expectation   10
    1.3.2   Food as theatre: the multisensory experience economy
          meets cuisine   11
1.4   The brain on flavour   12
    1.4.1   Neurogastronomy   12
    1.4.2   Do neurogastronomists make great-tasting food?   14
1.5   Food and the perception of everything else   16
1.6   Gastrophysics: the new science of the table   18
1.7   Food perception is fundamentally multisensory   21
1.8   Isn't modernist cuisine only for the lucky few?   22
1.9   *Amuse bouche*   25
    References   27

2   Let the Show Commence: On the Start of the Perfect Meal   **37**

2.1   Introduction   37
    2.1.1   Mood food   38
2.2   On the social aspects of dining   42
    2.2.1   Waiting staff   42
    2.2.2   The company   44
2.3   On the design of the menu   47
    2.3.1   Scanning the menu: 'Oysters, steak frites, field greens, oysters…'   48

2.3.2	'This dessert is literally calling me'	50
2.3.3	Images on the menu	52
2.3.4	On a diet? Does nutritional information help?	54
2.3.5	Price and behaviour	55
2.3.6	On the format of the menu	59
2.4	Conclusions	62
	References	62

3   Tastes Great, But What do We Call It? The Art and Science of Food Description   **71**

3.1	Introduction	71
3.2	Snail porridge	73
3.3	Can labelling enhance the taste and/or flavour of food?	77
3.4	Interim summary	81
3.5	On the neuroscience of naming food	81
3.6	Naming names	84
3.7	Does food labelling influence the perceived ethnicity of a dish?	85
3.8	Natural and organic labels	87
3.9	Health/ingredient labels	88
3.10	Local labels	90
3.11	Descriptive food labelling	91
3.12	Labelling culinary techniques	92
3.13	Surprise!	95
3.14	Expectations and reactions	96
3.15	Conclusions	98
	References	100

4   Plating and Plateware: On the Multisensory Presentation of Food   **109**

4.1	Introduction	109
4.2	A potted history of food presentation	111
4.3	The plate: the essential element of our everyday meal	115
4.3.1	On the colour of the plate	115
4.3.2	The shape of the plate	119
4.3.3	The size of the plate	121
4.3.4	On the haptic aspects of the plateware	122
4.4	Interim summary	128
4.5	The plate that is not a plate	128
4.5.1	Reaching new heights	129
4.5.2	On the smell and sound of the plateware	130
4.5.3	Camouflage	132
4.5.4	Improvised plateware	133
4.5.5	Purpose-made plateware	134
4.6	On the multiple contributions of the visual appearance of a dish	135
4.6.1	On the importance of harmony on the plate	136

4.7 Individual diner responses to the visual presentation of food 141
4.8 Conclusions 143
References 143

5 Getting Your Hands on the Food: Cutlery 151

5.1 Introduction 151
5.2 The story of cutlery 153
5.3 The material qualities of the cutlery 159
    5.3.1 The quality of the cutlery 160
    5.3.2 Tasting the cutlery 161
5.4 Size matters 165
5.5 On the texture/feel of the cutlery 166
5.6 Colourful cutlery 168
5.7 Cutlery that is not 169
5.8 Finger food 174
5.9 Eating without hands 175
5.10 Conclusions 177
References 177

6 The Multisensory Perception of Flavour 183

6.1 Introduction 183
6.2 Perceiving flavours 183
6.3 Taste 186
    6.3.1 Are you a supertaster? 187
6.4 Olfactory–gustatory interactions 188
    6.4.1 Cross-cultural differences in multisensory flavour perception 190
6.5 Oral-somatosensory contributions to multisensory flavour perception 191
    6.5.1 Are you a thermal taster? 193
6.6 Interim summary 193
6.7 The sound of food 194
6.8 Visual flavour 196
    6.8.1 How does colour influence flavour perception? 196
    6.8.2 Summary of research on visual flavour 200
6.9 The cognitive neuroscience of multisensory flavour perception 200
6.10 Conclusions 202
References 204

7 Using Surprise and Sensory Incongruity in a Meal 215

7.1 Introduction 215
7.2 How did sensory incongruity become so popular
and why is it so exciting? 216
    7.2.1 The search for novelty 216
    7.2.2 The rise of molecular gastronomy/modernist cuisine 216
    7.2.3 The rise of sensory marketing and multisensory design 217
    7.2.4 Globalization 217

7.3	Defining sensory incongruity	218
7.4	Noticing sensory incongruity	219
	7.4.1 Disconfirmed expectations	221
	7.4.2 Hidden and visible incongruity	222
7.5	A brief history of sensory incongruity at the dinner table	224
7.6	Colour–flavour incongruity	226
7.7	Format–flavour incongruity	227
7.8	Smell–flavour incongruity	230
7.9	Interim summary	231
7.10	The diner's response to sensory incongruity	232
	7.10.1 Attentional capture	232
	7.10.2 Surprise	232
	7.10.3 Memorability	232
7.11	Molecular gastronomy and surprise	233
7.12	Sensory incongruity and the concept of 'naturalness'	236
7.13	Individual differences in the response of diners to sensory incongruity	239
7.14	Conclusions	241
	References	242
8	**Looking for Your Perfect Meal in the Dark**	**249**
8.1	Introduction	249
8.2	The social aspects of dining in the dark	251
8.3	Why are dining in the dark restaurants so popular nowadays?	252
8.4	Seeing or not seeing (correctly) the food	255
	8.4.1 The importance of colour to food	255
	8.4.2 Do our other senses really become more acute in the dark?	260
8.5	Does dining in the dark really capture how the blind experience food?	264
8.6	Cooking in the dark	265
8.7	Conclusions	265
	References	266
9	**How Important is Atmosphere to the Perfect Meal?**	**271**
9.1	Introduction	271
9.2	Atmospherics and the experience economy	275
9.3	The Provencal Rose paradox	278
9.4	Does the atmosphere really influence our appraisal of the meal?	280
9.5	On the ethnicity of the meal	280
9.6	Tuning up how much money and time we spend at the restaurant	282
	9.6.1 The style and volume of the music	282
	9.6.2 The tempo of the music	284
	9.6.3 "Pardon?"	285
9.7	Context and expectation	286
9.8	The lighting	287
9.9	The olfactory atmosphere	288
9.10	On the feel of the restaurant	291

9.11   Atmospheric contributions to taste and flavour perception   294
9.12   Multisensory atmospherics   298
9.13   Conclusions   299
       References   301

10   Technology at the Dining Table   311

10.1   Introduction   311
10.2   Technology on the dining table   312
10.3   Transforming the dining experience by means of technology   315
10.4   Augmented Reality (AR) food: A case of technology for
       technology's sake?   317
10.5   Using QR codes to change our interaction with food   319
10.6   Fostering healthy eating through the incorporation
       of technology   320
10.7   Technology and distraction   322
10.8   Using technology to control the multisensory atmosphere   323
10.9   On the neuroscience of matching sound to food
       (and how technology might help)   324
10.10 On the future of technology at the table: digital artefacts   326
10.11 The SmartPlate   328
10.12 Anyone for a 'Gin & Sonic'?   328
10.13 The tablet as twenty-first century plateware?   329
10.14 Tips from the chef at the tips of your fingers   331
10.15 Conclusions   331
       References   333

11   On the Future of the Perfect Meal   339

11.1   Introduction   339
11.2   On the history of predicting the future of food   341
       11.2.1 A meal (or even a day's food) in a single dose   341
       11.2.2 On the mechanization of feeding   345
       11.2.3 Air 'food'   346
       11.2.4 Artificial flavours   348
11.3   From the past to the future of food   351
       11.3.1 *Sous vide* as the twenty-first century microwave   351
       11.3.2 3D printed food: an astronomical idea   353
       11.3.3 On the future of cultured meat   355
       11.3.4 Note-by-note cuisine   356
       11.3.5 Eating insects for pleasure: bug burger with
               insect paste, anyone?   358
       11.3.6 The new algal cuisine   362
11.4   Anyone for a spot of neo-Futurist cuisine?   363
       11.4.1 Food theatre: food as entertainment   364
       11.4.2 Plating art   365
11.5   Interim summary   366

11.6   Acknowledging our differences   367
11.7   The meal as catalyst for social exchange   367
11.8   Is it a restaurant or is it a science laboratory?   369
11.9   Pop-up dining, story telling and the joys of situated eating   371
11.10 Conclusions   372
       References   374

Index   **383**

Foreword

The day I fell down the enchanting rabbit hole into the magical wonderland of the senses and began exploring their influence on our culinary likes and dis-likes, I encountered a great deal of scepticism and resistance from chefs, diners, and food writers alike. In their eyes, the only thing of real importance was the food on the plate. The idea that the senses might influence our perception of flavour and help generate the pleasure and emotion that can accompany a meal was dismissed by some as nonsense, that reduced cooking and eating to mathematical formulae devoid of emotion. How wrong they were!

Although I only realised it later, my interest in the interplay of the senses and their influence on cuisine must stem from the event that originally made me want to be a chef when I was just 16 years old: a meal on the terrace of a three-star Michelin restaurant in Provence where the smell of the lavender bushes, the sound of the cicadas and the visual splendour of the setting almost seemed to eclipse the food and sent me down the rabbit hole. However, I date my conscious realisation of the culinary importance of the senses to 1997, when I created a dish that featured a crab-flavoured ice cream. The notion of a crab ice cream put some people off because ice cream is sweet, right? This associa-tion prevented them from enjoying a savoury version. However, I discovered that if I simply changed the name from 'crab ice cream' to 'frozen crab bisque', most people totally got it– even though it was the same ice cream! The idea that the actual name of a dish could change its whole context and enjoyment was, for me, a total eye opener. (As mentioned in the pages that follow, this dish was the inspiration for a paper from Sussex University on how the name of a dish can even change its perceived saltiness)

From then on, I researched whatever I could find on the incredible complex-ity of multisensory flavour perception, and began developing dishes that drew on what I was discovering. At the start of 2004, I unveiled my multisensory approach to cooking at one of the world's foremost gastronomic congresses in a presentation entitled 'Eating is a multisensory experience', and the rest, as they say, is history.

I first met Charles back in 2002 through a mutual friend and mentor of mine, Professor Tony Blake. I still have vivid memories of my first visit to his Cross-modal Research Laboratory in Oxford. He showed me the fascinating Sonic Chips experiment described in Chapter 6, and the idea that sound could radically affect our perception and enjoyment of food started my mind racing, like a kid in a sweet shop. I returned to Bray, got hold of a sound box, and started trying things out for myself.

Since then, Charles and I have worked together on a number of sonic experiments. For one of them we fed test-participants (actually members of the audience at the Art and the Senses conference held in Oxford in 2006) a scoop of bacon and egg ice cream. One group of participants ate the ice cream while listening to the sounds of bacon sizzling in the pan. The others tasted the ice cream while listening to the sounds of chickens clucking in the farmyard. In each case the sound appeared to intensify the relevant flavour. In another experiment, we fed participants oysters while listening either to the sound of the sea (think waves crashing gently on the beach), or to the sounds of farmyard animals, after which we asked them to rate how pleasant the oysters tasted. (Listening to the sound of the sea resulted in people rating the oyster as tasting significantly more enjoyable, but no more salty when compared to the farmyard soundtrack. Such results giving further support to the notion that sound can indeed influence our emotional response to food.) It was this last experiment (the results you'll find described within the pages that follow) that inspired what is now a classic on the menu at The Fat Duck, "Sound of the Sea", in which seafood, seaweed and edible "sand" are used to create what looks like the edge of the seashore, all of which is accompanied by an iPod and earphones so that the diner can hear the sounds of the waves lapping up against the shore while eating.

There are all sorts of other sensory questions that Charles and I have explored over the years, like whether listening to a low-pitched sound while eating a bitter, crunchy caramel would emphasize bitterness and whether listening to a sharp sound while tasting an acidulated toffee sauce would accentuate its acidity. We even investigated whether listening to a synaesthetically soft sound could enhance the richer, sweeter notes of the sauce. We've used jellies and *pâtes de fruit* in which the colour misleads you into expecting, say, a particular fruit when it is in fact a vegetable (i.e., blackcurrant that is in reality beetroot; lime that is fennel and pumpkin that tastes like apricot … adding fruit acids can flip the mind's interpretation of the colour of a food to blackcurrant) in order to probe the ways in which the senses can nudge us to a different place in terms of our perception of flavour. The senses of sight and taste have nudged the vegetable to a fruit of the same colour, leaving the smell saying little about the matter. And we've pursued ideas based on the early research from Köhler on sound symbolism in which people were shown a pair of two-dimensional shapes, a spiky one and an amoeba-like one, and asked which of the two was a "kiki" and which a "bouba". Despite both names being meaningless nonsense words, there was an overwhelming conviction that bouba was the rounded amoeba shape and kiki the pointed one. We tried this out with various foods and discovered that a similar correspondence between flavour and sound seemed to exist: for example, milk chocolate – even when brittle from a stint in the fridge – was generally considered more "bouba", while dark chocolate, even in the form of a light and airy mousse, was overwhelmingly "kiki".

As will by now be apparent, Charles has been one of my biggest inspirations. He is one of the world's leading researchers in the field of multisensory perception and together with Betina they have opened up new ways of experiencing food by focusing on everything that surrounds it. So it's very exciting that they have turned their ground-breaking fundamental research into a book so that you, too, can be inspired in much the same way that I have been – and still am. The pages that follow will open your eyes to new worlds and new ideas. Charles and Betina are the perfect guides for such a journey given their wide-ranging curiosity, great clarity of thought, and lively minds that are forever spotting connections that illuminate how the world of food and drink really works. If you're at all interested in food and the effect it has on our bodies and, more significantly, on our minds, then *The Perfect Meal* can't fail to entertain, inform and ultimately to dazzle.

Heston Blumenthal
The Fat Duck

Preface

Why is it that what you like I detest? How can it be that what we touch affects what we taste? Can people be nudged towards healthier food choices simply by incorporating a few psychological illusions and neuroscience insights into their cuisine? These are just a few of the intriguing questions that we address in the pages that follow. This book aims to provide the reader with the facts and figures needed to grasp what is it that makes them perceive and experience the food (one of life's greatest pleasures!) in the way that they do and how to improve upon it. Each of the chapters covers a number of the key factors that influence the diner's experience. Our interests lay in understanding from the fancy meals nowadays being served in modernist restaurants through to the family meal served in the comfort of our own home. Throughout, we highlight what we see as some of the most intriguing possible future trends when it comes to food and multisensory dining.

Our goal in writing this book has been to bring together and critically evaluate the large body of empirical research that has emerged in recent years documenting the profound effect that each one of our senses has on our perception and enjoyment of food. The focus, though, is not on the neuroscience of flavour, but rather on all of the other (non-food) factors that influence our overall multisensory experience of food. We outline the transition in research practice from the more traditional approaches to the study of flavour perception founded in the field of sensory science through to the emergence of a number of novel methods to understanding the diner's experience based on areas of research as diverse as cognitive and decisional neuroscience, marketing, design, and psychology. When taken together, these new ways of thinking about people's response to food give rise to a neuroscience-inspired approach to multisensory design. In this book, we describe all this exciting research in an accessible style for the general readership.

In this volume, we use both the latest research as well as relevant historical examples to illustrate how much more there is to the diner's experience than merely what is sitting there on the plate (if indeed there is a plate – nowadays you can't always be too sure). Indeed, there are researchers out there who are convinced that as much as half of the pleasure in a plate of food (or meal) actually comes from the "everything else"! In the pages that follow, we will highlight what we see as the most relevant and exciting findings to have emerged from the latest studies to have been conducted by sensory scientists, psychologists, neuroscientists, oenologists, and even economists, investigating how important each and every element of the meal (focusing especially on those cues that are extrinsic to the food itself) is to the diner's overall experience. This, then, is *the new science of the table* that we want to share our own passion for with the reader: A new field of research that is referred to as gastrophysics.

Many of the chefs whom we have had the good fortune to speak to over the years, and this includes a number with Michelin-stars under their belts (or should that be toques), are convinced that the meal is all, and only, about sourcing the right ingredients, preparing them correctly, and how the food is ultimately presented on the plate. Oftentimes, these chefs spend so much time thinking about the food itself that they forget to give adequate consideration to the mise-en-scene, for example, paying no attention to the music that happens to be playing in the background. As we will see later, such oversights can have a much greater impact on our dining experiences than might be expected.

Who is this book for? It is primarily aimed at all those people out there who are interested in food and the factors that influence our experience of it. That includes those working in the world of food, or writing about it, that is, the chefs, cooks, marketers, large food companies, research scientists, gourmets, and food lovers (or "foodies"), or simply the curious lay reader, who wants to know more about the key drivers underlying our experience of food. This book highlights a number of the most important influences that distinguish the wonderful, perhaps even that once-in-a-lifetime 'perfect meal' from the mundane fare of everyday life. We illustrate the central themes with dishes taken from the tables of some of the world's top restaurants. That said, it is our firm belief that many of the insights can easily be adapted to enhance the home dining experience as well.

Health researchers involved in trying to tackle the current global obesity crisis should also find a number of the findings summarized here to be relevant: So, for example, we provide numerous suggestions concerning ways in which the diner's mind can be tricked into thinking that more food has been consumed that is actually the case. We will also highlight a number of most innovative methods for making food taste sweeter which don't rely on adding a grain of sugar to our pudding. How? The tips here include everything from changing the colour of the plate to adding a little digital seasoning in the form of some sweet-sounding music. This book also contains a number of actionable insights for those working with the aging and hospital populations, where profound nutritional problems abound and are likely to increase unless we do something about it.

Many wonderfully creative minds have accompanied us on the journey that was writing this book. Our special thanks go to Heston Blumenthal and all of the research team at The Fat Duck in Bray for their ongoing interest and support at the frontiers of scientific and creative cuisine. We would also like to thank the many other chefs and culinary artists that we have had the great good fortune to collaborate or share all manner of outlandish ideas with: So, in no particular order our heartfelt thanks go out to Denis Martin (Restaurant Denis Martin); Ben Reade (Nordic Food Lab); Jozef Youssef (Kitchen Theory); Charles Michel (401B), Caroline Hobkinson (Stirring with Knives); Louise Bloor (Fragrant Supper Club); María José San Román (Monastrell Restaurant); Sriram Aylur (The Quilon Restaurant); Wylie Dufresne (WD~50); Blanch & Shock; Bompas & Parr: the chefs at Casa Mia in Bristol; and the Institut Paul Bocuse. We have also benefited greatly from the support of those working in the food science industry, particular thanks going to Francis McGlone, Tony Blake, Barry Smith, John Prescott, Rupert Ponsonby, Michael Bom Frøst, Line Holler Mielby, Ophelia Deroy; and Susana Fiszman. You have inspired us greatly and this book has been possible thanks to your generosity of spirit and ideas.

We would also like to thank all our contributors for having provided us with their images. They range from professional photographers through science researchers, to designers and architecture agencies and all share a passion and curiosity for food and eating experiences. We are also grateful to all those friends and family who have been kind enough to review some of the material that follows; Barbara and Thierry, thank you so much for going beyond the line of duty. Needless to say, the blame for any remaining inaccuracies lies squarely with us.

So, without further ado, let the meal begin …

1

Introducing the Perfect Meal

"Once at least in the life of every human, whether he be brute or trembling daffodil, comes a moment of complete gastronomic satisfaction. It is, I am sure, as much a matter of spirit as of body. Everything is right; nothing jars. There is a kind of harmony, with every sensation and emotion melted into one chord of well-being."
(Fisher 2005, p. 325)

1.1 Introduction

This is a book about the perfect meal and how to get it, or at least how to get closer to it: not in the sense of the chef travelling to the furthest corners of the globe in the search for the über-unusual and extreme of culinary delights (Bourdain 2002)[1]; nor in the behavioural economist's sense of trying to optimize the benefits, while minimizing the costs, of the financial transaction that is dining out (Cowen 2012); and nor does this book offer a chef's guide to, or search for, perfection as seen through the lens of molecular gastronomy or (better said) modernist cuisine (Blumenthal 2007; see also Rayner 2008). Rather, this is a book about how the latest insights from a diverse range of fields of research that include experimental psychology, design, neuroscience, sensory marketing, behavioural economics and the culinary and sensory sciences can, and in some cases already are, being used by a number of the

[1] Note that this interest in the unusual extends all the way from the celebrity chef though to the home dining setting. For example, Baumann (1996, p. 121) talks about the popular desire for *"not just ordinary cookbooks, but collections of ever more refined, exotic, out-of-this-world, … recipes; promises of taste-bud delights never experienced before."*

world's top chefs in order to deliver multisensory dining experiences that are more sensational, more enjoyable and consequently more memorable than anything that has ever gone before.

> *"What is 'real'? How do you define 'real'? If you are talking about what you can feel, what you can smell, what you can taste and see then 'real' is simply electrical signals interpreted by your brain."* (Morpheus in *The Matrix*; see Haden 2005, p. 354)

Here we are talking about experiences that are based on the emerging insights concerning the mind of the diner and not just on the whims and intuitions of the chef, or increasingly the culinary team, beavering away behind the scenes in many of the world's top restaurants (Spence 2013). It is our contention that, in the years to come, the search for the perfect meal will be facilitated as much by knowing about the mind of the diner and what makes it tick as it will by gaining further insights into the physiology of the human flavour system or by sourcing the most seasonal of ingredients and knowing how best to prepare (and present) them on the plate (Pollan 2006). The revolutionary new approach to the science of the perfect meal that we wish to showcase here is called 'gastrophysics'. Before immersing ourselves in it, let's take a look back over the evolution of gastronomic movements and trends that has led to our current culinary practices and food knowledge.

1.2   A brief history of culinary movements

Over the last half century or so there have been a couple of major culinary movements that have left their indelible mark on the way in which we think about food today. The first of these was *Nouvelle Cuisine* which emerged in France during the 1960s. In the early 1990s, molecular gastronomy arrived with a bang (often literally). Let's take a brief look at these movements in order to get a better sense of the culinary landscape in which we find ourselves today.

> *"Periods of gastronomic change are inevitably periods of gastronomic controversy. When there is no controversy, there is no inventiveness, because controversy of course doesn't appear if there is no tension between tradition and innovation, or the other way, between innovation and academic conventions."* (Revel 1985, on the introduction of the *Nouvelle Cuisine*)

1.2.1   Nouvelle Cuisine

The term itself dates from the 1730s–1740s when French writers used it to describe a break with the traditional way of cooking and presenting foods (Hyman and Hyman 1999). However, the culinary movement that now bears the name really took on a life of its own in the 1960s when the French food critics Christian Millau and Henri Gault used the term to describe the new culinary style that was then just starting to make its appearance in the kitchens of some of France's top chefs. Nowadays, the term *nouvelle cuisine* is used to refer to the use of seasonal ingredients with a focus on natural flavours, light textures (e.g. sauces that have not been thickened by the addition of flour and fat) together with a visual aesthetic that focuses on a presentation that is both simple and elegant (see Chapter 4). The French chefs who were instrumental in developing this new type of cuisine, including Paul Bocuse and Jean and Pierre Troisgros, were undoubtedly influenced by the minimalist Japanese style that placed a value on serving smaller portions. Indeed, the opening of the first French culinary school in Japan in 1960 by chef Shizuo Tsuji resulted in a much greater cultural exchange between Japanese and leading French chefs, including Paul Bocuse and Alain Chapel. The latter also embraced the use of ingredients sourced from many different parts of the world. In fact, this is also why it was so natural for *nouvelle cuisine* to morph seamlessly into 'fusion' food.

> "*Really, the concern with how the food looked can be traced back to the emergence of nouvelle cuisine. The pictures of these dishes have set themselves in the mind of the public. Nouvelle cuisine was essentially photogenic … Think of the glorious coloured photographs of these dishes, which have become eponymous with the purveying of recipes.*" (Halligan 1990, p. 121)

It was precisely this emphasis on the visual appearance of food that led Alexander Cockburn, in a 1977 article that appeared in the *New York Review of Books*, to introduce the term 'gastroporn'.[2] This term, which has now made it into the Collins English Dictionary, is defined as 'the representation of food in a highly sensual manner'. It should therefore be noted that even food writing can qualify for this epithet.

1.2.2   The rise of molecular gastronomy

There can be no doubt that the fusion of the physical sciences with culinary artistry has fundamentally changed the fine dining landscape over the last couple of decades or so (Belasco 2006; Roosth 2013) and has been enthusiastically covered in the press under the title of 'molecular gastronomy'. This revolutionary new approach to cuisine is one that has attracted a phenomenal amount of media interest from pretty much every corner of the developed world (see Barham *et al.* 2010). The term itself was first coined by the Oxford-based Hungarian physicist Nicholas Kurti (who back in 1969 gave a

---

[2] Cockburn memorably described one of Paul Bocuse's cookbooks as a '*costly […] exercise in gastro-porn*' (cited in Poole 2012, p. 59).

presentation at the Royal Institute in London entitled *The Physicist in the Kitchen*; see Kurti 1969; Kurti and Kurti 1988). Particularly influential here was also a paper that Kurti wrote together with the French physical chemist Hervé This in the popular science magazine *Scientific American* (Kurti and This-Benckhard 1994a, b).

But what exactly is molecular gastronomy? McGee (1984) talks in terms of *"the scientific study of deliciousness"*. Perhaps a more precise, albeit less grammatical, definition comes from Roosth (2013, p. 4) who describes it as *"a food movement whose practitioners – chemists who study food and chefs who apply their results – define [sic] as the application of the scientific method and laboratory apparatuses [sic] to further cooking."*

Nowadays, there is certainly a bewildering array of new techniques and ingredients, some natural, others much more artificial/processed,[3] available to the budding modernist chef, no matter whether operating in the restaurant or home environment (e.g. see Blumenthal 2008; Myhrvold and Young 2011; Youssef 2013). Harold McGee, the brilliant North American author on kitchen science, has written a number of influential books in which he explores the science underpinning the practice of molecular gastronomy (McGee 1984; 1990). There he investigates such things as culinary proverbs, sayings and old wives' tales. He has done more than perhaps anyone else to explore the physics and chemistry that lie behind a host of everyday culinary phenomena such as, for example, the Maillard reaction (McGee 1990).[4]

Fortunately for us there are already many great chefs and eminent scientists, not to mention flavour houses, working on the physics and chemistry of flavour (e.g. Barham 2000; Alícia and elBullitaller 2006; Konings 2009; Barham *et al.* 2010; Chartier 2012; Humphries 2012). We are therefore not going to cover these aspects of molecular gastronomy in any detail in this book (see McGee 1990; This 2005, 2012, 2013, for detailed coverage of this theme). We will, however, be taking a closer look at some of the most intriguing dishes to have emerged from these modernist kitchens over the last couple of decades. We will discuss some of the legendary dishes from the elBulli restaurant in Spain and The Fat Duck in Bray (UK). We're going to dissect a number of the dishes from the Chicago School of Restaurants; think Grant Achatz's Alinea and Homaro Cantu's Moto. We'll also be taking a look at a few of the dishes championed by those innovative new restaurants that have sprung up across Spain in recent years (part of *la nueva cocina* movement; Lubow 2003; Steinberger 2010). However, our interest in discussing many of these amazing dishes will not be the culinary magic underlying the preparation of the ingredients on the plate, but rather to try and understand some of the key psychological

---

[3] In fact, it is all those unnatural ingredients, all the colorants, gelling agents, emulsifiers, acidifiers and taste enhancers that has led some authors to suggest, not entirely ironically, that the menus at molecular gastronomy restaurants ought to carry a health warning because of all the additives that they contain (Campbell 2009).

[4] Named after the French doctor Louis-Camille Maillard who *"discovered that when amino acids are heated in the company of sugar, the reaction produces hundreds of new molecules that give cooked food its characteristic color and much of its smell."* (Pollan 2013, p. 88).

and neuroscientific principles that lie behind the wonderful experience of eating them. And having got a handle on these fundamental insights, the challenge will then be to demonstrate how they can be used in everyday life, for example, to provide tips to help any one of us eat a little more healthily without having to compromise on the sensory pleasure of the experience.

1.2.3 Molecular gastronomy or modernist cuisine?

A number of the chefs with whom we collaborate most closely have something of a love/hate relationship with the term 'molecular gastronomy' (e.g. Blumenthal and McGee 2006; McGee 2006; Rayner 2006; Blumenthal 2008; Gopnik 2011). In fact, many of those working in the field would much rather have you refer to what they do as 'modernist cuisine'. There are a number of reasons behind this terminological debate that are perhaps worth mentioning here. First, many chefs object to the term 'molecular gastronomy' because they feel that what has been happening in the kitchen in recent years is about so much more than merely playing with molecules, films, foams (or *espumas* as the Spanish like to call them) and gels, etc. In the pages that follow, you'll see this is a view with which we most wholeheartedly agree.

What is more, many of those working in this area are also sensitive to the criticism that what they deliver can be seen as nothing more than a form of elitist cuisine. This notion, at least to those who worry about such things, is strengthened by the term 'gastronomy'.[5] As Heston Blumenthal put it in an interview back in 2006:

> "*Molecular makes it sound complicated … and gastronomy makes it sound elitist … We may use modern thickeners, sugar substitutes, enzymes, liquid nitrogen, sous vide, dehydration and other non-traditional means but these do not define our cooking. They are a few of the many tools that we are fortunate to have available as we strive to make delicious and stimulating dishes*" (Rayner 2006)

The preference among many of those practitioners working in the kitchen is therefore for the more inclusive and less overtly chemical label of 'modernist cuisine'.

What with so much baggage associated with the term 'molecular gastronomy', it should perhaps come as little surprise that Myhrvold and Young (2011), in what *The Independent* newspaper described as "*the most spectacular cookbook the world has ever seen*" (Walsh 2011, p. 11), chose to title their 3000-page masterpiece *Modernist Cuisine*. This 5-volume shelf-filler is undoubtedly a veritable feast for the eyes, detailing with absolutely stunning photography pretty much every tool and technique of the new art and science of the table (those with an addiction to gastroporn take note). That said, 'molecular gastronomy' would appear to be the term that has stuck in the

---

[5] In his classic volume *The Physiology of Taste*, the French polymath Jean Anthelme Brillat-Savarin (1835) defined gastronomy as "*the reasoned knowledge concerning all aspects of food*".

public consciousness. Indeed, a quick search on Google Scholar on 24 August 2013 brought up 1080 hits for the term 'molecular gastronomy' as compared to just 123 for 'modernist cuisine'. Furthermore, many other up-and-coming young chefs such as Josef Youssef (who like many others trained for a while in the kitchens of The Fat Duck) appear to have no qualms about using the term 'molecular' (as Youssef himself does in the title for his new book; see Youssef 2013).

Deciding on the right name for this global culinary movement would seem to be a debate that is going to run and run. As such, we trust that you will forgive us for using the two terms fairly interchangeably in this book, although we also acknowledge the fact that 'molecular gastronomy' fails to capture many of the most important innovations that have permeated the research kitchens of some of the world's top restaurants over the last few years (see also McGee 2006; Schira 2011).[6]

In the pages that follow, we will repeatedly see how many of the most interesting things that have been going on recently in the world's top restaurants are about so much more than merely innovative food chemistry (especially in the area of novel gelling agents such as methylcellulose, xanthan gum and alginate) and kitchen technology (here we are thinking of devices such as the RotoVap, Pacojet, Thermomix and Gastrovac). Rather, the table of the future will likely involve the delivery of marketable (and hence branded)[7] multisensory dining experiences: experiences that are as much about theatrical performance, entertainment and, increasingly, interaction as they are about the delivery of nutritious and filling food to the hungry and soon-to-be rather poorer diner (Berghaus 2001). In addition, as far as we can tell, technology is also going to be an ubiquitous feature of our fine (and possibly also home) dining in the years to come.

> "*They work on extracting the essence of the ingredient, and they play with the sense and textures,*" Remolina says. "*All the senses are involved. Now food is a show.*" (Park 2013 interviewing Remolina)

Of course, not everyone is convinced by the turn that so many top-end dining experiences are taking (e.g. Gill 2007; Poole 2012). And that's fine too (to be expected, even; see the earlier quote from Revel). As we hope to show in the pages that follow, even if you plan never to set foot in a modernist

---

[6] What is more, Hervé This – one of the scientists credited with coining the term 'molecular gastronomy' – has already pronounced the movement dead (see Ashley 2013)! For This (2012), the future is all about note-by-note cuisine, which he defines as: "*a culinary trend in which no plant (vegetables, fruits) or animal (meat, fish) tissues are used, because these traditional food ingredients are mixtures of compounds giving poor control to the cook. Instead, note-by-note cuisine makes use of "pure" compounds in order to build all aspects of dishes: taste, odor, color, texture, and so on.*" (This 2012, p. 243). Sounds tasty? You can read more about this new approach to cuisine in Chapter 11.

[7] As Visser (1991, p. 124) presciently notes, when it comes to cuisine the contemporary taste for novelty offers "*a wonderful marketing milieu*".

restaurant, there are still insights to be gained from studying the food that is being served in such venues nowadays – insights that can be applied no matter your favourite food or style of cuisine. Even the slowest of slow food (see Petrini 2007) still has to be served somewhere, and will most likely be eaten with the aid of some sort of cutlery. It is crucial to remember, then, that the atmosphere affects what we think about the food no matter where we happen to be or what we happen to be eating (slow food or modernist cuisine). The same applies when we start to think about the cutlery, the company and even the naming of the dishes that we order. The key point to note here is that while our growing understanding of the new sciences of the table may well be best advanced by looking at what is being served at the top modernist restaurants, the insights that will be uncovered there can hopefully be applied wherever we happen to eat and no matter what we happen to be eating.

1.2.4   On the rise of the celebrity chef

While nouvelle cuisine and molecular gastronomy have swept the world stage, another profound change in the balance of power within the restaurant sector has also taken place. Traditionally, all of the activity in a fancy restaurant would revolve around the front of house. Just think back to the time when the omnipotent restaurateur would meet and greet his guests by name as they arrived, wielding the power to decide who would get to sit at the best tables (Steinberger 2010). Meanwhile, the anonymous chef would normally keep a low profile out back doing exactly as he or she was told. In fact, should the chef in one of these restaurants change, the diner might well not know about it; even if they did, they likely wouldn't care too much. However, the last couple of decades have seen a fundamental shift of power from the front of house to the back (which is no longer always to be found out back).

The rise of the glass-screened kitchen, which has become such a signature feature of so many restaurants nowadays, can be seen as an architectural acknowledgement of this transition. For those who have had the opportunity to dine there, think of the glass-screened kitchen that forms the centrepiece of Heston Blumenthal's Dinner restaurant in the Mandarin Oriental Hotel in London. There is simply no way that the diner can get to their table without catching an eyeful of the action taking place in the kitchen (including all of those pineapples slowly spit-roasting). It is certainly hard to imagine that there has ever been a time previously when anyone would have thought it worthwhile to beam the action live from the kitchen direct to the diners' table (as Daniel Facen now does in his Italian restaurant; see Schira 2011). And never before has the celebrity chef been guaranteed to pack out stadium after stadium (as happened to Heston Blumenthal during his recent tour of Australia) while talking about and demonstrating the latest culinary creations from their kitchens.

1.3   The search for novelty and surprise

Before taking a look at the relevant science underlying the field of gastrophysics, it is perhaps worth dwelling for a moment on the search for novelty that is such a signature feature of so much of contemporary cuisine (and that includes, obviously, *nouvelle cuisine* but also modernist cuisine). This search very often seems as if it were a recent phenomenon. However, Beaugé (2012) makes the case that diners have actually been interested in all that is new for well over a century now. As proof, just take the following: "*It is an exceedingly common mania among people of inordinate wealth to exact incessantly new or so-called new dishes ... Novelty! It is the prevailing cry; it is imperiously demanded by everyone. ... What feats of ingenuity have we not been forced to perform, at times, in order to meet our customer's wishes? Personally, I have ceased counting the nights spent in the attempt to discover new combinations.*" While this might well sound like something that came from the keyboard of one of today's overworked celebrity chefs, the words were actually penned more than a century ago by Auguste Escoffier, head cook of the Paris Ritz and London Savoy (Escoffier 1907, p. vii).

The key point, then, is that we shouldn't think of the search for novelty as being a late twentieth century phenomenon. The desire, at least at the top end of cuisine, has been with us for a very long time. That said, an argument can be made that there probably hasn't been a time previously when the appetite for anything and everything new was quite as strong as it is today, nor found across such a broad section of the dining public. But where exactly does this overriding search for novelty, for the unusual, for the surprising and for the latest 'new thing' come from? According to Baumann (1996, pp. 116–121), contemporary dining can be seen in terms of the post-modern 'consuming body': the modernist diner as the receiver of sensations. In fact, in his book *Life in Fragments*, Baumann stresses how we currently live in a period of uncertainty: we live in a world where we are unsure if what we are getting is really the best of all possible sensations. The problem for the diner, then, is that it simply isn't possible to measure those sensations and experiences objectively in order to know whether or not they really are the very best.[8]

"*Novel or strange edibles are no longer scorned but prized, dinner-party fare is judged according to its surprise value.*" (MacClancy 1992, p. 209)

This uncertainty, then, leads the diner – and the modernist chef preparing the food for that diner – to search for the new products and improved food

---

[8] Peter Barham, professor of physics and active molecular gastronomist at the University of Bristol, and a number of his esteemed scientific colleagues have gone so far as to suggest that it may, in fact, soon be possible to "*give some quantitative measure of just how delicious a particular dish will be to a particular individual*" (Barham *et al.* 2010, p. 2361). We have to say that we don't yet share their optimism in this regard (but see Savage 2012).

experiences that just might live up to the promise of delivering heightened sensory pleasures at the table (Baumann, 1996, pp. 116–117, as cited in Sutton, 2001, pp. 117-119).[9] Notice here how novelty comes in many forms: from sourcing the most unusual and/or exotic of ingredients or outré vegetables from the very furthest corners of the globe (MacClancy 1992; Baumann 1996, p. 121; Sutton 2001; Bourdain 2002); from presenting familiar ingredients and flavours in formats that are entirely unfamiliar (see Chapter 7); or from the introduction of unusual new elements into the dining experience, be it technology at the dining table (Chapter 10), dining-in-the-dark (Chapter 8) or the addition of elements of theatre or magic to the gastronomic proceedings (Chapter 11). We believe that the delivery of novel culinary experiences that diners find both satisfying and multisensorially stimulating is increasingly going to be facilitated by our rapidly growing knowledge about how the diner's brain integrates the various sensory and conceptual elements in a dish, by understanding that taste and flavour resides in the mind (and not the mouth) and, of course, by taking this science to the table. As Gill (2007, p. 119) notes "… *taste is something that happens in your head and not, as you might imagine, on your tongue.*" Marion Halligan (1990, p. 209) makes a similar point: "*Chefs, whose livelihood is others' eating, know that the best food begins in the mind*" … to which we would like to add that that is where the best food experiences end up as well!

At the same time, however, it is worth remembering that the search for novelty can have some unexpected consequences. Although we may be willing to try anything once (Abrahams 1984, p. 23), as least if we happen to be a neophile (Rozin 1999), much of contemporary cuisine cannot really be described as comfort food (Rayner 2008, p. 193; Stuckey 2012, p. 65). What is undoubtedly also the case is that culinary surprise never tastes as sweet the second time around. In fact, we may find ourselves in the bizarre position of having a truly wonderful meal at the hands of a modernist chef (who knows, perhaps even *the* perfect meal), while at the same time having absolutely no desire to want to repeat the experience ever again (cf. Stuckey 2012, on this theme). Take the following from a recent review of the London eatery Restaurant Story:

> "*Still, Sellers is a serious talent, and his achievement in launching a restaurant this fine at the age of 26 is worth celebrating. Like a good book, Restaurant Story left me feeling stimulated, satisfied and wanting to tell my friends about it. It also left me with a suspicion that, much as I'd enjoyed it, I would probably never need to return.*" (MacLeod 2013)

---

[9] The search for novelty is "*not merely the treatment of the consumptive experience as an end-in-itself but the search for ever more novel and varied consumptive experiences as an end-in-itself. It is the desire to desire, the wanting to want which is its hallmark*" (Baumann 1996).

1.3.1   The taste of expectation

Expectations are a key point when talking about novelty and surprise. It has been demonstrated that, generally speaking, we tend to like food and drink more if they meet our expectations than if they do not (see Peterson and Ross 1972; Pinson 1986; Lee *et al.* 2006; but see also Garber *et al.* 2000). Whenever we eat and drink in fact, even before we have taken the first mouthful, our brains will have made a prediction about the likely taste/flavour of that which we are about to ingest (Small 2012). They will also have made a judgment call about how much we are going to like the experience (this is known as hedonic expectancy; Cardello and Sawyer 1992; Woods *et al.* 2011). Note also that the appearance sets up expectations regarding the likely satiating properties of a food too, which can also impact on a diner's subsequent feelings of satiety (Brunstrom and Wilkinson 2007; Brunstrom *et al.* 2010).

"A great deal of the pleasure of food is expectation." (Gill 2011, p. 13)

Food scientists have demonstrated that when a food or beverage item fails to meet our expectations we are likely to evaluate it, both immediately and for a long time thereafter, more negatively than if our expectations had been met (e.g. Cardello 1994; Deliza and MacFie 1996; Schifferstein 2001; Raudenbush *et al.* 2002; Deliza *et al.* 2003; Zellner *et al.* 2004; Yeomans *et al.* 2008). It turns out that we may be especially sensitive to disconfirmed expectation when it comes to our experience of food and drink, since these are the stimuli that we actually take into our mouths (Koza *et al.* 2005). As such, we need to take special care to avoid the risk of poisoning (see Chapter 7). Such findings are once again of fundamental importance to the modernist chef who may well be thinking about deliberately confounding his or her diners' expectations. Take the following example to illustrate the point: when Heston Blumenthal and his colleagues served a savoury ice-cream that looked like sweet strawberry to unsuspecting diners in the setting of the laboratory, those who hadn't been forewarned that it might be salty rather than sweet liked the dish far less both at the time and when tested several weeks thereafter than those who had been told (by the name of the dish) to expect a savoury flavour. In fact, simply giving the dish the name 'Food 386' helped to prepare diners for surprise, to expect the unexpected and so keep their mind open to new experiences. Just how many great-tasting dishes have been spoiled, one wonders, by the failure to get the name of the dish right (see Chapter 3 on the wonderful world of food naming).

"I watched the Blonde get her first course, a neat timbale of salmon hash, beet-cured salmon and sweet dill dressing (what's beet-cured salmon, please?). Her pretty face was a picture of serene expectation. Then, a moment later, it was as if she were [sic] sitting still, but her head were [sic] travelling at Mach three.

She let out a small, strangulated mew and coughed: 'Cat food.' What, it's like cat food? 'No, it is cat food. It's Rory Bremner beethinied salmon doing such a good impression of cat food, it's uncanny'." (Gill 2007, p. 108)

Of course, any self-respecting modernist chef wouldn't climb very far up the international San Pellegrino rankings if they were to listen to advice such as 'We really do think that you shouldn't surprise your diners! Laboratory-based research unequivocally suggests that people just don't like it.' This is one of the key areas where the results of laboratory research differ from what happens in many Michelin-starred restaurants. Now, when it comes to surprising the diner (which often involves trying to disconfirm or confound their expectations), this is something that the modernist chef excels at (in a positive way). Indeed, to be surprised is something that many diners have now come to expect when dining at one of the modernist temples to haute cuisine (Rayner 2008). However, our enjoyment of surprise, especially when it comes to food and drink – that is, the stuff that goes into our mouths and that we may swallow and which, as was just mentioned, has the potential to poison us – is going to be very much context dependent.

While surprise can undoubtedly be a very enjoyable and exciting experience if the diner knows that they are safe in the hands of one of the growing number of culinary artists who has specifically designed the experience to be 'just so', it can be far less pleasant when dining at a friend's house or if you find yourself taking part in a culinary experiment in the context of the research laboratory. Understanding the role of expectations in our dining experiences is therefore going to be absolutely crucial to approaching the perfect meal, as we will see in Chapter 7.

"Standing in Ferran Adrià's kitchen at elBulli, it is easy to believe that you have slipped down the rabbit hole. Adrià, who would have been the caterer of choice for the Mad Hatter, invents food that provokes all the senses, including the sense of disbelief. His success is almost as amazing as his food." (Lubow 2003)

1.3.2 Food as theatre: the multisensory experience economy meets cuisine

In the pages that follow, we are going to see how the new art of the table is increasingly as much about the theatre of the overall experience as it is about the taste of the food on the plate (in a way, building on Pine and Gilmore's 1998, 1999 influential work on 'the experience economy'; see also Kotler 1974; Hanefors and Mossberg 2003). At this point in history and for the foreseeable future, should we be lucky enough to stumble across it or search it out (as one of the growing number of food tourists; Boniface 2003; Hall *et al.* 2003; Rayner 2008), the perfect meal will likely involve some combination of great (and probably novel) culinary sensations together with a healthy dose of theatre/story-telling in what will be a truly immersive multisensory dining experience (Blumenthal 2013).

> "*It is food as theatre.*" (Elizabeth Carter, Good Food Guide editor, cited in BBC News story 'Fat Duck wins award despite scare')

1.4   The brain on flavour

At this stage in the proceedings, it should be clear that the perfect meal involves so much more than merely how the food on the plate tastes. As such, it suddenly becomes clear that we need to draw on a whole new range of scientific disciplines/insights in order to really understand what is going on in the diner's mind in response to the all-new multisensory experiences that they find themselves exposed to.

Now, it isn't strictly true to say that scientists have *not* been studying the experience of flavour; they have. More often than not however, this study is carried out in a very basic way typically at the behest of one of the large food or drink companies (Meiselman 2013). The results that emerge from such research may well have been of interest to the company who wants to know how to reduce the salt in their breakfast cereal without the consumer detecting it (Stuckey 2012), or else answering a company's queries about exactly how much fish meal you can feed a chicken before the average supermarket consumer will taste it in the breast meat.[10] However, while such research is undoubtedly worthy, it fails to address many of the most pressing questions about how to deliver the most stimulating and memorable multisensory dining experiences with which we are concerned in this book. We are fortunate here that our understanding of how the brain experiences flavour have benefited greatly from the recent emergence of a new field of research that goes by the name of 'neurogastronomy'.

1.4.1   Neurogastronomy

Neurogastronomy – the study of the complex brain processes that give rise to the flavours that we all experience when eating or drinking – really emerged as a scientific discipline in the first years of the twenty-first century. The term itself was first coined by Gordon Shepherd, a distinguished professor at Yale School of Medicine (Shepherd 2006, 2012). We certainly believe that a number of the studies that have investigated which parts of the brain light up when

---

[10] This isn't the kind of tasting panel that you want to get stuck on for a year, although we have friends who have been (and they were scarred by the experience, never being able to look at a chicken breast in quite the same way ever again). Note that things can get a whole lot worse; see Pickering (2008) if you don't believe us!

a participant, lying in the brain scanner, is fed something or other (often some liquid or purified foodstuff delivered by means of a tube inserted into their mouth) have generated some fascinating results (e.g. St-Onge *et al.* 2005). Neuroimaging studies have, for example, enabled researchers to understand why exactly it is that people think that a drink tastes better when they have been told that it costs more (Plassman *et al.* 2008; Spence 2010). They have also highlighted the way in which different brands of soft drink (e.g. Coke vs Pepsi) can end up recruiting different brain networks (McClure *et al.* 2004; see also Kühn and Gallinat 2013).

Neuroimaging has also been used to investigate whether wine experts use more of their brain when tasting than the rest of us do; the answer, it turns out, depends on which study you read (Castriota-Scanderbeg *et al.* 2005; Pazart *et al.* 2011). Furthermore, surprising though it may seem, more of our brain lights up when we merely think about (or anticipate) food than when we actually get to taste it (O'Doherty *et al.* 2002; see also Pelchat *et al.* 2004).[11] Researchers have even started to delve into the question of which parts of the brain become more active when we decide whether or not we would like to taste a particular novel combination of ingredients (i.e. something that we have never eaten or come across before; Van der Laan *et al.* 2011; Barron *et al.* 2013). For example, do you think that you would like the taste of a raspberry and avocado smoothie? Or how about a green tea jelly, or beetroot custard? Only future research will tell whether today's modernist chefs exhibit increased neural activation in areas such as the medial prefrontal cortex (mPFC) that have been shown to light up when we perform such a task, given all the practice they have undoubtedly had in terms of imagining weird and wonderful combinations of ingredients with which to assault their diners' senses (Maguire *et al.* 2000).

It turns out that food really is one of the most effective stimuli in terms of modulating brain activity. This is especially true if we happen to be hungry. For example, in one neuroimaging study, a 24% increase in whole brain metabolism was observed when a group of hungry participants were shown, and allowed to smell, their favourite foods (e.g. a bacon, egg, cheese sandwich or cinnamon buns; see Wang *et al.* 2004). This is a massive change in brain activity in what is by far the body's most blood-thirsty organ (e.g. Wrangham 2010; Allen 2012), especially when compared to the 1–2% signal changes that are typically reported in the literature.[12]

---

[11] This presumably explains why, when offered a great meal at a fancy French restaurant, people typically choose to delay the pleasure of actually consuming it (Loewenstein and Prelec 1993). Gilbert (2006, p. 17) summarizes the underlying premise here: "*Forestalling pleasure is an inventive technique for getting double the juice from half the fruit.*" Seen in this light, the interminable waiting list for a table at many of the world's top restaurants could actually work to enhance the diner's experience when they eventually get to eat there.

[12] According to Allen (2012, pp. 51–52), "*The human brain accounts for only 2% of the body's mass, but a whopping 20–25% of the body's resting metabolic rate.*"

"... *on a day-to-day basis, from the moment we are born until the moment we die, there is nothing that concerns us more than food.*" (Allen 2012, p. 180)

At this point, we can only speculate as to whether there might be a link between the profound neural and physiological changes that can be triggered when a person looks at (and/or smells) an appetizing plate of food and the recent growth of gastroporn.[13] Indeed, the growing importance of the visual appearance of food, a trend that as we have seen already was really promoted by the emergence of the nouvelle cuisine movement, seems to make perfect sense once it is realized that 'eye appeal' really is half the meal (or as Apicius, the first century Roman gourmet is purported to have said: "*The first taste is always with the eyes*").[14] Given just how important the sight of food is, we are clearly going to need to learn as much as we can about the visual aesthetics of plating (see Chapter 4).

In fact, one of the most fascinating examples of the way in which our brain controls our food behaviours actually comes not from neuroimaging research but rather from neuropsychology (that is, from the study of patients suffering from brain damage). Take the bizarre case of those patients afflicted by Gourmand Syndrome (Regard and Landis 1997; Steingarten 2002). This is a rare neurological condition in which a stroke (one that typically affects the insula) results in an individual suddenly acquiring a profound and all-consuming interest in fine food! This can sometimes happen to those who previously expressed no interest in food whatsoever (i.e. those would eat to live rather than vice versa). Seemingly overnight, these patients develop an overriding passion for fine gastronomic cuisine. Such curious examples left Jeffrey Steingarten (2002), the famous North American food critic, to ponder: "*With nearly every bite I take, in the back of my mind there looms the same nagging question: Who is having all the fun? Is it my brain or is it really me?*"

1.4.2   Do neurogastronomists make great-tasting food?

Given the importance of the brain to multisensory flavour perception, one question that would likely spring to mind here is whether you are likely to have your perfect meal while sitting in a restaurant serviced by a chef practicing neurogastronomy. This is no longer a purely hypothetical question. For while he may not have come up with the term, the credit for first combining culinary science with brain science should probably go to Miguel Sánchez Romera, a friend of Ferran Adrià. For a while, Sánchez Romera combined two careers, one as a neurologist by day and the other as a

---

[13] As Allen (2012, p. 74) notes, there certainly needs to be some account of why there are now so many more cookbooks out there than anyone could ever manage to cook from over an entire lifetime.

[14] Although given how much our brains appear to like looking at appetizing foods, it is somewhat surprising that most Western diners dislike the idea of menus that contain full-colour pictures of the dishes being served (see Chapter 2).

practicing chef by night. Somehow, he even found time to write the intriguing book *La Cocina de los Sentidos* in which he combines his two passions (Sánchez Romera 2003). He eventually closed his Spanish restaurant situated close to Barcelona, L'Esguard de Sant Andreu de Llavaneres, and moved to New York City's Chelsea district to open another one named Romera (McLaughlin 2011). Miraculously, it looks like the restaurant has managed to survive the excoriating review it received from Frank Bruni in *The New York Times* (Bruni 2011).

> "*Its chef, Miguel Sánchez Romera, is a doctor who worked for years as a neurologist. He has coined a whole new genre for his cooking, which favors squishy textures, kaleidoscopic mosaics of vegetable powders, and a wedding's worth of edible flowers. He calls it neurogastronomy, which "embodies a holistic approach to food by means of a thoughtful study of the organoleptic properties of each ingredient," or so says the restaurant's Web site. Organoleptic means 'perceived by a sense organ'. I looked it up.*" (Frank Bruni on Romero, one of the world's first neurogastronomy restaurants in Chelsea, New York City; Bruni 2011)

Of course, it is unlikely that the neurogastronomy movement will lose momentum simply because of the activities of any one of its practitioners (or because of a negative review, no matter how bad it might be). We would, however, argue that this example helps to illustrate the more fundamental point that neurogastronomy –understanding the brain on flavour – provides insights about only a small part of what makes a wonderful meal truly great. It should always be remembered that sticking people into the noisy claustrophobic coffin that is the contemporary functional magnetic resonance imaging (fMRI) brain scanner is a most unnatural activity. And when it comes to the study of flavour, things rapidly get much worse (see Spence 2012a). To get a sense of the gulf that separates neurogastronomy research from the real world of dining experiences that we are trying to understand in this book, just imagine yourself lying with your head clamped absolutely still. You have a tube inserted into your mouth pumping in who knows what liquid or puréed concoction as you lie flat on your back.[15] Worse still, each squirt of real flavour is typically washed down with a gob of artificial saliva. (OK, that may not be what the scientists conducting these studies call it, but that is essentially what it is – the most neutral of mouth washes!)

Can such research really provide useful insights about the organization of the flavour perceptual system in the human brain? Absolutely! Just see Small (2012) for a summary of the current state of the art in this regard. That said, it is important not to lose sight of the fact that the situation of the isolated participant being scanned in a noisy neuroimaging machine in a science faculty

---

[15] In the near future, the introduction of vertical bore scanners (that will allow the participant to sit upright while the scanning apparatus is lowered over their head) will allow for more naturalistic consumption behaviours while brain images are acquired.

is very far removed from the social interaction of eating a great meal in a wonderful location surrounded by your close friends (see Chapter 2). As in so many other areas, one needs to be cautious about the 'neuromania' (the term coined by Legrenzi and Umiltà 2011) that has swept the cognitive neurosciences (and many other fields of research) in recent years.[16]

As the results reviewed above have shown, research in the field of neurogastronomy is really starting to help researchers understand a little more about the fundamentally important role of food in the organization and responsiveness of the human brain. However, while knowing more about how the brain processes flavour is one thing; understanding the key factors contributing to the perfect meal is quite another. With that clearly in mind, let us then move on to look at the other new sciences of the table that will make their appearance in the pages ahead.

1.5   Food and the perception of everything else

How much of our pleasure in savouring a great meal resides in the quality, freshness and seasonality of the ingredients and how they have been prepared, and how much depends on 'the everything else', that is, the tablecloths, the feel of the cutlery, the name of the dish and the atmosphere and ambiance? This is a debate that we have had with a number of chefs. Every one of us, whether a chef, a diner or even a food critic, likes to think that we can taste the quality of the food. That is, we all (and this includes experimental psychologists and budding gastrophysicists alike) believe that we can evaluate the merits of a food or dish and ignore the 'everything else'. However, a very large body of empirical evidence suggests that this is simply an illusion: a convincing one, granted, but an illusion nonetheless. In fact, the field of experimental psychology research is filled with exposing just such misperceptions that permeate so many aspects of our daily lives (e.g. Chabris and Simons 2011; Kahneman 2011). Certainly, when the scientists investigate what happens to people's ratings of food and drink when they change the colour of the plate, the ambient lighting, the music, the cutlery, etc., they often find that those ratings change significantly. This is not only true for the sensory-discriminative qualities of what a diner happens to be eating (e.g. what it tastes of and how intense the flavour is), but also for their hedonic responses (i.e. how much do they like the experience). But when you ask people do you think that the colour of the plate or the weight of the cutlery had any influence on your experience of the dish, we all say "*Of course not. Are you crazy?*"

---

[16] It sometimes seems like everyone wants to stick the 'neuro-' prefix in front of whatever discipline it is they happen to be interested in, giving it an air of scientific respectability that simply may not be warranted (see also Tallis 2008; Poole 2012, pp. 71–72; Spence 2012a). One needs to be especially cautious here given McCabe and Castel's (2008) claim that people are more likely to be convinced by an argument if it is accompanied by a colourful brain image; the more colourful the picture, the more convincing the argument appears to be (but see also Michael *et al.* 2013 for the latest take on an ever-changing story).

"How much of our enjoyment of a great meal originates in the food and drink itself and how much comes from the 'everything else'?" is a question that we are frequently asked by journalists hungry for a figure or better still a percentage to put in their columns. Now the serious scientist is loath to provide such a number; it obviously depends on so many different factors (and whatever number you give, you will undoubtedly be criticized by your academic colleagues for having simplified matters too much or for having failed to consider some or other factor or issue). Nevertheless, when you combine all of the evidence outlined in the pages that follow, it's hard not to come away from the research convinced that a 'good half' of our experience of food and drink is determined by the 'everything else'. We are going to come across a lot of research showing how pretty much every conceivable factor can make a difference to the way in which we perceive, respond to and remember food (and drink).

Of course, the food itself is absolutely critical. No one can argue with the claim that sourcing the best, the freshest and the most flavourful ingredients and having the culinary skills to allow those components to show their full potential and to harmonize them with whatever else happens to be on the plate is going to be a necessary precondition for the perfect meal. However, if that wonderful food is served up at 35,000 ft in an airplane or in a grotty work canteen, it simply will not taste the same. The profound impact of the atmospherics of the environments in which we choose to eat and drink on our dining experience will be covered in Chapter 9.

We can't deny that the claim that so much of the experience of a great meal depends on the 'everything else' comes as anathema to many of the chefs we work with. In fact, we have known some of them to get more than a little agitated when we start to talk to them about how important the 'everything else' really is. Many chefs, especially those of a more traditional persuasion, will tell you that great food is nothing more than simply the freshest, tastiest of ingredients, skilfully prepared and beautifully presented. For many of them, there really is nothing else. Very often however, these are the very same chefs who have their restaurant situated in a converted knitting museum or who are happy to let the duty manager make the musical selection by blasting every diner in the restaurant with their own personal iPod selection. Others that we have spoken to start muttering something like *"You mean that you can serve dog food, and people will like it if you just play the right music?"* [17], the blood rushing to their faces. We honestly believe that that is unlikely to be the case (Chossat and Gergaud 2003). What we are really much more interested in is making sure that wonderful food is really shown at its best.

---

[17] We are reminded of the infamous dog food study (Bohannon *et al.* 2009). A group of researchers in the US invited people over to taste a range of pâtés, one of which was actually made from blended dog food (top quality, of course). Remarkably, the participants were unable to pick out the pet food from the other four store-bought pâtés. But before you decide to save a little money next time you invite your friends round for dinner, it should be noted that the blended dog food didn't score particularly well in terms of liking.

1.6   Gastrophysics: the new science of the table

We have hopefully convinced you by now that understanding the perfect meal requires us to know about much more than merely the preparation and presentation of the food on the plate and more than just which parts of the brain light up. The question then becomes one of which other sciences are going to be relevant in developing a further understanding of what is really driving the diner's experience. Another important question that emerges, and one that we hope we'll find an answer to, is: "Why do you love foods that I hate?" (Lauden 2001). Why is there seemingly just as much chance of finding people who love the new concepts of dining as there is of finding people who detest them?

In the pages that follow, you will be hearing a lot more about 'gastrophysics', a term that first appeared in a 2005 article in *New Scientist* magazine. The year 2012 saw the first international symposium devoted specifically to the topic of 'The Emerging Science of Gastrophysics'. This ground-breaking meeting was held at the Royal Danish Academy of Sciences and Letters in Copenhagen (see Mouritsen and Risbo 2013). By now, the meaning of the 'gastro-' part of this term should be clear. The 'physics' in the title, meanwhile, we take to refer to the science of psychophysics. This is a field of research in which scientists investigate the way in which people (formerly subjects, although we now prefer to call them participants; but whatever you call them, they are mostly still WEIRD[18]) respond to sensory stimuli; essentially treating the human as if they were a physical detector that responds in a highly predictable manner to a given set of parameterized sensory inputs (Gregory 1987).[19] Gastrophysics isn't a culinary movement as such. Rather, it is simply the name given to a range of tools, techniques and ways of thinking about the diner's response by means of assessing the impact of various factors – that are both internal and external to the food and drink itself – on the multisensory dining experience.[20]

So what are the tools, techniques and approaches that lie at the heart of the new discipline of gastrophysics? We believe that well-controlled experiments, conducted both in the laboratory and out there in the real world, are absolutely central to this nascent field of research: experiments that involve the rigorous measurement of the diner's experience using carefully designed questionnaires, response scales and behavioural tests (some of them involving implicit measures). As psychologists, we are highly reticent about putting

---

[18] That is, Western, Educated, Intelligent, Rich (relatively speaking) and Democratic (see Henrich *et al.* 2010).

[19] The curious thing is that few traditional psychophysicists, not to mention experimental psychologists, have seemingly ever wanted to get their hands dirty messing around in the world of taste and flavour.

[20] Some of the other terms that people have come up with over the years include "gastrophy", as suggested by the nineteenth century French Utopian socialist Charles Fourier (Ferguson 2004, p. 100; Cowen 2012, p. 13), and "gustemology", a set of *"approaches that organize their understanding of a wide spectrum of cultural issues around taste and other sensory aspects of food"* (Sutton 2010, p. 215). Neither of these terms has caught on however, so we'll stick with gastrophysics here.

too much weight on the unconstrained self-report of participants. Why? Time and again research has shown that relying on such reports can often paint a misleading picture concerning the critical factors that are actually driving perception and behaviour (Martin 2013; see also North *et al.* 1997 for a particularly nice example of this dissociation).

It is important to note that many of the laboratory studies in this area can be criticized for their lack of ecological validity. Indeed, one important limitation associated with trying to interpret the results of many studies that have been published to date is that the scientists concerned have typically tended to eliminate all of the extraneous variables in order to focus on just one (or at most a couple) of factor(s) that are of particular interest to them. As such, a participant may find him- or herself sitting alone in a dark science lab[21] with no background noise (or perhaps some white noise playing over headphones), nor any other distractions, focusing their attention squarely (or so the experimenter hopes) on the task at hand. Rating the sourness of 20 samples of yoghurt on a 7-point sweetness scale, for example, would not be an unusual test.

A second problem that one should be aware of when interpreting the results of food research that has been conducted in the science lab is that, in the majority of cases, the scientists running the studies will typically have utilized what is known as a within-participants experimental design. What this means in practice is that each participant is exposed to each and every one of the conditions of interest (often several times in quick succession). This kind of experimental design is favoured because it enables the researchers to rule out any differences between participants as the cause of any effects that they observe; these differences could hinder the interpretation of a between-participants experimental design. In this case, we can never really be sure whether the participants in each group were really matched in all possible regards. However, one unintended consequence of the between-participants experimental design is that it can serve to emphasize any differences between the various conditions in a manner that is rather unnatural.

As a consequence of such limitations, the gold standard in terms of ecological validity is when the scientist is able to test his/her hypotheses in an actual restaurant (or cafeteria) setting. However, the challenge for the budding gastrophysicist is that it can be very difficult to find a restaurateur who is willing to have his/her venue taken over in the name of science. (And those who do allow it may not want the results to be made public). Using such realistic settings to conduct behavioural experiments can however be criticized for lack of control over a number of other key variables that might be expected to impact on a person's responses. For instance, consider the fact that the diners are eating at different speeds while chatting among themselves; they potentially may

---

[21] The walls of all the best psychophysics labs are painted with matt black paint. This helps to avoid any unwanted reflections.

be distracted from the food and the experimental variables of interest (e.g. Meiselman 1992; Köster 2003).

In the pages that follow, we are ideally looking for those factors that contribute to delivering the perfect (or at least enhanced) meal that have been backed up by the results of both types of research, that is, where the weaknesses of laboratory studies are made up for by the strengths of the real-world tests and vice versa (see also de Graaf *et al.* 2005). At present however, having such converging evidence is a rare luxury. What is more, the results of these different types of study can sometimes actually deliver different conclusions (e.g. see Chapter 5 for an example when psychologists investigated the effects of changing the size of the plate on people's consumption behaviours). When this happens, it is obviously going to be that much more difficult to know quite what conclusion to draw. Perhaps the best that one can say is that more research is needed, and leave it at that.[22]

One final issue to bear in mind here is that, as yet, there is a real paucity of long-term follow-up studies. While many of the studies reviewed in this book provide relatively convincing evidence concerning the short-term consequences of this or that intervention – whether it be changing the colour of the plates, the weight of the cutlery, or the name of the dish, etc. – what we really need to see much more of in the years to come are longer-term follow-up studies investigating whether a given intervention continues to influence people's performance over weeks, months and possibly even years after it has first been introduced.

That said, and despite these various limitations, it is very exciting to see how many of the current generation of young chefs are interested in going beyond the basic culinary science that they have been taught in cookery schools such as the Institut Paul Bocuse and the Cordon Bleu in order to learn more about the key insights from gastrophysics (and even computational gastronomy; see Ahn *et al.* 2011) that are most relevant to delivering great-tasting food experiences. Many of these chefs are really curious to learn more about the minds, and not just the palates, of their diners. Increasingly, we also see chefs at all levels of fame and fortune working with designers to create custom menus, plateware, cutlery and dining spaces.[23] These chefs are creating experiences that build on all that contemporary design and technology has to offer paired with the latest findings from the field of gastrophysics.

> "'We're quite close to throwing out the theory of five tastes,' [Heston] says. 'Researchers have found 21 receptors for bitterness on the tongue. There is a growing argument that fat is a taste.' All of this will change the way chefs flavour their dishes." (Heston Blumenthal being interviewed by Jay Rayner 2006)

---

[22] Note that a number of scientific journals no longer allow their authors to make such statements. But are there any areas of science where such a claim doesn't apply?

[23] In fact, the last few years have seen an exciting shift from straight sensory science to a more integrated approach to culinary arts and meal science (Gustafsson 2004).

A growing number of chefs, spearheaded by the likes of Heston Blumenthal and Andoni Aduriz in San Sebastian, are now spending time visiting the psychology, physiology and/or neuroscience labs in order to gain whatever scientific insights they can to enable them to deliver differentiated culinary experiences to their diners. In some cases, diners may never have had such experiences before.[24] Many chefs are now increasingly coming to realize that what they put on the plate is only a part of the diner's overall experience. We have all had the experience: we all know that food and drink taste different depending on where we happen to be eating or drinking (see Chapter 9), not to mention with whom (see Chapter 2).[25] The challenge is therefore how to bring the science of the diner together with the rapidly evolving science of the kitchen into an all-new culinary experience that can really blow a diner's mind.[26] There is also a growing interest here in taking the diner on some kind of emotional journey.

1.7 Food perception is fundamentally multisensory

What is particularly exciting at the present time is that we currently know far more about the principles of multisensory integration giving rise to flavour perception than ever before (Stevenson 2009; Dijksterhuis 2012; Spence 2012c, 2013). What is more, a number of the world's top chefs have started to wake up to the importance of stimulating all the senses to deliver multisensory dining experiences that are more engaging, more exciting and ultimately more memorable than ever before. Just take the following quotes:

> *"Eating is the only thing we do that involves all the senses. I don't think that we realize just how much influence the senses actually have on the way that we process information from mouth to brain."* (Heston Blumenthal, Tasting menu from 2004, The Fat Duck restaurant, Bray, UK)

> *"Cooking is the most multisensual art. I try to stimulate all the senses."* (Ferran Adrià, elBulli, quoted in Anonymous 2007, p. 19)

Delivering great-tasting food is about more than merely stimulating each of the diner's senses individually (Dornenburg and Page 1996). Rather, it is

---

[24] Unfortunately, while many chefs and culinary teams are generally interested in the new sciences of the kitchen, they rarely have the funds to support such research.

[25] This phenomenon has started to be studied under the evocative title of the Provencal Rose Paradox.

[26] Note that many of the top restaurants actually lose money, and lots of it. But this isn't always as problematic as it sounds because very often the flagship restaurants act as the loss-leaders for the chef's brand, whether selling cookbooks or salad dressing, TV shows or stadium-filling tours (see Gill 2011).

a matter of knowing how one sense affects another. Chefs will likely benefit from learning a little more about the latest in scientific discoveries as they relate to the world of cuisine. For example, once you come to understand that many scientists believe that around 80–95% of what you think of as flavour actually comes from your nose (Martin 2004; Rosenblum 2010; Ge 2012), you might start thinking rather differently about the aromatic element of your dishes. Furthermore, the chef's ability to deliver great-tasting food can't be harmed by the knowledge that there may be as many as 20 different basic tastes according to some researchers, not just the 4–5 that most people can name (sweet, sour, bitter, salty and umami; Stuckey 2012). We'd also hope that chefs would want to find out more about how changing the aroma of a food (by adding the aroma of strawberry or vanilla, say) can change its perceived sweetness, and how changing the colour of a food or beverage can send a very powerful signal to the diner's brain about the likely taste and flavour that they are about to experience. Every food producer would benefit from knowing how to make the food and drink taste 10% sweeter without the addition of any extra calories or artificial sweeteners (Spence *et al.* 2010). Over the last decade or so, neuroscientists have started to uncover a number of the key rules that are used by the diner's brain (in fact, used by every one of our brains) to combine the information from the tongue with that from the nose, eyes and ears: rules such as superadditivity, subadditivity and sensory dominance. In Chapter 6 we'll take a closer look at the latest evidence concerning the multisensory perception of flavour and the rules governing how all of our brains combine the evidence from each of their senses.

> *"But the biggest development will be in what [Heston] calls 'sensory design'. No longer will eating out be just about putting stuff in our mouths and deciding whether it's nice. 'Eating is a multisensory experience.'"* (Heston Blumenthal being interviewed by Jay Rayner in 2006)

1.8   Isn't modernist cuisine only for the lucky few?

Journalists sometimes question what we do: *"Isn't it the case,"* they ask, *"that what you do is all very esoteric. Something that only a moneyed minority can ever enjoy? Surely, there is nothing in all of this that has any relevance whatsoever to the everyday person preparing a meal for their friends at home, say?"* We would counter such a question by pointing to the fact that the modernist restaurant provides the perfect verve for new culinary ideas coming from the field of gastrophysics. What happens in the top-end restaurant provides the ideal test-bed for culinary innovation. In fact, the overarching idea behind much of our own research is that the best of what starts in the modernist restaurant will eventually feed down to the high street and home dining environment.

The top modernist chefs can be thought of as equivalent to the F1 drivers of the motor racing world. Just think of the many millions that are spent on technological innovations directed at saving the Lewis Hamiltons of this world one-hundredth of a second on the Grand Prix circuit. It is hard to justify such expense in itself, but the fact is a number of the most successful of the innovations that start out in such cars will sooner or later find their way into the cars coming off the production line. It may take 5 years or even longer, but sooner or later it will happen and that is a key part of what makes the whole endeavour worthwhile. We believe that exactly the same is true for the top modernist chefs. Indeed, a number of the ideas that could only ever have been trialled in a modernist restaurant (or perhaps in a Futurist dinner party; see Poole 2012) are now starting to percolate down to society at large. Some of the latest techniques and inventions that high-end chefs have been perfecting over the last decade or so have now become standard practice in the catering industry. The modernist restaurant can therefore be seen as a veritable hotbed for gastronomic invention, and hence the ideal venue in which to study gastrophysics. Take for instance the technique of *sous-vide* cooking that has been popularized and developed by chefs such as Heston Blumenthal and Thomas Keller. This technique is likely to become increasingly popular across the commercial restaurant sector for the cost saving (relative to other cooking techniques) it offers, if not for the differentiated flavour experience and enhanced flavour retention that this technique can provide.

For those with an eye on the marketplace, one can also see that certain modernist dishes that were once available only at the world's most exclusive tables (foods like the mustard ice cream served at Heston Blumenthal's The Fat Duck) are now being served more and more in down-to-earth venues[27] or found on supermarket shelves or kitchen stores. In fact, a growing number of the world's most famous practising molecular gastronomists are now teaming up with supermarket chains or else selling their own product lines direct to market. For example, think of all those spherification (or molecular gastronomy) kits that are available to chefs (e.g. http://www.albertyferranadria.com) or celebrity chef's ranges of food (http://www.telegraph.co.uk/foodanddrink /foodanddrinknews/7873046/Heston-Blumenthal-launches-range-of-food-at-Waitrose.html). The modernist or molecular approach can also be seen making its way into other sectors. For instance, the molecular mixologists are increasingly following ever more closely on the heels of the molecular gastronomists (e.g. Sherman 2008; http://en.wikipedia.org/wiki/Molecular _mixology; http://www.thecocktaillovers.com/tag/professor-charles-spence/), and they aren't the only ones. Visit one of the modernist ice cream parlours (Schlack 2011, http://www.humphryslocombe.com) and even modernist fish

---

[27] Of course, whether one thinks this is a good thing is a very different matter (see Chapter 8 for some of the pitfalls of applying modernist principles in the home setting). We don't know about you, but we are certainly starting to tire of seeing 'the triple-cooked chip' appearing on menus in restaurants and gastropubs across the globe.

and chips (take a walk down Upper Street in Islington, North London if you don't believe us).

In line with the idea of broadening the reach of modernist cuisine, a number of the molecular gastronomy chefs have now attempted to open restaurants that are more affordable to the general population. This is not to say that all of the wonderfully wacky culinary ideas that have been tried and tested in the confines of the modernist restaurant will necessarily work outside in the wider world of cuisine, as we'll see in the last chapter. While popping candy has certainly helped to enliven the menu at The Fat Duck, this proved to be just a step too far when Heston Blumenthal added a shake of this noisy food additive to the food on the menu at the UK's Little Chef chain of motorway pit stops (Fleming 2013).

So will the latest discoveries from the emerging field of gastrophysics only benefit the culinary elite? Absolutely not. For us, the challenge in this area is to take a number of the top insights gleaned from studying what goes on at some of the world's top restaurants and translate them so that they can be utilized by diners at every table. We are firmly of the belief that many of the insights and discoveries reported in this book (a number of which have come from the modernist restaurant) are actually things that any one of us can utilize at home in order to improve our everyday dining experiences.

If food really does seem to be of higher quality when eaten with the aid of heavy cutlery (Piqueras-Fiszman and Spence 2011), if desserts really do taste sweeter when served on round white plates (Piqueras-Fiszman *et al.* 2012), if playing Italian opera really can make your pizza and pasta sauce taste more authentically Mediterranean (see Spence 2012 for a review) and if we all eat less if served from smaller bowls onto smaller plates with the aid of smaller cutlery (or, better still, chopsticks for the unacquainted; Wansink 2006), the implications are right there for each and every one of us to incorporate into our daily routines. For one very simple trick here, just try holding the bowl that you eat from in your hands, rather than letting it rest on the table, assuming it is suitably heavy (and the heavier the better in this regard). The weight in your hand will likely make you feel more satiated with however much food you eat (Piqueras-Fiszman and Spence 2012).[28] The hope is that the insights discussed in the chapters that follow can be used as the basis for delivering culinary experiences and guidelines that can make everyone's life better.[29]

---

[28] It has been estimated that in the UK 30% of the money spent on food is on that which we eat away from home (Binkley *et al.* 2000).

[29] The economist Cowen (2012, pp. 2–3) makes a similar point at the beginning of his recent behavioural economics book *An Economist gets Lunch*. There he notes that understanding where the quality of the experience lies has huge implications, not just for the world of fine dining; it may also constitute an important step towards feeding the world's 7 billion people – a grand and worthy aim we're sure you will agree. However, we must confess that do not see quite how or why? The 'percolating down' argument unfortunately does not mean that those suffering from food shortages in Africa will necessarily be fed.

Last, but by no means least, is the fundamental question of whether our search for the basic science underpinning the perfect meal can help to combat the global obesity crisis (e.g. Caballero 2007). We, along with many others, are certain that one part of the solution here will have to come from our growing understanding of the neuroscience and behavioural economics of flavour (Lau *et al.* 1995). We passionately believe that gastrophysics has its part to play here; it can provide the evidence to back up the guidelines and practical suggestions concerning the subtle nudges (Thaler and Sunstein 2008) that might help to move people towards healthier eating behaviours (Marteau *et al.* 2012). Each and every one of these strategies should not only help those with nutritional issues to eat more healthily, but also to feel more rewarded (and hopefully satiated) after the meal.

> "*A bad or mediocre meal is more than just an unpleasant taste; it is an unnecessary negation of life's pleasures. It is a wasted chance to refine our tastes, learn about the world, and share a rewarding experience.*" (Cowen 2012, p. 11)

In fact, the happiness experts tell us that blowing our hard-earned cash on experiences such as a great meal is the most rewarding path to happiness (Dunn and Norton 2013). However, the pleasure may soon diminish after any more than one or two delicious meals as Jay Rayner (British food critic and broadcaster) discovered; his wonderful book recounting his 7-day sojourn to a number of the world's top French Michelin-starred restaurants describes just how quickly the pleasure of fine dining can wear off (Rayner 2008).[30]

1.9   *Amuse bouche*

So, hearing your stomach start to rumble, we would like to end this chapter by serving you a few *amuse bouche* concerning some of the insights that you are going to come across in this study of the perfect meal. First, let us tell you that nothing will be left to chance, from the mood you're in when your first course arrives[31] to the position on the menu of that oh-so-tempting starter (or *hors d'oeuvre* if you will), to how much thinner your wallet is going to be when you eventually leave (Chapter 2). After reading this book, you will hopefully understand a little more about why it is that most people are willing to pay twice as much for an '*Omelette à la Norvégienne*' than for a 'Baked Alaska', being one and the same dessert. You will also learn why some people may end

---

[30] One can think of this as the gastronomic equivalent of Morgan Spurlock's heroic attempt to eat nothing but McDonalds for a month, as portrayed in the memorable 2004 documentary *Super Size Me*.

[31] As we'll see in the next chapter, the waiters at The Fat Duck carefully watch their diners to know whether they are left or right handed. If you are a leftie, when you sit down after a visit to the bathroom the cutlery will have been reset to match your personal proclivities. Nothing said; an absolutely seamless performance.

up feeling a little nostalgic while eating a 'Grandma's chicken pie' dish from the work canteen (Chapter 3).

In addition, and as a palate cleanser, we are going to see how the visual appearance of a dish can be just as important as its flavour and that, on occasion, the canvas will turn out to be just as important as the painting itself (Chapter 4). The canvas, or in this context where and how the food is presented to the diner sitting at the table, can play all manner of tricks on you to the extent of modulating how much you eat and how much you (or your brain) thinks that you enjoyed the experience afterwards. We will be serving up some of the dishes that will help you to understand what we mean here; we might even make you salivate in the process.

What about the cutlery? Bigger, smaller, textured or smooth, warm even (no, really)? The everyday utensils that we instinctively grab for when dining have evolved and modelled the ways in which we interact with food; as we will see later, they may even have changed the shape of our jaws (Chapter 5). Once the table has been cleared away, we bet that you will not be able to believe the many creative new ways in which cutlery (or its absence) is being utilized by the world's top chefs today. During the *intermezzo* of our feast, you'll be served a shot of cognitive neuroscience so that you will know a little more about how your brain integrates all the sensory information that it receives while your stomach is filling up, from eye and ear to nose and mouth (Chapter 6). (We hope to have served this somewhat complex material in a format that you will find at least reasonably easy to digest!) You'll then taste some sensorially incongruent dishes, a perfect means of achieving and hopefully delighting the exigent diners; the only essential prerequisite to enjoy it is that you are something of a neophile when it comes to food (Chapter 7).

For dessert, we'll be walking you through the experience of dining in the dark (Chapter 8). You will soon discover just how important vision is in terms of perceiving what it is that we are eating; perhaps this will help to explain why this isn't everyone's cup of tea. On the other hand, we will also be highlighting a number of other alternative approaches including those who have chosen to play with the lighting, the music and all the other atmospheric elements that are capable of immersing the diner in a truly engaging multisensory experience (Chapter 9).

As a *digestif*, we'll be serving up the latest trends in digital technology and showing how it is being brought to the table (or at least to the dining room) by the most adventurous of modernist chefs (Chapter 10). We promise to show you some of the ways in which the latest technologies are currently being developed to enhance our eating experiences as you prepare for your dessert. At the end of the perfect meal we hope that you still have a little space left to hear about some of the amazing predictions regarding the future of food, based on what's happening in the top restaurants today (Chapter 11). Without wanting to advance any particular agenda, we simply hope to enthuse the open-minded reader about some of the fabulous possibilities lying just around the corner. Ultimately, we hope to share a little of our passion and excitement for these food experiences.

In the pages that follow we hope to convince you that gastrophysics, the new science of the table, really does hold the potential to enhance everyone's dining experience; the results of the research in this area can bring each and every one of us a little closer to that perfect meal. As you will see, achieving that goal is going to require us to bring together the evidence that is now emerging from a diverse range of research fields including experimental psychology, design, neuroscience, sensory marketing, behavioural economics and the culinary and sensory sciences.

And with that, let us wish you *"Bon appétit!"*

References

Abrahams, R. (1984) Equal opportunity eating: A structural excursis on things of the mouth. In *Ethnic and Regional Foodways in the United States: The Performance Group Identity* (eds L. Brown and K. Mussell), pp. 19–36. University of Tennessee Press, Knoxville, TN.

Ahn, Y.-Y., Ahnert, S. E., Bagrow, J. P. and Barabási, A.-L. (2011) Flavor network and the principles of food pairing. *Scientific Reports*, **1(196)**, 1–6.

Alícia and elBullitaller (2006) *Modern Gastronomy A to Z: A Scientific and Gastronomic Lexicón*. CRC Press, Boca Raton, FL (originally published as *Léxico Científico Gastronómico*, Editorial Planeta, Barcelona).

Allen, J. S. (2012) *The Omnivorous Mind: Our Evolving Relationship with Food*. Harvard University Press, London.

Anonymous (2007) Does it make sense? *Contact: Royal Mail's Magazine for Marketers* (Sensory Marketing Special Edition). Redwood, London.

Ashley, S. (2013) *Synthetic food: Better cooking through chemistry*. Available at http://www.pbs.org/wgbh/nova/next/physics/synthetic-food-better-cooking-through-chemistry/ (accessed January 2014).

Barham, P. (2000) *The Science of Cooking*. Springer-Verlag, Berlin.

Barham, P., Skibsted, L. H., Bredie, W. L. P., Bom Frøst, M., Møller, P., Risbo, J., Snitkjaer, P. and Mortensen, L. M. (2010) Molecular gastronomy: A new emerging scientific discipline. *Chemical Reviews*, **110**, 2313–2365.

Barron, H. C., Dolan, R. J. and Behrens, T. E. J. (2013) Online evaluation of novel choices by simultaneous representation of multiple memories. *Nature Neuroscience*, **16**, 1492–1498.

Baumann, Z. (1996) *Life in Fragments: Essays in Postmodern Morality*. Blackwell, Oxford.

Beaugé, B. (2012) On the idea of novelty in cuisine: A brief historical insight. *International Journal of Gastronomy and Food Science*, **1**, 5–14.

Belasco, W. J. (2006) *Meals to Come: A History of the Future of Food*. University of California Press, Berkeley, CA.

Berghaus, G. (2001) The futurist banquet: Nouvelle Cuisine or performance art? *New Theatre Quarterly*, **17(1)**, 3–17.

Binkley, J. K., Eales, J. and Jekanowski, M. (2000) The relation between dietary change and rising US obesity. *International Journal of Obesity and Related Mental Disorder*, **24**, 1032–1039.

Blumenthal, H. and McGee, H. (2006) Statement on the 'new cookery'. *The Observer*, December 10. Available at http://www.guardian.co.uk/uk/2006/dec/10 /foodanddrink.obsfoodmonthly (accessed January 2014).

Blumenthal, H. (2007) *Further Adventures in Search of Perfection: Reinventing Kitchen Classics*. Bloomsbury, London.

Blumenthal, H. (2008) *The Big Fat Duck Cookbook*. Bloomsbury, London.

Blumenthal, H. (2013) *Historic Heston Blumenthal*. Bloomsbury, London.

Bohannon, J., Goldstein, R. and Herschowitsch, A. (2009) Can people distinguish pâté from dog food? American Association of Wine Economics (AAWE) Working Paper No. 36.

Boniface, P. (2003) *Tasting Tourism: Travelling for Food and Drink*. Ashgate, Aldershot.

Bourdain, A. (2002) *A Cook's Tour: In Search of the Perfect Meal*. Bloomsbury, London.

Brillat-Savarin, J. A. (1835) *Physiologie du goût* (The philosopher in the kitchen/The physiology of taste). J. P. Meline, Bruxelles. Translated by A. Lalauze (1884) *A Handbook of Gastronomy*. Nimmo and Bain, London.

Bruni, F. (2011) Dinner and derangement. *New York Times*. Available at http://www .nytimes.com/2011/10/18/opinion/bruni-dinner-and-derangement.html?_r=0 (accessed January 2014).

Brunstrom, J. M. and Wilkinson, L. L. (2007) Conditioning expectations about the satiating quality of food. *Appetite*, **49**, 281.

Brunstrom, J. M., Rogers, P. J., Burn, J. F., Collingwood, J. M., Maynard, O. M., Brown, S. D. and Sell, N. R. (2010) Expected satiety influences actual satiety. *Appetite*, **54**, 631–683.

Caballero, B. (2007) The global epidemic of obesity: An overview. *Epidemiologic Reviews*, **29**, 1–5.

Campbell, M. (2009) World's top chef 'poisons' diners with additives. *The Sunday Times*, 11 September, p. 29.

Cardello, A. V. (1994) Consumer expectations and their role in food acceptance. In *Measurement of Food Preferences* (eds H. J. H. MacFie and D. M. H. Thomson), pp. 253–297. Blackie Academic and Professional, London.

Cardello, A. V. and Sawyer, F. M. (1992) Effects of disconfirmed consumer expectations on food acceptability. *Journal of Sensory Studies*, **7**, 253–277.

Castriota-Scanderbeg, A., Hagberg, G. E., Cerasa, A., Committeri, G., Galati, G., Patria, F., Pitzalis, S., Caltagirone, C. and Frackowiak, R. (2005) The appreciation of wine by sommeliers: A functional magnetic resonance study of sensory integration. *NeuroImage*, **25**, 570–578.

Chabris, C. and Simons, D. (2011) *The Invisible Gorilla and Other Ways our Intuition Deceives Us.* Harper, London.

Chartier, F. (2012) *Taste Buds and Molecules: The Art and Science of Food, Wine, and Flavor* (translated by Levi Reiss). John Wiley and Sons, Hoboken, NJ.

Chossat, V. and Gergaud, O. (2003) Expert opinion and gastronomy: The recipe for success. *Journal of Cultural Economics*, **27**, 127-141.

Cowen, T. (2012) *An Economist Gets Lunch: New Rules for Everyday Foodies.* Plume, New York.

de Graaf, C., Cardello, A. V., Kramer, F. M., Lesher, L. L., Meiselman, H. L. and Schutz, H. G. (2005) A comparison between liking ratings obtained under laboratory and field conditions: The role of choice. *Appetite*, **44**, 15–22.

Deliza, R. and MacFie, H. J. H. (1996) The generation of sensory expectation by external cues and its effect on sensory perception and hedonic ratings: A review. *Journal of Sensory Studies*, **2**, 103–128.

Deliza, R., Macfie, H. J. H and Hedderley, D. (2003) Use of computer-generated images and conjoint analysis to investigate sensory expectations. *Journal of Sensory Studies*, **18**, 465–486.

Dijksterhuis, G. (2012) The total product experience and the position of the sensory and consumer sciences: More than meets the tongue. *New Food Magazine*, **15(1)**, 38–41.

Dornenburg, A. and Page, K. (1996) *Culinary Artistry.* John Wiley and Sons, New York.

Dunn, E. and Norton, M. (2013) *Happy Money: The Science of Smarter Spending.* Simon and Schuster, New York.

Escoffier, A. (1907) *A Guide to Modern Cookery.* Heinemann, London.

Ferguson, P. P. (2004) *Accounting for Taste: The Triumph of French Cuisine.* University of Chicago Press, Chicago.

Fisher, M. F. K. (2005) The pale yellow glove. In *The Taste Culture Reader: Experiencing Food and Drink* (ed. C. Korsmeyer), pp. 325–329. Bloomsbury, Oxford.

Fleming, A. (2013) What makes eating so satisfying? The Guardian, 25 April, available at http://www.guardian.co.uk/lifeandstyle/wordofmouth/2013/apr/23/what-makes-eating-so-satisfying (accessed January 2014).

Garber Jr, L. L., Hyatt, E. M. and Starr Jr, R. G. (2000) The effects of food color on perceived flavor. *Journal of Marketing Theory and Practice*, **8**, 59–72.

Ge, L. (2012). Why coffee can be bittersweet. *FT Weekend Magazine*, October 13/14, 50.

Gilbert, D. (2006) *Stumbling on Happiness.* Alfred A. Knopf, New York.

Gill, A. A. (2007) *Table Talk: Sweet and Sour, Salt and Bitter.* Weidenfeld and Nicolson, London.

Gill, A. A. (2011) The last supper: 30 courses down, 20 to go. *The Sunday Times*, 31 July, p. 13.

Gopnik, A. (2011) Sweet revolution. *The New Yorker*, 3 January, available at http://www.newyorker.com/reporting/2011/01/03/110103fa_fact_gopnik (accessed January 2014).

Gregory, R. L. (1987) *The Oxford Companion to the Mind*. Oxford University Press, Oxford.

Gustafsson, I.-B. (2004) Culinary arts and meal science – a new scientific research discipline. *Food Service Technology*, **4**, 9–20.

Haden, R. (2005) Taste in an age of convenience. In: *The Taste Culture Reader: Experiencing Food and Drink* (ed. C. Korsmeyer), pp. 344–358. Bloomsbury, Oxford.

Hall, C. M., Sharples, E., Mitchell, R., Macionis, N. and Cambourne, B. (2003) *Food Tourism Around the World: Development, Management and Markets*. Butterworth-Heinemann, Oxford.

Halligan, M. (1990) *Eat my Words*. Angus and Robertson, London.

Hanefors, M. and Mossberg, L. (2003) Searching for the extraordinary meal experience. *Journal of Business and Management*, **9**, 249–270.

Henrich, J., Heine, S. J. and Norenzayan, A. (2010) The weirdest people in the world? *Behavioral and Brain Sciences*, **33**, 61–135.

Humphries, C. (2012) Delicious science. Chefs are teaming up with researchers to create avant-garde dishes. Is 'molecular gastronomy' more than a fad? *Nature*, **486**, 10–11.

Hyman, P. and Hyman, M. (1999) Printing the kitchen: French cookbooks, 1480–1800. In *Food: A Culinary History from Antiquity to the Present* (eds J.-L. Flandrin and M. Montanari), pp. 394–402. Columbia University Press, New York.

Kahneman, D. (2011) *Thinking, Fast and Slow*. Penguin Books, London.

Konings, H. (2009) *Latte macchiato: Trends voor het volgende decennium* (Latte macchiato: Trends for the next decade). Available at http://www.sensefortaste.com/ (accessed January 2014).

Köster, E. P. (2003) The psychology of food choice: Some often encountered fallacies. *Food Quality and Preference*, **14**, 359–373.

Kotler, P. (1974) Atmospherics as a marketing tool. *Journal of Retailing*, **49**, 48–64.

Koza, B. J., Cilmi, A., Dolese, M. and Zellner, D. A. (2005) Color enhances orthonasal olfactory intensity and reduces retronasal olfactory intensity. *Chemical Senses*, **30**, 643–649.

Kühn, S. and Gallinat, J. (2013) Does taste matter? How anticipation of cola brands influences gustatory processing in the brain. *PLoS ONE*, **8(4)**, e61569.

Kurti, N. (1969) The physicist in the kitchen. A transcript from the weekly Evening Meeting of the Royal Society London Friday 14th March. *Proceedings of the Royal Institution of Great Britain*, **42**, 451–467.

Kurti, N. and Kurti, G. (eds) (1988) *But the Crackling is Superb: An Anthology on Food and Drink by Fellows and Foreign Members of the Royal Society*. Institute of Physics Publications, Bristol.

Kurti, N. and This-Benckhard, H. (1994a) Chemistry and physics in the kitchen. *Scientific American*, **270(4)**, 66–71.

Kurti, N. and This-Benckhard, H. (1994b) The amateur scientist: The kitchen as a lab. *Scientific American*, **270(4)**, 120–123.

Lau, K.-N., Post, G. and Kagan, A. (1995) Using economic incentives to distinguish perception bias from discrimination ability in taste tests. *Journal of Marketing Research*, **32(May)**, 140–151.

Lauden, R. (2001) A plea for culinary modernism: Why we should love new, *fast, processed food. Gastronomica*, **1(1)**, February, 36–44.

Lee, L., Frederick, S. and Ariely, D. (2006) Try it, you'll like it: The influence of expectation, consumption, and revelation on preferences for beer. *Psychological Science*, **17**, 1054–1058.

Legrenzi, P. and Umiltà, C. (2011) *Neuromania: On the Limits of Brain Science* (translated by F. Anderson). Oxford University Press, Oxford.

Loewenstein, G. F. and Prelec, D. (1993) Preferences for sequences of outcomes. *Psychological Review*, **100**, 91–108.

Lubow, A. (2003) A laboratory of taste. *The New York Times*, 10 August, available at http://www.nytimes.com/2003/08/10/magazine/a-laboratory-of-taste.html?pagewanted=all&src=pm (accessed January 2014).

MacClancy, J. (1992) *Consuming Culture: Why You Eat What You Eat*. Henry Holt, New York.

MacLeod, T. (2013) Restaurant Story, 201 Tooley Street, London SE1. *The Independent*, 11 May, available at http://www.independent.co.uk/life-style/food-and-drink/reviews/restaurant-story-201-tooley-street-london-se1-8607805.html (accessed January 2014).

Maguire, E. A., Gadian, N. G., Johnsrude, I. S., Good, C. D., Ashburner, J., Frackowiak, R. S. *et al.* (2000) Navigation-related structural changes in the hippocampi of taxi drivers. *Proceedings of the National Academy of Sciences of the USA*, **97**, 4398–4403.

Marteau, T. M., Hollands, G. J. and Fletcher, P. C. (2012) Changing human behaviour to prevent disease: The importance of targeting automatic processes. *Science*, **337**, 1492–1495.

Martin, G. N. (2004) A neuroanatomy of flavour. *Petits Propos Culinaires*, **76**, 58–82.

Martin, N. (2013) Seeing the future through the eyes of your consumers: New perspectives from consumer and sensory sciences. *New Food Magazine*, **16(2)**, 37–42.

McCabe, D. and Castel, A. (2008) Seeing is believing: The effect of brain images on judgments of scientific reasoning. *Cognition*, **107**, 343–352.

McClure, S. M., Li, J., Tomlin, D., Cypert, K. S., Montague, L. M. and Montague, P. R. (2004) Neural correlates of behavioral preference for culturally familiar drinks. *Neuron*, **44**, 379–387.

McGee, H. (1984) *On Food and Cooking: The Science and Lore of the Kitchen* (revised edition, republished 2004). Scribner, New York.

McGee, H. (1990) *The Curious Cook*. Collier Books, New York.

McGee, H. (2006) Foreward. In *Modern Gastronomy A to Z: A Scientific and Gastronomic Lexicón* (originally published as *Léxico Científico Gastronómico*, Editorial Planeta, Barcelona). CRC Press, Boca Raton, FL.

McLaughlin, K. (2011) Cerebral palate. *Wall Street Journal*, available at http://online.wsj.com/article/SB10001424052748703421204576329403800844910.html#ixzz1NYmoWdTK (accessed January 2014).

Meiselman, H. L. (1992) Obstacles to studying real people eating real meals in real situations. *Appetite*, **19**, 84–86.

Meiselman, H. L. (2013) The future in sensory/consumer research: Evolving to a better science. *Food Quality and Preference*, **27**, 208–214.

Michael, R. B., Newman, E. J., Vuorre, M., Cumming, G. and Garry, M. (2013) On the (non)persuasive power of a brain image. *Psychonomic Bulletin and Review*, **20**, 720–725.

Mouritsen, O. G. and Risbo, J. (2013) Gastrophysics - do we need it? *Flavour*, **2**, 3.

Myhrvold, N. and Young, C. (2011) *Modernist Cuisine. The Art and Science of Cooking*. Ingram Publisher Services, USA.

North, A. C., Hargreaves, D. J. and McKendrick, J. (1997) In-store music affects product choice. *Nature*, **390**, 132.

O'Doherty, J., Deichmann, R., Critchley, H. D. and Dolan, R. J. (2002) Neural responses during anticipation of a primary taste reward. *Neuron*, **33**, 815–826.

Park, M. Y. (2013) *A history of how food is plated, from medieval bread bowls to Noma*. Available at http://www.bonappetit.com/trends/article/a-history-of-how-food-is-plated-from-medieval-bread-bowls-to-noma (accessed February 2014).

Pazart, L., Menozzi, C., Comte, A., Andrieu, P. and Vidal, C. (2011) La dégustation "en aveugle", décryptée en IRM fonctionnelle? (Blind tasting, decrypted in fMRI?) *Revue des Oenologues et des Techniques Vitivinicoles et Oenologicques: Magazine Trimestriel D'information Professionnelle*, **38(139)**, 43–44.

Pelchat, M. L., Johnson, A., Chan, R., Valdez, J. and Ragland, J. D. (2004) Images of desire: Food-craving activation during fMRI. *NeuroImage*, **23**, 1486–1493.

Peterson, R. A. and Ross, I. (1972) How to name new brands. *Journal of Advertising Research*, **12 (6)**, 29–34.

Petrini, C. (2007) *Slow Food: The Case for Taste* (translated W. McCuaig). Columbia University Press, New York.

Pickering, G. J. (2008) Optimizing the sensory characteristics and acceptance of canned cat food: Use of a human taste panel. *Journal of Animal Physiology and Animal Nutrition*, **93**, 52–60.

Pine II, B. J. and Gilmore, J. H. (1998) Welcome to the experience economy. *Harvard Business Review*, **76(4)**, 97–105.

Pine II, B. J. and Gilmore, J. H. (1999) *The Experience Economy: Work is Theatre and Every Business is a Stage*. Harvard Business Review Press, Boston, MA.

Pinson, C. (1986) An implicit product theory approach to consumers' inferential judgments about products. *International Journal of Research in Marketing*, **3(1)**, 19–38.

Piqueras-Fiszman, B. and Spence, C. (2011) Do the material properties of cutlery affect the perception of the food you eat? An exploratory study. *Journal of Sensory Studies*, **26**, 358–362.

Piqueras-Fiszman, B. and Spence, C. (2012) The weight of the container influences expected satiety, perceived density, and subsequent expected fullness. *Appetite*, **58**, 559–562.

Piqueras-Fiszman, B., Alcaide, J., Roura, E. and Spence, C. (2012) Is it the plate or is it the food? Assessing the influence of the color (black or white) and shape of the plate on the perception of the food placed on it. *Food Quality and Preference*, **24**, 205–208.

Plassmann, H., O'Doherty, J., Shiv, B. and Rangel, A. (2008) Marketing actions can modulate neural representations of experienced pleasantness. *Proceedings of the National Academy of Sciences USA*, **105**, 1050–1054.

Pollan, M. (2006) *The Omnivore's Dilemma: A Natural History of Four Meals*. Penguin Press, New York.

Pollan, M. (2013) *Cooked: A Natural History of Transformation*. Penguin Books, London.

Poole, S. (2012) *You Aren't What You Eat: Fed Up With Gastroculture*. Union Books, London.

Raudenbush, B., Meyer, B., Eppich, W., Corley, N. and Petterson, S. (2002) Ratings of pleasantness and intensity for beverages served in containers congruent and incongruent with expectancy. *Perceptual and Motor Skills*, **94**, 671–674.

Rayner, J. (2006) 'Molecular gastronomy is dead.' Heston speaks out. Observer Food Monthly, 17 December, available at http://observer.guardian.co.uk/foodmonthly/futureoffood/story/0,,1969722,00.html (accessed January 2014).

Rayner, J. (2008) *The Man Who Ate the World: In Search of the Perfect Dinner*. Headline Publishing Group, London.

Regard, M. and Landis, T. (1997) "Gourmand syndrome": Eating passion associated with right anterior lesions. *Neurology*, **48**, 1185–1190.

Revel, J.-F. (1985) *Un festin en paroles/Histoire littéraire de la sensibilité gastronomiuque de l'Antiquité à nos jours* (A feast in words/Literary history of culinary sensibility from antiquity to today). Éditions Suger, Paris (1st edition 1979).

Roosth, S. (2013) Of foams and formalisms: Scientific expertise and craft practice in molecular gastronomy. *American Anthropologist*, **115**, 4–16.

Rosenblum, L. D. (2010) *See What I am Saying: The Extraordinary Powers of our Five Senses*. W. W. Norton and Company Inc., New York.

Rozin, P. (1999) Food is fundamental, fun, frightening, and far-reaching. *Social Research*, **66**, 9–30.

Sánchez Romera, M. (2003) *La Cocina de los Sentidos* (The Kitchen of the Senses). Editorial Planeta, Barcelona.

Savage, N. (2012) Technology: The taste of things to come. *Nature*, **486**, 18–19.

Schifferstein, H. N. J. (2001) Effects of product beliefs on product perception and liking. In *Food, People and Society: A European Perspective of Consumers' Food Choices* (eds L. Frewer, E. Risvik and H. Schifferstein), pp. 73–96. Springer-Verlag, Berlin.

Schira, R. (2011) Daniel Facen, the scientific chef. Available at http://www.finedininglovers.com/stories/molecular-cuisine-science-kitchen/ (accessed January 2014).

Schlack, L. (2011) Catch cold: London's Chin Chin Laboratorists puts the cool factor back into ice cream. *Hemispheres Magazine*, **July**, 29–30.

Shepherd, G. M. (2006) Smell images and the flavour system in the human brain. *Nature*, **444**, 316–321.

Shepherd, G. M. (2012) *Neurogastronomy: How the Brain Creates Flavor and Why it Matters*. Columbia University Press, New York.

Sherman, L. (2008) Molecular mixology. Available at http://www.forbes.com/2008/07/01/molecular-mixology-cocktails-forbeslife-drink08-cx_ls_0701science.html (accessed February 2014).

Small, D. M. (2012) Flavor is in the brain. *Physiology and Behavior*, **107**, 540–552.

Spence, C. (2010) The price of everything – the value of nothing? *The World of Fine Wine*, **30**, 114–120.

Spence, C. (2012a) Book review of *Neurogastronomy: How the Brain Creates Flavor and Why it Matters* by Gordon M. Shepherd. *Flavour*, **1**, 21.

Spence, C. (2012c) Multi-sensory integration and the psychophysics of flavour perception. In Food Oral *Processing: Fundamentals of Eating and Sensory Perception* (eds J. Chen and L. Engelen), pp. 203–219. Blackwell Publishing, Oxford.

Spence, C. (2013) Multisensory flavour perception. *Current Biology*, **23**, R365–R369.

Steinberger, M. (2010) *Au Revoir To All That: The Rise and Fall of French Cuisine*. Bloomsbury, London.

Steingarten, J. (2002) *It Must've Been Something I Ate*. Knopf, New York.

Stevenson, R. J. (2009) *The Psychology of Flavour*. Oxford University Press, Oxford.

St-Onge, M.-P., Sy, M., Heymsfield, S. B. and Hirsh, J. (2005) Human cortical specialization for food: A functional magnetic resonance imaging investigation. *Journal of Nutrition*, **135**, 1014–1018.

Stuckey, B. (2012) *Taste What You're Missing: The Passionate Eater's Guide to why Good Food Tastes Good*. Free Press, London.

Sutton, D. E. (2001) *Remembrance of Repasts: An Anthropology of Food and Memory*. Oxford, Berg.

Sutton, D. E. (2010) Food and the senses. *Annual Review of Anthropology*, **39**, 209–223.

Tallis, R. (2008) The neuroscience delusion. *The Times Literary Supplement*, 9 April, available at http://tomraworth.com/talls.pdf (accessed January 2014).

Thaler, R. H. and Sunstein, C. R. (2008) *Nudge: Improving Decisions about Health, Wealth and Happiness*. Penguin, London.

This, H. (2005) *Molecular Gastronomy: Exploring the Science of Flavor* (translated M. B. Debevoise). Columbia University Press, New York.

This, H. (2012) Molecular gastronomy is a scientific activity. In *The Kitchen as Laboratory: Reflections on the Science of Food and Cooking* (eds C. Vega, J. Ubbink and E. van der Linden), pp. 242–253. Columbia University Press, New York.

This, H. (2013) Molecular gastronomy is a scientific discipline, and note by note cuisine is the next culinary trend. *Flavour*, **2**, 1.

Van der Laan, L. N., de Ridder, D. T. D., Viergever, M. A. and Smeets, P. A. M. (2011) The first taste is always with the eyes: A meta-analysis on the neural correlates of processing visual food cues. *NeuroImage*, **55**, 296–303.

Visser, M. (1991) *The Rituals of Dinner: The Origins, Evolution, Eccentricities, and Meaning of Table Manners*. London: Penguin Books.

Walsh, J. (2011) Dinner, deconstructed. *The Independent*, 8 April, 11–13.

Wang, G.-J., Volkow, N. D., Telang, F., Jayne, M., Ma, J., Rao, M. *et al.* (2004) Exposure to appetitive food stimuli markedly activates the human brain. *NeuroImage*, **212**, 1790–1797.

Woods, A. T., Lloyd, D. M., Kuenzel, J., Poliakoff, E., Dijksterhuis, G. B. and Thomas, A. (2011) Expected taste intensity affects response to sweet drinks in primary taste cortex. *Neuroreport*, **22**, 365–369.

Wrangham, R. (2010). *Catching Fire: How Cooking made us Human*. Profile Books, London.

Yeomans, M., Chambers, L., Blumenthal, H. and Blake, A. (2008) The role of expectancy in sensory and hedonic evaluation: The case of smoked salmon ice-cream. *Food Quality and Preference*, **19**, 565–573.

Youssef, J. (2013) *Molecular Cooking at Home: Taking Culinary Physics out of the Lab and into your Kitchen*. Quarto, London.

Zellner, D., Strickhouser, D. and Tornow, C. (2004) Disconfirmed hedonic expectations produce perceptual contrast, not assimilation. *American Journal of Psychology*, **117**, 363–387.

2
Let the Show Commence: On the Start of the Perfect Meal

2.1 Introduction

In this chapter, we take a look at just how important the opening phases are to our overall enjoyment of the meal. Surely everyone has been in the situation in which those at one's table are (still) strangers, and you're desperately trying to think of something to say to your dining companions in order to engage them in conversation? At that moment, everyone's looking around, here and there, saying something like: "Yes, this *is* a nice place." If the evening continues like this, you might sneak a look at your watch before any of the food has even arrived at the table.

In order to try and avoid such uncomfortable situations, we are going to take a closer look at some of the innovative strategies that some of the world's top chefs and restaurateurs have introduced into their dining establishments to help their diners break the ice (laugh, even) and consequently enjoy their meal that much more. While great company is certainly no guarantee of a perfect meal, it is definitely an important step in the right direction. As we will see below, the company we keep affects the food choices that we make. Here, one can also ask whether groups or couples enjoy a meal more than a lone diner sitting by themselves (perhaps one of the growing number of people who are nowadays to be seen keeping themselves entertained with their phones; see Anonymous 2013)? More generally, we will look at the effect that our mood has on our perception of food.

The Perfect Meal: The Multisensory Science of Food and Dining, First Edition.
Charles Spence and Betina Piqueras-Fiszman.
© 2014 John Wiley & Sons, Ltd. Published 2014 by John Wiley & Sons, Ltd.

In this chapter, we are going to cover everything from the way in which the serving staff greet you on arrival at the restaurant[1] through to the design of the menu itself. Apart from reflecting the style of the restaurant, these elements can play a surprisingly important role in influencing what we decide to eat, how much we (or someone else, our host?) ends up paying for the pleasure and how much any one of us enjoys the overall dining experience. As surprising as the effects of layout, price and images on the menu on the food and drink choices that we make is the effect of attention from serving staff. Just as surprising, but not in a good way, is how inattentive some servers can be, even in the most formal of dining establishments; their attitude can sometimes end up spoiling your meal.

So, let's start this chapter at the moment at which the expectant (not to mention hungry) diners have just been seated at their table.

2.1.1   Mood food

Should you have the good fortune to dine at Denis Martin's namesake restaurant in Vevey, Switzerland (see www.denismartin.ch; Martin 2007), the first thing that you will notice when you take your seat is the colourful plastic cow sitting centre stage on the starched white linen tablecloth (see Figure 2.1). 'It must be a salt or pepper shaker' is the thought going through the minds of most diners at this two-Michelin-starred restaurant situated on the edge of Lake Geneva (just down the road from Nestlé's Global HQ, not to mention a couple of miles from the house in which Charlie Chaplin lived out his final years). Your coats have been checked, you have been seated and then you wait. And at some point it will likely strike you as odd that there isn't a single piece of cutlery to be seen on the table. Eventually, if you wait long enough – and Denis himself sees this happen most nights – some inquisitive soul will pick the cow up and turn it over in order to investigate matters further. As soon as this happens, the plastic cow will make a low-pitched mooing sound that breaks the hushed and respectful silence that has descended over the expectant roomful of diners. And then the diners at the table laugh. It turns out that what sits there centre stage on each and every table in this temple to the art of 'molecular gastronomy' is nothing other than a cheap children's toy.

You can guarantee that, as soon as the first diner has made their cow moo, those sitting at the other tables won't be able to keep themselves from joining in. In no time at all, and completely unexpectedly, you have a chorus of mooing farm animals and a roomful of laughter. As a diner, you just can't help but chuckle with childish delight. It is an incredibly simple, yet supremely effective, means of breaking the ice. Once Denis hears that chorus

---

[1] Assuming, that is, you have managed to avoid the clutches of the waiter standing outside, trying desperately to lure you in. According to Coren (2012, p. 181), the warmth of the waiter's entreaty for you to step inside is inversely related to the standard of the service that you can expect once seated!

Figure 2.1 The only tableware to greet the expectant diner at Denis Martin's Vevey restaurant in Switzerland. *Source*: Reproduced with permission of Denis Martin.

of laughter, hovering out of sight in the kitchens, he knows that it is time to send out the first dish.

But what is really going on here? What is the underlying rationale behind the plastic cow? Denis tells how, a few years ago, stiff businessmen and -women would be sitting in hushed and respectful silence at the tables of his restaurant.[2] The problem was that they did not seem to be savouring the food as much as he felt sure that they should. It was as if those who arrived in an uptight and formal state of mind couldn't really relax and appreciate either the food or the overall dining experience. At some point, Denis realized that the diners' perception of his food was being coloured, at least in part, by their mood on arrival at his restaurant. At some level, we all know this to be true. After all, how can one enjoy a great meal while arguing with one's partner? Doesn't the food and wine often taste so much better when we are relaxed and happy sitting by the Mediterranean on our summer holidays than when we are back home on a cold winter's evening sitting in front of the fire. (This phenomenon is known as 'Provencal Rose Paradox' and is covered in more detail in Chapter 9.)

---

[2] Of course, the fact that the restaurant is housed in what looks like a museum charting the history of the sewing machine probably doesn't help much with the atmosphere.

Having identified the problem, what was to be done about it? Denis knew that he had to do something to break the ice and improve the mood of his diners before they took their first bite of his delicious food. The answer, one that presumably works best in Switzerland, was the black and white spotted cow standing daintily on a bright red plastic pedestal. Most diners probably don't think too much about what is going on here. It just seems like a playful gesture, nothing more. But from a scientific point of view, this casual intervention raises the following intriguing question: does our mood really exert that much of an effect on how we perceive the food that we eat? Apart from breaking the ice at the table, and hopefully bringing a smile to the diners' faces, what effect does that plastic moo-cow (and the positive mood that it is designed to induce) actually have on a diner's perception of the food, not to mention on their eating behaviours?

Rather than using a noisy plastic farm animal, researchers in the laboratory setting often compare people's perception of food or drink after artificially inducing either a positive or negative mood. This is normally achieved by one of several means. Some researchers have their participants watch short film clips (Macht and Mueller 2007; Platte *et al.* 2013). For example, in Macht and Mueller's study as in many others, a 3-minute-long video sequence taken from the movie *The Champ* was used. In the researchers' favourite scene, a boy is seen crying at the death of his father. Another short clip from *When Harry Met Sally*, specifically the moment when Harry and Sally discuss an orgasm at a restaurant, was used as the uplifting clip.[3]

Once the mood has been set (or set up), the question has to be asked: does the food and drink really taste better if we happen to be in a good mood? Seo and Hummel (2012) examined this very question using two odours as stimuli (one a pleasant floral odour and the other which had an unpleasant vinegary smell). They found that the experimentally induced mood had a significant effect on people's ratings of the odours. Both were rated as smelling significantly better after (and while) listening to the sound of a baby's laughter than when listening to a baby crying. In fact, by now, several studies have independently arrived at a similar conclusion regarding the effect of mood on the rating of food, as we'll see below. Platte *et al.* (2013) recently demonstrated that those individuals who could be characterized as slightly more depressive found it harder to discriminate the fat content of a food after watching an emotional movie clip (once again, they used the scene from *The Champ* as the sad clip and a little something from *An Officer and a Gentleman*, specifically the

---

[3] These movie fragments have been shown to reliably elicit sadness and amusement in viewers (Gross and Levenson 1995). Other researchers have chosen to use the sounds of babies laughing or crying (hasn't everyone experienced how annoying the sound of a crying baby can be at a restaurant or in a plane?) or else positive and negative sound clips or pieces of music. For anyone who wants to try this kind of mood induction at home, we'd recommend Samuel Barber's *Adagio for Strings* as a supremely depressing sonic selection. According to Huron (2007), and he should know as he has been around the world testing all sorts of music on anyone he could find to listen to it, this is the universally most depressing piece of music known to mankind.

scene where the protagonist gets together with his girlfriend, for the uplifting clip) compared to after watching a neutral emotion clip.

So, a diner's mood really can influence his/her perception of an aroma, but does it also influence their ratings of real food and drink items, not to mention how much they ultimately decide to consume? Haven't we all watched the stereotypical movie scene of a woman slowly emptying the tub of ice cream in her lap in order to try and drown her sorrows? To answer this question, Yeomans and Coughlan (2009) gave a group of women snack foods to try while they watched 20-minute film segments that had been pre-selected to elicit a mood that was positive, negative or neutral. While no significant effects were found in terms of the participants' hedonic ratings of the food as a function of the film clip that they had been watching, most of the participants ate (snacked) more after watching the positive film clip and less after watching the negative clip. Those who ate more following the negative film clip tended to score higher on measures of both restraint and disinhibition (that is, people who try to restrict themselves but who, most of the time, end up eating without inhibition).

In a similar study it was found that male participants consumed more chocolate, and evaluated it more favourably, after watching a scene from a happy as opposed to a sad movie (Macht *et al.* 2002; see also Garg *et al.* 2007). It would therefore appear that, in the context of a restaurant, restaurateurs ought to do whatever they can in order to ensure that the diners are in as good a mood as possible to keep the plates returning to the kitchen clean and the orders rolling in![4]

Researchers have argued that one explanation for the effects of a diner's mood on his/her perception of food and drink might be the feelings-as-information theory and what is known as the phenomenon of mood misattribution (e.g. Greifeneder *et al.* 2011; Schwarz 2012; compare this to earlier research on halo effects, e.g. Thorndike 1920). In a series of elegant experiments, Schwarz and Clore (1983) demonstrated that people use their mood as a source of information (or cue) when making judgments, even if these judgments have nothing to do with the source of their mood. For example, people tend to rate their general long-term life satisfaction more positively when they are interviewed on a sunny as opposed to a rainy day. Similarly, the positive mood induced by the background music (or by the cow sitting on the table at Denis Martin's restaurant), the attitude of the serving staff or the company we keep at the dinner table (as we'll see next) can all influence a diner's perceptions and evaluations of food and drink. Having briefly covered the effect of mood on food, let us now move on to look at the social aspects of the meal.

---

[4] Here, it is worth noting that extremes of mood can also affect a person's ability to detect certain basic tastes.

2.2   On the social aspects of dining

2.2.1   Waiting staff

The way in which a diner is greeted and served by the waiting/serving staff can have a significant impact on their mood and on their overall experience of a meal. The wait staff ideally represent something of the style of the restaurant and will not only answer any questions that the diner may have about the menu or where the toilets are located,[5] but will also present the specials of the day and answer questions in a manner that may nudge us more towards choosing one menu item versus another. After all, who hasn't been in the situation in which the waiter says with a big smile: "Allow me to tell you that we also serve our award-winning pastry chef's special, which isn't on the menu. It's an Italian cassata made with gold leaf and Irish cream, served with a mango and pomegranate compote, a champagne sabayon and a chocolate sculpture of a stilt fisherman perched on top." (Note that the prices of those 'special' dishes that we hear about from the waiter's mouth are often not mentioned.) Or they may finish by saying: "This is his signature dessert."[6] (Very likely one of the most expensive items on the dessert menu.) And we timidly say without looking him/her in the eyes: "Hmm, it sounds really tempting... but no, I think this time I'll go for the molten almond cake instead... Yes. Thanks." We even ask ourselves why we feel a bit uncomfortable about having chosen another item instead of the one that was so earnestly recommended to us. We sincerely hope then, after all, that at least one of the others dining at our table will choose it or, if not the special, then at least the chef's signature dish. After all, the waiter really ought to be compensated for the effort that went into learning such long descriptions of the dishes by heart and having to repeat it for the benefit of the diners sitting at each and every one of the tables that they cover!

However, often the diner will be tempted to follow the waiter's recommendations. But is the dish that is recommended in such a positive manner liked more? At the training restaurant in Bournemouth University, Edwards and Meiselman (2005) tested the effect of positive and negative verbal recommendations on the diners' rating of the food. Imagine the waiter coming to your table and saying something like: "To assist you in your selection, could I just

---

[5] For example, the waiting staff at Per Se in New York are specifically instructed not to point diners in the direction of the toilet if asked, but instead to lead them there in person; every detail is designed to add that personal touch to the diner's meal experience (Damrosch 2008).

[6] While it can feel constraining for a chef to always have to include their signature dish (that is, the dish for which they are most famous) on the menu, it is important to note another of the roles that are played by the items of the menu: to give the diner a feeling of familiarity. This crucial element can be lost in those restaurants that change everything on the menu too often (or in those restaurants like The House of Wolf in London, where the pop-up chef changes on a more-or-less monthly basis), but may be an important trigger in determining whether diners really feel like going back to a particular restaurant or not (see Dornenburg and Page 1996, pp. 257–259). Having the chef's signature dish on the menu can certainly help in terms of promoting just such a sense of familiarity.

say that 'dish X' has been particularly popular this week?" That's the positive condition. In the negative condition, the waiter would instead say something like: " … has not been particularly popular … " There was also a neutral condition, where the waiter made no specific comment regarding the popularity of the dish. Perhaps unsurprisingly, twice as many of the diners selected 'dish X' in the positive condition than in the negative condition (40%, 41% and 19% of the diners in the positive, neutral and negative conditions, respectively). However, when the diners had to rate how much they liked that dish on a 9-point scale, the average was the same (around 7.6) in all three cases. While the results of this study would appear to suggest that the wait staff can certainly discourage a diner from choosing a specific menu option, their positive praise doesn't seem to enhance the taste of the food.

At Dinner, Heston Blumenthal's London restaurant, the dishes on the menu have the date of when the dish was created; next to the name of each dish one sees something like "(c. 1660)". If the curious diners ask what exactly the number means, they will be told that it is the approximate date of when the dish first appeared. This will also provide the opportunity for the waiter to explain to the diner a little more about how the dish was created back in the day and how Blumenthal and his team have reinterpreted the original recipe.[7]

The serving staff also have a number of other subtle strategies up their sleeves to please the diner, even if it's not suggesting which dish you should order. For instance, they will sometimes approach the diner with a phrase like: "Hi, my name is John, I'll be taking care of you this evening". This makes them appear friendly and polite and makes the diner feel more empathetic towards them. There have even been reports of the staff at one restaurant chain in the UK being explicitly taught the art of 'subtle flirtation' to try and ensure a better interaction with the diner (Rowley 2010). In a study conducted at a branch of the Charlie Brown's casual dining restaurant chain in Southern California, adopting such a strategy increased the average size of the tips by a not-insignificant 55% (Engle 2004; see also McDermott 2013). Sometimes, the wait staff may even greet you by your first name (as they make sure to do at Union Square Cafe, Gramercy Tavern, Eleven Madison Park, among other restaurants in New York run by the über-successful restaurateur Danny Meyer; Meyer 2010). The idea here is to make the diner feel special, and some diners might think that they are paying for the privilege of having the wait staff pretend that they are a big shot. Of course, there are others who might find it grating, seeing it instead as an act of fake hospitality (and, on occasion, even fake friendship). There are undoubtedly important cross-cultural differences here (Goodwin and Verhage 1989). However, it is of course possible that the waiter calls us by our name because (s)he recognizes us (and supposedly we should recognize them too!).

---

[7] When we were last in the restaurant, the waitress joked: "Don't worry, it's not the number of calories!"

That said, there can be a fine line between the maître d' recognizing you, and perhaps remembering your table preferences, and a kind of creepy sense that you might get should the staff appear to know too much about you (see Feiler 2002; Meyer 2010). At many a top restaurant, the waiting staff are taught to be especially observant and attentive without ever being invasive of a diner's privacy. Imagine yourself at The Fat Duck, for example. You order some wine (nowadays there is only a tasting menu, so no food decisions have to be taken), take a nibble of your bread roll, and then decide to take a little trip to check out the washrooms. (How exactly does a modernist washroom look, you wonder? Do they pump out the smell of roast coffee and the sound of kitchen noises, as Heston tried when he was parachuted in to help save the Little Chef roadside restaurant chain in the UK? see 'Big chef takes on Little Chef', http://www.channel4.com/programmes/big-chef-takes-on-little-chef, 2013.) When you come back, you might notice that the waiter has refolded your napkin (or perhaps changed it for a new one). Danny Meyer takes it one step further. He says that he worries if ever he notices that a diner's gaze isn't focused squarely on the table (see Meyer 2010).

Wurgaft (2008) describes a trip to another of the world's top restaurants, The French Laundry in California. There, the waiters are trained to replace the warm brioche on your side plate should you let it get too cold before eating it (perhaps because you have been recounting some witty anecdote). Meanwhile, in a number of restaurants such as Ricard Camarena's Arrop in Valencia, Spain, when the waiter comes around a second time with the assorted basket of freshly baked breads, they will have remembered exactly which type of roll you had last time.[8] The waiter deliberately asks you whether you want the same one (pointing with the tongs at the type you had before) or another one for a change (and all that without you having left even a crumb on the plate).[9]

> *"Heston's grub is great – but so what if your date is ugly? … My point, I suppose, is this. Food is only a small part of what makes a dining experience great. Acoustics are just as important. So is lighting, especially if you have an ugly date. But by far and away the most important thing is the company."* (Jeremy Clarkson 2012)

2.2.2 The company

Eating and drinking are fundamentally social activities, and always have been, at least according to Martin Jones (2007), an archaeologist based at

---

[8] Of course, anyone who is being offered a second bread roll obviously either hasn't read, or perhaps simply doesn't agree with, Coren's (2012, pp. 72–73) line on this subject that, whatever you do, you should never eat the bread when you go out for dinner.

[9] What's perhaps a little ironic here is that Wansink and Linder (2003) have reported that within 5 minutes of leaving an Italian restaurant, more than 30% of the diners they questioned were unable to remember how much bread they had eaten. More surprising still, one in 10 of the diners didn't even remember that they had eaten any bread at all! All the more reason to skip the bread.

Cambridge University. Certainly, the food never seems to taste as good when dining alone at a restaurant. The rise of both fast food culture and specialty coffee shops in recent years has, in part, been down to the kinds of social interactions that such venues facilitate: a family meal (of sorts) in the former case (James 2005, p. 378; Finkelstein 2008) and a relaxed place to chat and socialize in the latter (Luttinger and Dicum 2006). Even the field of gastronomy itself may have developed, at least in part, as a means of bringing people together (see Petrini 2007, p. 8). As Alice Waters, the famous American cook, once put it: *"Eating is something we all have in common. It's something we all have to do every day and it's something we can all share"* (cited in Sonnefeld 2003, p. xii).

Researchers have now started to uncover some of the psychological principles that underlie the more social aspects of the meal (see Mennell *et al.* 1992; Sobal 2000; Logue 2004). Of course, social dining throws up its own set of social/etiquette problems. Who gets to order the wine? Who gets to choose their entrée first? And what happens if that delectable tasting menu is only available for the whole table, and not for individual diners? US researcher Dan Ariely has started to assess how these decisions affect a diner's subsequent enjoyment. Apparently those who get to order first when dining at the restaurant tend to enjoy their food more than those who choose later (Ariely 2008; see also Nowlis *et al.* 2004; Tanner *et al.* 2008). In one representative study, Ariely and Levav (2000) went to a bar and offered groups of drinkers a taste of one of four new beers (Copperline Amber Ale, Franklin Street Lager, India Pale Ale and Summer Wheat Ale). Each of those sitting at the table was allowed to choose one of the beers to taste. The drinkers then had to rate how good it tasted. The results showed that whoever chose first liked their beer more, on average, than the rest of the group (amounting to a difference of around 10%). Interestingly, no such difference was observed when the drinkers placed their orders by marking a piece of paper instead (i.e. when the normally public act of selection was turned into a very private one). Ariely explains this result in terms of individuals within a group 'taking the road less travelled', sacrificing the chance of choosing their preferred beer in favour of a less-liked beer and thus leading to increased variety among the group.

Intriguingly, subsequent research has shown that the tendency to order something different from those who have ordered already correlates with a personality dimension known as the 'need for uniqueness' (Ariely and Levav 2000). While it is apparently the case that diners generally like to choose something different from those who have already ordered, adopting this strategy can sometimes lead an individual to select something that they might not otherwise have chosen alone (or when ordering in secret/private), and hence potentially end up enjoying the experience a little less. There seems no reason to think the same sequence effects wouldn't also affect a diner's enjoyment of their meal when dining with friends or colleagues.

It has been well established that the presence of other people influences not only what we eat, but also how much we consume. Furthermore, those around us at the dinner table can also influence how much we like whatever it is that we are eating, should they start pulling some extreme facial expressions such as that of pleasure or disgust (see Barthomeuf *et al.* 2009). People also adapt to what, and how much, others at the table are eating (Polivy *et al.* 1979; Goldman *et al.* 1991). If we see that others are only nibbling on a couple of canapés or sandwich triangles at a social event, very likely we'll refrain from devouring the whole tray even if we happen to be ravenous (but that isn't to say that we won't keep a close watch on the food out of the corner of our eye). People also exhibit a tendency to mimic the food choices of others (Tanner *et al.* 2008). As one would expect, eating with those who are familiar to us changes the dynamics of the dining situation and can lead to an extended meal (Bell and Pliner 2003). The more people there are at the dinner table, the longer we stay (Sommer and Steele 1997). de Castro (2000) reported that we eat around 35% more with one other person present, 75% more when dining with three others and we eat nearly twice as much when the number at the table reaches 7![10]

Elsewhere, it has been shown that how much more we eat depends on who exactly it is that we happen to be dining with (de Castro 1994). For example, there were significant differences between the mean meal sizes (measured in terms of the number of calories consumed) for both males and females. For all comparisons, meals eaten alone were significantly smaller than those eaten with others (no matter whether the others were of the same or opposite sex; de Castro 1994). On average, male diners were seen to eat 36% more in social conditions than when eating alone, whereas the female diners ate 40% more. That said, the two sexes responded somewhat differently to the presence of companions of the same and opposite sex. While meal sizes were equivalent for males eating with either males or females (though this apparently depends on whether they are in their first date or not; see Mori, Chaiken and Pliner 1987), females ate significantly more when eating with males than with other females. In addition, in comparison to the meals eaten with co-workers or other people whom we do not know very well, the meals eaten with one's spouse and family tended to be larger and eaten more rapidly, while meals eaten with friends were larger but tended to be of a longer duration.

While some may want to defend the notion that "a 'proper' or 'ideal' meal has to be eaten with others" and that "eating alone is devalued and is not considered a 'real' meal" (Sobal and Nelson 2003), there is actually little evidence to support the claim that the food tastes any better when we dine in company. Social interaction can certainly make the dining experience more enjoyable

---

[10] That said, for those women going on a first date it might be worth bearing in mind that your date will tend to be attracted to heavier women at the start of the meal (i.e. when he is hungry). However, during the course of the meal his preferences will start to shift towards lighter women as his stomach fills up, at least if research by Swami and Tovée (2006) is anything to go by.

(Sobal 2000), but we would rather say that it depends on whether we happen to enjoy the company (after all, isn't bad company sometimes worse than eating alone?). To conclude: while we all tend to enjoy food more when dining in company (than when dining alone), it can also throw up its own challenges.

> "*The mind knows not what the tongue wants.*" (Howard Moskowitz, quoted by Gladwell 2004)

2.3 On the design of the menu

The menu can be considered as the '*carte de présentation*' for any restaurant: either explicitly or implicitly, the menu announces the style of the restaurant, its taste, its origins (or ethnicity) and its status (not to mention its pretensions). But the function of the menu goes much further than that (Kelson 1994). Have you ever thought about what cues the weight of the menu may subliminally be sending to your brain? How about the price of the dishes themselves, or the font in which the menu items are printed? Do any of these factors contribute to a sense of quality or, perhaps, luxury? What is certain is that they help to set the diner's expectations. As we have seen already (see Chapter 1), managing the diner's expectations plays a major role in terms of delivering on the promise of the perfect meal.

With such a long list of items, we can spend a while looking at the menu in our hands. Did you ever stop to wonder what the ideal number of items on a restaurant menu might be? (Even if you haven't, you can be rest assured that the psychologists out there have.) According to the results of one study by a Professor John Edwards of Bournemouth University, it turns out that there is an optimum number of items per section, at least for diners, when it comes to the listing of the dishes on the menu. In fast-food establishments, most people apparently want to have a choice of six items per category (the typical categories being starters, chicken, fish, vegetarian and pasta dishes, grills and classic meat dishes, steaks and burgers, and desserts). By contrast, in fine dining restaurants the ideal for diners is apparently seven starters, ten main courses and seven desserts. Having fewer items on the menu can leave the diner feeling that there simply isn't enough choice (Fleming 2013). Any more, and there is a danger that the diner will start to become disconcerted with all the difficult decisions that they are being forced to make (and perhaps also demotivated; see Iyengar and Lepper 2000; Iyengar 2010).[11] That said, researchers have been working hard to figure out how to group items so that it doesn't feel like there are quite so many options to choose from, preventing the diner from feeling overburdened.

---

[11] Forcing people to make such difficult choices between tasty-looking menu options has been shown to result in increased brain activity in the frontal parts of the diner's brain; for more on the neuroscience of menu choice, see Chapter 3.

In this section, we will take a look at those functionally subliminal elements in the design of the menu that can help to influence a diners' expectations, their food choice (unless, of course, one happens to be dining at a restaurant where there is a fixed tasting menu) and general appraisal of a restaurant, all without them without even being aware of it and long before any food is tasted. We will delve into the fascinating science of menu design. Does the justification (e.g. left, right or centred) of the prices on the menu affect what a diner chooses? We will look at the latest research suggesting that there are hotspots on the menu (that is, positions where the diner's eyes are more likely to come to rest). We are uncertain whether, when one has a rumbling stomach and one's mind remains undecided, such functionally subliminal suggestions designed to make us narrow down the array of mouth-watering options that we consider really helps; true, we might save ourselves a few moments of indecision, but then again we might well end up choosing something that we wouldn't otherwise have ordered had we gone on anything other than gut feel. Note that there is a lot more to be said about the names and sensory descriptions on the menu, but that is a topic that deserves its own chapter (see Chapter 3).

> "*A star is a popular, high-profit item; in other words, an item for which customers are willing to pay a good deal more than it costs to make. A puzzle is high-profit, but unpopular. A plowhorse is the opposite – popular yet unprofitable. Marketing consultants employed to assemble many of the most high-profile restaurants' menus try to turn puzzles into stars, nudge customers away from plowhorses, and convince everyone that the prices on the menu are more reasonable than they look.*" (Poundstone 2010a, p. 160)

2.3.1 Scanning the menu: 'Oysters, steak frites, field greens, oysters ... '

Most mainstream strategies for menu design focus on the presentation of the content, that is, how to draw a diner's attention towards targeted items or categories on the menu. While it is undoubtedly the case that diners will not order what they cannot see, it is still presumptuous to assume that simply by increasing the amount of visual attention that a diner pays to a particular menu item will significantly increase the chances of their ordering it (see Carmin and Norkus 1990; Reynolds *et al.* 2005). Note that it has been estimated that diners spend only around three minutes looking at the menu (Hedden 1997).[12] Nevertheless, it can still be worth ensuring that the diner notices the existence of specific items by drawing their attention to them or by making them

---

[12] Although perhaps nowadays diners take longer in venues where the menus are more complex/descriptive.

more prominent on the menu. This is precisely what popular menu design recommendations often focus on. For example, items earmarked for increased promotion through visual design are usually placed at the top or bottom of a category list, or else placed in one of those 'sweet spots' on the menu where a diner's eyes[13] pass through most frequently (see von Keitz 1988).

These design recommendations are based on two well-known effects in the field of experimental psychology: one is the serial position effect, commonly referred to as the rule of primacy and recency (e.g. Miller 1992; Ditmer and Griffin 1994). This refers to a person's overall ability to recall the first (primary) and last (most recent) items in a list more easily than any of the others. Often, the literature on menu design would appear to suggest that the most common scanpath in a two-page menu is that shown in Figure 2.2. According to industry convention (Main 1994), the most desirable locations on the menu for those items that the restaurateur really, really wants to shift lie at positions 1 (primacy), 7 (recency) and perhaps position 5 (where the diner's gaze pattern would be expected to pass through most frequently).

Another oft-cited gaze pattern among menu design practitioners is that proposed by graphic designer William Doerfler (see McVety *et al.* 2009). He believes that the area where most people will end up focusing their attention on a two-page document is anywhere in the upper-right section as marked by a diagonal line from the top-left corner through to the bottom right. That is, the area just above the middle of the right-hand page. One could easily imagine Keith McNally having been familiar with such research when he was planning the design of the menu for his New York *brasserie* Balthazar (http://balthazarny.com/menus/lunch.pdf), since the Oyster bar section is placed on the upper-right of the menu. Just mere coincidence you think?

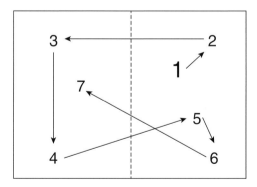

Figure 2.2 Diagram showing what is, supposedly, the most common scanpath when it comes to the two-page restaurant menu. *Source*: Yang, 2012

---

[13] A 'scanpath' is the name given to a series of eye movements as a person shifts his/her fixation between various points during the viewing of a stimulus.

"Menus, he says, employ a litany of psychological tricks to ensnare the unwary diner. With a combination of pictures, bold fonts and careful positioning of items, the savvy restaurant owner will have you parting with your cash faster than you can say 'Hey! It's still 10 days until pay-day!'" (William Poundstone, quoted in Sharp 2009)

However, it is important to bear in mind that there is something of a controversy concerning the results of eye-tracking studies in this area that have sometimes revealed different pattern of results. The Gallup Organisation (1987) published a study in which it was reported that a diner's scanpath when inspecting a restaurant menu is more like what is seen when a person reads a book (that is, they go from top to bottom on the left page, and then do the same thing on the right page; see Yang 2012). Yang recently obtained a very similar pattern of results (that is, a book-like pattern of gaze shifts) when using the latest in eye-tracking technology.[14]

By now, you are probably asking yourself: 'but doesn't the scanpath depend on many other factors as well, such as the order in which the menu categories (breads, salads, pastas, entrées, etc.) are organized?' That is undoubtedly something to bear in mind. In fact, most of the research that has been conducted to date has been based on menu items ordered from lighter dishes (starters, salads, etc.) through to heavier courses (pasta, meats). This leaves us wondering whether, if the order of these categories were to be altered on the menu, the pattern of eye movement would still be the same. Or do diners always systematically go through the menu in the order in which courses are normally presented, and only then move their gaze around here and there? The answer to this question is still being debated by practitioners in the field. However, findings regarding the effect of item positioning would appear to be somewhat controversial. For example, a couple of studies that directly tested such effects suggested that there was no statistical relationship between the position of the items and their selection. As Bowen and Morris (1995, p. 9) conclude *"This study … provided little support for the proposition that the menu design can influence the sale of selected items"* (see also Kincaid and Cursun 2003).

2.3.2   'This dessert is literally calling me'

The other psychological effect on which menu design recommendations tend to be based is the Von Restorff effect. This refers to a person's ability to recall distinctive items from a list, namely those that are salient or distinctive with respect to the rest of the presented items (Hunt 1995). As applied to

---

[14] Of course, if a person knows that their eye movements are being monitored, there is always the worry that that knowledge may change how they inspect whatever it is that the psychologists happen to have them looking at.

items on the menu or even to entire menu categories, some of the tactics that have been suggested to increase distinctiveness include the use of salient font colour and size, texture, the inclusion of photographs of the food items, boxes and highlighting (Stoner 1986; Hug and Warfel 1991), among a host of other techniques (Panitz 2000). As an example (or paradigm) of menu design, let's take the *à la carte* menu at The Delaunay, one of the latest restaurants from Corbin and King[15] in London's Covent Garden (http://www.thedelaunay.com/menus). A good half-dozen boxes highlight the options that the restaurateurs presumably want you to order, from (at the time of writing) some eggs starting at £6.75, cheese at £9.75 or a comforting banana split at £7.75. Note the signs of the chef (most probably Austrian or German) imploring you to order the dish of the day (*tagesteller*, if you will) or some sausage of breaded meat; if you have the menu in front of you it's impossible not to notice this. Perhaps, if you're there on a romantic date, you might be tempted to order the suggested dessert Salzburg Soufflé[16] which comes in at "£8.00 per person" for two (any reason not to quote a simple £16?). However, going back to the theory, whatever the source(s) of distinctiveness might be von Restorff demonstrated that vividness or perceptual salience was not a prerequisite for improved memorability (Hunt 1995).

While the highlighted items on the menu will certainly draw our attention and, on occasion, seem to call out to us to be chosen as if we were Alice in Wonderland (Carroll 1865), what is not so clear at the present time is what effect they actually have on the food choices that we make. In their study, Bowen and Morris (1995) redesigned the menu of Eric's, a restaurant at the Hilton Hotel, located on the University of Houston campus. The restaurant concerned had recently introduced a salad bar and a corresponding section in the menu. However, the salad bar was not as popular as the manager had hoped. Bowen and Morris focused their efforts on trying to enhance the visibility of this particular menu item in order to test whether this change could be used to nudge diners toward the salad. The location was changed from the top of the left-hand page to the top of the right-hand page, descriptive information was added, a double-line box was drawn around the item, a small illustration was included in the corner and the font size of the title was increased. However, after five weeks of the revised menu being handed out to diners, the sale of salads had not increased significantly. While these researchers had previously observed that this kind of menu treatment might have some effect in a fast-food setting (Lorenzini 1992), they ended up concluding that their disappointing results in the restaurant setting might have been attributable to either one of two reasons: (1) menus in full-service restaurants are somewhat more complex than in fast-food joints and diners are not in such a rush to choose; and (2) since the specials sold extremely well, and these items are not included

---

[15] The partnership of Chris Corbin and Jeremy King has resulted in some of London's most iconic restaurants such as Brasserie Zédel, Colbert and The Wolseley.

[16] What is more, all these suggestions are highlighted in red.

in the menu but only described by the service staff, the layout of the menus in full-service restaurants may have less persuasive power than the waiter. Our suggestion here would be that the menu should be used as a complimentary tool by the waiter in order to be maximally effective.

Elsewhere, Reynolds *et al.* (2005) designed a menu boxing two of the items (while at the same time placing them in the supposed sweet spots on the menu) to check whether they would increase the number of diners choosing those items during 11 sittings at Rhapsody, a restaurant on a university campus. The result? This strategy had no effect whatsoever on the sales of the boxed items! We guess that further research is needed here in order to confirm the robustness of this null effect; perhaps the boxed dish (or its name) was simply not appealing enough (see the next chapter).[17] What does seem clear, though, is that the persuasive abilities of the waiting staff are much more powerful (if used wisely) than any other strategy that the restaurateur can implement on a piece of paper, as we'll see below.

> "*There's a big taboo about pictures in classy restaurants – you wouldn't normally get them in the classy places.*" (Poundstone 2010b)

2.3.3   Images on the menu

One might also think that our desire for one or other item on the menu, and our ultimate choice, can be modulated by including a photo of the dish or some other related pictorial representation. Indeed, some menus have drawings, whether the restaurant logo itself (check out the happy flying pig logo of the St. John restaurant in London, https://www.stjohngroup.uk.com/) or as watermarks, while others include pictures of the food items themselves. (Note that in fast-food joints, "*the more they want you to buy something, the bigger they make the image on the menu board*"; Freedman 2013). But do diners need (or even want) to see the images of the items on the menu itself? Sometimes it is undoubtedly helpful if one wants to avoid ordering something that happens to sound much nicer than it looks (think of sweetbreads; we'll discuss this in the next chapter). On other occasions, however, most probably at high-end restaurants, you might rather prefer to be surprised by the dish than by first seeing it in all its technicolour glory on the menu.[18] Showing pictures of the meals on the menu tends to make a restaurant look cheap (strange, given how

---

[17] Yang (2012) takes the available data to suggest that while there might not be any especially 'sweet spots' on a restaurant menu, there may be what she calls 'sour spots', that is, places where the diner's eyes don't loiter. She suggests these can be found at the bottom of the menu page.

[18] Of course, that said, you might well have seen the images online many times before or else been 'instagrammed' by friends and colleagues.

much our brains seem to like looking at food and how they increase the blood flow to certain parts of the brain).[19]

Verma *et al.* (1999) carried out a field study at Chicago's O'Hare International airport with groups of consumers from four different nationalities. Those who were cajoled into taking part were encouraged to look at a card containing five different restaurant options. The options provided information about the names of the restaurants, the type of food that they served and, among other details, whether or not their menus included pictures. While the overall results were somewhat mixed, it did seem as though the English speakers didn't like the idea of restaurants incorporating pictures of food on their menus (certainly in the case of burgers or pizza). That said, they didn't seem to mind it so much when the food served at the restaurant consisted of hot dogs or deli items instead. By contrast, the Japanese-speaking participants liked the idea of menus containing food pictures. The latter results might be attributable to the fact that the Japanese, or Asians in general, are simply more used to having pictures of the food displayed on their menus. Indeed, any *gaijin* who has had to face an Asian menu by themselves is only too glad of pictures, avoiding the embarrassing situation of having to go outside the restaurant with the waiter and point out what they want from the window display.[20]

Another interesting question to ask here is whether the decorative pictorial elements on the menu have any effect on the food choices that we make? Guéguen *et al.* (2012) conducted an intriguing study in a restaurant to figure this out. The researchers designed three menus, all identical except for the fact that they incorporated differently themed watermarks: one related to the seaside, another to the countryside and the third was neutral. On the menu that had been designed to prime the sea, different fish appeared on the front and a boat on the back of the menu. The countryside menu included various farm animals (cows, pigs and sheep) on the front of the menu and a country landscape on the back. In the neutral condition, tables and chairs appeared on the front and a kitchen cabinet on the back. Of the possible choices (salads, meat, fish and dessert), 69% of those in the countryside condition selected meat, while only 14% selected fish. Compare this with what happened when the diners ordered from a seaside-themed menu: only 42% of the diners selected meat while the percentage of diners choosing fish rose to 43% (crucially, the results of these two conditions were significantly different). Meanwhile, the percentages of diners ordering meat and fish from

---

[19] It doesn't seem that far-fetched, then, to think about the possibility of having QR codes embedded in the menu (perhaps at the back, since at the moment they are not very appealing to look at). This would provide the curious diner with the opportunity to scan the codes and see what the dish they are thinking of ordering actually looks like to avoid disappointment, while keeping the menu itself clear of images.

[20] The alternative is to visit a Yo Sushi-type restaurants that are popular in Japan, where you simply grab the food from the conveyor belt as it passes by in front of your very eyes.

the neutral menu were 65% and 22%, respectively (not significantly different from the countryside-themed menu). Taken together, these results hint at the efficiency of certain primes in terms of influencing the behaviour of diners in a real-life restaurant setting. This confirms the presumption that the subtle sensory cues present in the customer's immediate environment can influence their behaviour, even if the diner is not necessarily aware of them.

2.3.4   On a diet? Does nutritional information help?

Another sort of information that is increasingly being provided in food descriptions on restaurant menus these days (mainly in fast-food joints), and which can bias the food choices that the diner makes, relates to the nutritional information and general healthiness of a dish (although it can often be difficult to figure out quite what counts as healthy in a food context; Leake 2009; Shooter 2011; Blythman 2012) as well as the list of ingredients (see Kahkonen and Tuorila 1998; Goerlitz and Delwiche 2004). These kinds of labels, at least if they are noticed,[21] can influence people's food choices, how they rate the consumption experience and even how much they end up eating. Such results have been obtained from multiple studies, conducted both in the laboratory and, more importantly, in a variety of real-world settings (Kozup *et al.* 2003; Wansink *et al.* 2004). Indeed, a growing worry nowadays among many diners concerns the number of calories that may be packed into the various dishes on a restaurant's menu (especially if the calorie counts in the famous chefs' cookery books are anything to go by; see Howard *et al.* 2012). It turns out that consumers apparently have surprisingly little idea about how many calories they're putting into their mouths with each forkful.

In a set of state-wide surveys carried out in the US before obligatory menu labelling was introduced, Burton *et al.* (2006) discovered that consumers tended to underestimate the fat and calorie content of restaurant menu items by as much as 50% (see also Howlett *et al.* 2009). These researchers then went on to look at whether providing nutritional information would alter consumers' attitudes and/or food choices. As one might have guessed, they found that when people were made aware that food items were much 'worse' for them (for instance, higher in calories) than expected, they were more likely to change their choices as compared to when their expectations were closer to the actual calorie content of the food (see also Downs *et al.* 2013). So, it could be said that having this information displayed on the menu at least allows the diner to make more informed decisions (hopefully nudging them toward healthier options, though perhaps having a deleterious impact on the restaurant's profit; cf. Campbell-Arvai *et al.* 2012). However, Downs

---

[21] Bear in mind here that only 50–70% of customers tend to notice such information if it is included in the menu (Driskell *et al.* 2008; Hammond 2012).

et al. (2009) reported that displaying the calorie count on a sandwich menu (in a study where the participants believed that the sandwiches were being given out as a reward for having completed a survey) had little impact on their food choices. Surprisingly, displaying this information made dieting participants less likely to choose a low-calorie sandwich than when no such information was provided (see also Roberto *et al.* 2009).

On the other hand, it should also be taken into account that the healthier options are not always the cheaper ones. Further, with one having to consider many numbers in the menu (the amount of fat, the salt content, how many calories and even, in some cases, the fibre content, oh and not forgetting the price), there is a very real danger that the diner might simply become overwhelmed by too much information. The result? Apart from the obvious increase in time spent taking the decision of what to order in the first place, the price itself may end up taking a less important place in the diner's decision making (Poundstone 2010a).

2.3.5 Price and behaviour

Many diners make their food selections based on the price of the items that they see displayed on the menu; at the very least, let's say that it is a factor often considered when having the first-world dilemma of having to choose between two or more equally tempting options. But how are prices strategically displayed on the menu so that you find yourself inclining towards the more expensive items? Most menus will bundle expensive items together with cheaper ones: more specifically, they price some items relatively high for what is on offer and list them next to the most expensive items of all, such as the oyster bar mentioned in Section 2.3.1. This approach to menu design can certainly make the other items look like a genuine bargain! In addition, as mentioned earlier, the most expensive items would normally be found in the upper right-hand corner of the menu where, supposedly, most of us look first. What this strategy is designed to do is to boost the so-called 'cheaper neighbour' effect.

Think here only of the now-famous second-cheapest wine phenomenon. The behavioural economists are fond of telling us that most people shun the cheapest wine on the menu; many people (your authors included) are afraid that ordering that would make them look cheap. So, what do we do instead? Often, in order to keep the price of the alcohol down while at the same time looking at least somewhat discriminating (and not like an utter cheapskate), we order the second-cheapest wine. Well, some of us do (e.g. Poundstone 2010a; Cowen 2012). Of course, the cheapest option isn't necessarily the worst one.[22]

---

[22] It is worth remembering that, even if you can't pick out the price in a blind wine tasting, if you happen to know how much you are going to have to pay you will end up liking the taste of whatever goes in your mouth more (Plassmann *et al.* 2008; see Spence 2010 and Spence in press for reviews).

Bear in mind here that the price of the exact same bottle of wine or champagne can differ enormously from one restaurant to the next. For instance, Chung (2008) reported that a bottle of 1999 Dom Pérignon Champagne cost $155 at Legal Sea Foods in Washington; at another seafood restaurant a few hundred metres away, McCormick and Schmick's, the same wine was offered at $250. Meanwhile, at Carnevino in Las Vegas, a bottle could have been yours for $450. Finally, at Per Se in New York, it came in at an eye-watering $595. In the latter case, the argument is presumably that the price reflects the top level of wine service that one would expect to receive at such a famous restaurant. *"An inexpensive bottle might be priced three to four times its wholesale cost, while a pricey wine may be marked up only 1.5 times. This so-called progressive mark-up helps sell more expensive wines"* (Chung 2008).[23] According to Randy Caparoso, restaurant wine consultant at Wine List Consulting Unlimited, purchasing wine is like buying a pair of shoes: the nicer the store, the higher the price! Given that the drinks are where most restaurants make their margins, the behavioural economists have even been known to advise tight-fisted diners to go light on the wine if they want to maximize the benefit while minimizing the cost of their gastronomic transaction (Cowen 2012).

> *"A neighbourhood restaurant where you're greeted at the door by the owner who also seats you, takes your order and cooks your food has two to three times lower overall expenses than a restaurant with fresh flowers, valets, five chefs and an army of waiters."* (Randy Caparoso, cited in Roberts 2010)

However, the prices of a restaurant menu (of both the food and drinks) also help the diner to get an idea of their status and of how much emphasis they put on profit. For instance, currency signs and zeroes are mostly out nowadays; '£15.00' looks much more money-focused and expensive than the nakedly trendy '15' and '£9.99' seems a little manipulative, at least as compared to '£9.50' or even '£9.95'. If you eat at one of Russell Norman's Polpo group of restaurants in London (http://www.polpo.co.uk/), you may even find the occasional "9$^1/_2$". Other restaurants simply spell the price out in words, as we see more often in house numbers (but why do they think we're in the mood to play 'Where's Wally?' with the prices?). With such a variety of ways in which to display the price on the menu, restaurateurs might be wondering which method would entice diners tend to spend the most.[24]

In a study conducted by Cornell University's Center for Hospitality Research at a restaurant of the Culinary Institute of America (that's CIA for short), from August to November 2007, three versions of the daily menu were created (Yang *et al.* 2009). They were identical except for the way in which

---

[23] Surely there will soon be an app (if not out there already) for wines that will allow the miserly diner to know how much mark-up their restaurant is charging relative to its nearby competitors.
[24] We love to find studies that answer these intriguing questions!

the prices were displayed, for instance: '20', '$20.00' or 'twenty dollars'.[25] Contrary to what the researchers had expected, the numerical pricing resulted in diners spending significantly more than either of the other two pricing formats. This surprising result may have been due to the fact that seeing dollar signs repeated in the menu (no matter whether it is the symbol or the word; Dehaene and Akhavein 1995) could have activated a diner's awareness of the price, and consequently resulted in their being more prudent. Our question here is what would have happened if only the value of the items had been worded, without the word 'dollars'?

Even the justification of the menu (whether it's aligned to the left or to the right margin) can also influence the kinds of choice we make, by apparently making us focus on either the food items themselves (if aligned to the left) or on the price, usually to the right side of the names (hence making more price-conscious choices; Poundstone 2010a).

> "I decide what I'm going to eat according to the name of the food, the sound of the word. It's silly to say mayonnaise tastes good. The z (as in the Russian spelling) ruins the taste – it's not an appealing sound ... " (Luria 1968, p. 82)

A nice example of menu design that brings together several of the topics covered in this section is that of Serendipity3 in New York (see http://www.serendipity3.com/food.htm). Just check out their menu. First, notice that the items are left justified and don't include any dollar signs. Now that we know some of the tricks of the trade, this makes us think that the restaurateurs concerned are hoping that the diner will focus on the names of the dishes and not on their price. Certain names on the menu are literally calling the diner as seen in the previous section, such as the '"Can't Say No" Sundae'. Second, the range of prices looks pretty crazy. Looking at the Sweets section of the menu, the desserts start at 7.00, passes through 22.50 for a 'Cheese Cake Vesuvius' and reaches all the way up to the dizzying heights of the 'Golden Opulence Sundae' at $1,000.00! Note that the menu includes the $ sign for this option, and for this option only. Obviously, the restaurateurs want to make sure that their diners notice the price of this particular sundae (apart from the fact that the font in which the name of this dessert is printed in bold and larger than the rest of the items, and placed at the very bottom of the three-column menu). So, will you ever order it? Not likely. Will anyone else? Unlikely too, unless perhaps as an ostentatious Veblen purchase (see Veblen 1899/1992). But does it make the price of the 'Cheese Cake Vesuvius' sound reasonable, and the other desserts

---

[25] Kim and Kachersky (2006) have proposed that Arabic numerals may draw more attention in those situations in which people have to compute totals, so this format could enable the diners to calculate the cost of the meal more easily and hence result in them making more price-sensitive choices instead.

just too commonly cheap? It most certainly does[26] and that, presumably, is the primary reason behind the inclusion of this outrageously priced menu item, as well as to get lots of free publicity (see also Wharton 2008 for other exorbitantly priced omelettes, baked potatoes, bagels and pizzas, topped with truffles, caviar and/or even gold flakes and see Padoa-Schioppa and Assad 2008 on the neural representation of desirable menu options as a function of what else happens to be on offer).

When offering a fixed-price menu where the diner has to choose one dish for each course the menu will, on occasion, include the price of the items as they appear on the *à la carte* menu. Normally, one would hypothesize that people would tend to choose the option that is most expensive since the final price of their meal is fixed. As a result, one strategy that devious restaurateurs might well think of using is to raise the price of the item on the *à la carte* menu that they most want to serve on the fixed-price menu (this could be one means of turning a 'puzzle' into a 'star'; Poundstone 2010a). However, contrary to this suggestion, researchers have reported that adopting such a mendacious strategy doesn't necessarily work, since diners seem to have stable preferences and do not always go for the most expensive item in such situations (Heffetz and Shayo 2009).[27]

Behavioural economists have been having a field day these last few years on the topic of pricing and its effect on consumer behaviour. However, some of the recommendations that they have come up with are, at least to our way of thinking, downright preposterous. Take, for example, Taylor Cowen's (2012) suggestion that we should all order the least-appealing item on the menu to get the best value for money; no, really (see the quote below).[28] When you combine these sage words with his other suggestion not to order the wine, since that is where most restaurants make their mark-up, it doesn't look like you are going to have much of a fun night ahead – better just order a take-away and a 6-pack and eat at home in front of the telly.

> *"At fancy and expensive restaurants ($50 and up for a dinner is an imperfect benchmark for this category) there is a simple procedure, which I outlined in my book* Discover Your Inner Economist. *Look at the menu and ask yourself: "Which of these items am I least likely to want to order?" Or "which of these*

---

[26] See Huber and Puto (1983) on the classic study showing how adding an item to a list will tend to draw people's preferences in the direction of that item.

[27] In contrast, a menu design that breaks all these rules is that of Eleven Madison Park in 2011, designed by Juliette Cezzar (see www.underconsideration.com/artofthemenu/archives/eleven _madison_park.php). All of the names of the items consist of single words displayed in a 4x4 grid (examples of items in each row are: 'langoustine', 'carrot', 'chicken', 'chocolate'), with no individual prices or any sort of highlighting strategy in the format. In this fixed-priced menu, the diner could choose any four items for a four-course meal; one could only guess from the names which were the entrées, mains and desserts. See Chapter 3 for further tips on naming strategies.

[28] That said, perhaps we shouldn't trust anyone's advice on what to do in a restaurant if they are unsure of how to spell restaurateur (see Cowen 2012, p. 207).

sounds the least appetizing"? Then order that item. The logic is simple. At a fancy restaurant the menu is well thought out. The time and attention of the kitchen are scare. An item won't be on the menu unless there is a good reason for its presence. If it sounds bad, it probably tastes especially good." (Cowen 2012, p. 71)

Finally, how many of you have gone to a restaurant and found no prices at all? It is not so long ago that when a gentleman took a lady out to dinner in a fancy restaurant, he would be given a menu with the prices while the woman's menu would merely list the names of the dishes. The idea was to let her order whatever she wanted without having to think about the price, but the man would know how much her dinner was going to cost and hence could adjust his order accordingly if he, for example, wanted to stay within a given budget. Nowadays, this idea is pretty much obsolete (offensive even) in the minds of most diners and restaurateurs; it's now much harder to spot who is the 'host' and who is/are the 'guest/s'. However, one can still find a few restaurants where this tradition continues (either as a matter of course or upon request)[29]. Other restaurants simply don't show the price to anyone,[30] especially when it comes to the drinks. Why? Many people find menus without prices create an uncomfortable situation for those who are dining unless, that is, you are one of those for whom money is no object.

2.3.6 On the format of the menu

What about the overall format of the menu? Consider those glossy or velvety papers, embossed silver letters spelling out the restaurant's name on the cover, menus that open in surprising ways or menus which include the history of each dish. Every detail in the format of the menu tells us something about the personality (or pretensions) of the restaurant. For instance, the daily special menu sheet not only tells us what is available today but also helps to convey a certain feeling, that is, whether the daily specials involve the use of fresh ingredients. Sophisticated, minimalistic, rough, pop or naturalistic; we believe that, as long as the menu presentation has been designed with good taste, every style is welcome.[31]

Given our own research on weight and its correlation with price (see Piqueras-Fiszman and Spence 2012) we would expect that the weight of

---

[29] The Fat Duck actually had a very interesting version of this menu idea, where one could keep folding over the pages indefinitely like in some sort of magic trick.

[30] The Hi Lo Jamaican Eating House restaurant down the Cowley Road in Oxford (http://www .hilojamaicaneatinghouse.co.uk/) used to be infamous for this when the first author was a student. As far as anyone could tell, the owner would simply pick a price that he thought that the diners could afford to pay.

[31] That said, you simply can't mount an expensive French wine list on a piece of plywood with a few sheets of tatty paper held together with nothing more than a rusty old bulldog clip on the upper left corner. Or, better said, you *shouldn't* (this is precisely what one of your authors came across in a tiny Paris restaurant serving fabulous food run by a pair of Japanese chefs recently).

menu will probably provide a good (albeit subtle) clue as to the quality of the food offering. The heavier the better or, rather, the heavier the more expensive. If our intuition is correct, then perhaps all those pop-up restaurants and gastropubs that offer their diners nothing more than a sheet of A4 paper with the menu of the day printed on are missing out on a trick or two here.

Chef Homaro Cantu, owner of the Moto restaurant in Chicago, stretches the imagination and takes you on a culinary adventure from the very first bite of your menu (see Figure 2.3). Any chapter on menu design wouldn't be complete without mentioning this chef's unique tasting menu. In this case, though, it *really* is a tasting menu. Chef Homaro prints the evening's offerings on edible paper and flavours each section according to the principle ingredients that the dishes include. You just break a piece off and place it in your mouth tasting, for instance, the tomato, basil and mozzarella that go into making an Italian entrée (Perez 2005). This edible paper can assume so many different forms on the plate such as a risotto, a lime margarita or a Vietnamese spring roll.[32] At least two or three food items made of paper are likely to be included in a meal at Moto, which might include 10 or more tasting courses.

In several restaurants with fixed tasting menus the physical paper menu is not given at the beginning but at the very end of the evening, more as a

Figure 2.3   The literally tasty 'tasting menu' served at Homaro Cantu's Moto restaurant printed with food-based ink. *Source*: Reproduced with permission of Homaro Cantu. *See colour plate section*

---

[32] Cantu prints on pieces of edible paper made of soybeans and cornstarch, using delicious organic food-based inks that he makes himself.

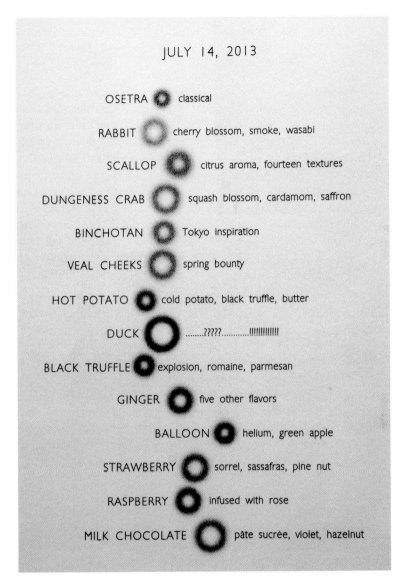

JULY 14, 2013

OSETRA classical

RABBIT cherry blossom, smoke, wasabi

SCALLOP citrus aroma, fourteen textures

DUNGENESS CRAB squash blossom, cardamom, saffron

BINCHOTAN Tokyo inspiration

VEAL CHEEKS spring bounty

HOT POTATO cold potato, black truffle, butter

DUCK ....?????...........!!!!!!!!!!!

BLACK TRUFFLE explosion, romaine, parmesan

GINGER five other flavors

BALLOON helium, green apple

STRAWBERRY sorrel, sassafras, pine nut

RASPBERRY infused with rose

MILK CHOCOLATE pâte sucrée, violet, hazelnut

Figure 2.4 The menu (given at the end of the meal) of Alinea. It uses the alignment of the items from the central axis and the diameter of the ring elements to make the diner remember their taste and the intensity of the flavour, respectively

souvenir than anything else. Here the word *souvenir* (meaning 'memories' in English) is well used since, without it, remembering each of the courses (and their order) is practically impossible. This is certainly the case at restaurants such as Alinea in Chicago and Eleven Madison Park in New York. In the former, the menu items are aligned along the centre of the sheet (of translucent paper) but those of sweeter items are shifted more towards the right while the more savoury options are shifted to the left. In addition, there are rings to indicate by means of their diameter the likely intensity of the flavour (see Figure 2.4).

Other restaurants use tablets in order to allow their diners to scroll up and down the menu. Some restaurants take these ideas to an extreme and use projectors to display their menus on the diners' tables or digital tables which themselves act as the menu. The tablet wine list certainly made for a welcome change on a recent visit to the Australasia restaurant in Manchester. How many times have you carefully selected a certain vintage of the wine from the list (a particularly good vintage, don't you know), only to find half-way through the bottle that they have actually given you a slightly different (and always inferior) vintage. "Oh, sorry sir, we haven't had time to reprint the menu" the waiter says apologetically. No such excuse needed once the wine list goes electronic. We cover the topic of digital technologies at the dining table in full in Chapter 10.

2.4   Conclusions

As we have seen in this chapter, the opening of the meal needs to be right including everything from the initial greeting, through the menu and pricing set expectations, and anticipation that will then determine how the rest of the meal unfolds. Moreover, the different elements that the diner interacts with during the start of the meal (the serving staff, the items on the menu, its design, not to mention the other diners) can modulate, to a lesser or greater extent, everything from the food choices and how much the diners spend to the overall amount that they consume.

Eventually, we'll see at the end of the meal that all of these elements contribute to the overall experience. Given that we very often perceive what we believe that we are going to experience (as we saw in Chapter 1), setting the right expectations turns out to be absolutely crucial. In the next chapter, we will see just how important the names and descriptions of the dishes on the menu, or as espoused by the wait staff, can be in delivering the perfect meal.

References

Anonymous (2013) Lone diners use their phones for company. *The Times*, 26 April, 31.
Ariely, D. (2008) *Predictably Irrational: The Hidden Forces that Shape our Decisions.* Harper Collins Publishers, London.

Ariely, D. and Levav, J. (2000) Sequential choice in group settings: Taking the road less traveled and less enjoyed. *Journal of Consumer Research*, **27**, 279–290.

Barthomeuf, L., Rousset, S. and Droit-Volet, S. (2009) Emotion and food. Do the emotions expressed on other people's faces affect the desire to eat liked and disliked food products? *Appetite*, **52**, 27–33.

Bell, R. and Pliner, P. L. (2003) Time to eat: The relationship between the number of people eating and meal duration in three lunch settings. *Appetite*, **41**, 215–218.

Blythman, J. (2012) Whole milk is healthier than skimmed and baked potatoes make you fat … Why what you thought you knew about healthy food is wrong. *Daily Mail Online*, 13 March, available at http://www.dailymail.co.uk/health/article-2114101/Whole-milk-healthier-skimmed-baked-potatoes-make-fat--Why-thought-knew-healthy-food-wrong.html (accessed January 2014).

Bowen, J. T. and Morris, A. J. (1995) Menu design: Can menus sell? *International Journal of Contemporary Hospitality Management*, **7**, 4–9.

Burton, S., Creyer, E. H., Kees, J. and Huggins, K. (2006) Attacking the obesity epidemic: The potential health benefits of providing nutrition information in restaurants. *American Journal of Public Health*, **96**(9), 1669–1675.

Campbell-Arvai, V., Arvai, J. and Kalof, L. (2012) Motivating sustainable food choices: The role of nudges, value orientation, and information provision. *Environment and Behavior*, published online 13 December 2012, doi: 10.1177/0013916512469099.

Carmin, J. and Norkus, G. X. (1990) Pricing strategies for menus: Magic or myth? *Cornell Hotel and Restaurant Administration Quarterly*, **31**, 44.

Carroll, L. (1865) *Alice's Adventures in Wonderland*. MacMillan Publishing Co., London.

Chung, J. (2008) Cracking the code of restaurant wine pricing. *Wall Street Journal*, 15 August, available at http://online.wsj.com/article/SB121875695594642607.html (accessed January 2014).

Clarkson, J. (2012) Heston's grub is great – but so what if your date is ugly? *The Sunday Times*, 6 May, available at http://www.thesundaytimes.co.uk/sto/comment/columns/jeremyclarkson/article1031654.ece (accessed January 2014).

Coren, G. (2012) *How to Eat Out: Lessons from a Life Lived Mostly in Restaurants*. Hodder and Stoughton, London.

Cowen, T. (2012) *An Economist Gets Lunch: New Rules for Everyday Foodies*. Plume, New York.

Damrosch, P. (2008) *Service Included: Four-star Secrets of an Eavesdropping Waiter*. William Morrow, New York.

de Castro, J. M. (1994) Family and friends produce greater social facilitation of food intake than other companions. *Physiology and Behavior*, **56**, 445–455.

de Castro, J. M. (2000) Eating behavior: Lessons from the real world of humans. *Ingestive Behavior and Obesity*, **16**, 800–813.

Dehaene, S. and Akhavein, R. (1995) Attention, automaticity, and levels of representation in number processing. *Journal of Experimental Psychology: Learning, Memory and Cognition*, **21**, 314–326.

Ditmer, P. R. and Griffin, G. (1994) *Principles of Food, Beverage and Labor Cost Control for Hotels and Restaurants.* Von Nostrand Rheinhold, New York.

Dornenburg, A. and Page, K. (1996) *Culinary Artistry.* John Wiley and Sons, New York.

Downs, J. S., Loewenstein, G. and Wisdom, J. (2009) Strategies for promoting healthier food choices. *American Economic Review,* **99,** 159–164.

Downs, J. S., Wisdom, J., Wansink, B. and Loewenstein, G. (2013) Supplementing menu labelling with calorie recommendations to test for facilitation effects. *American Journal of Public Health,* **103,** 1604–1609.

Driskell, J., Schake, M. and Detter, H. (2008) Using nutrition labeling as a potential tool for changing eating habits of university dining hall patrons. *Journal of the American Dietetic Association,* **108,** 2071–2076.

Edwards, J. S. A. and Meiselman, H. L. (2005) The influence of positive and negative cues on restaurant food choice and on food acceptance. *International Journal of Contemporary Hospitality Management,* **17,** 332–344.

Engle, J. (2004) Server tips: How to get more out of customers. Available at http://articles.chicagotribune.com/2004-04-25/travel/0404250508_1_people-tip-servers-diners (accessed January 2014).

Feiler, B. (2002) The therapist at the table. *Gourmet Magazine,* October, available at http://www.gourmet.com/magazine/2000s/2002/10/therapistatthetable (accessed January 2014).

Finkelstein, J. (1989) *Dining Out.* Polity Press, Cambridge.

Fleming, A. (2013) Restaurant menu psychology: Tricks to make us order more. *The Guardian,* 8 May, available at http://www.guardian.co.uk/lifeandstyle/wordofmouth/2013/may/08/restaurant-menu-psychology-tricks-order-more (accessed January 2014).

Freedman, D. H. (2013) How junk food can end obesity. Available at http://www.theatlantic.com/magazine/archive/2013/07/how-junk-food-can-end-obesity/309396/3/ (accessed January 2014).

Gallup Report (1987) Through the eyes of the customer. *The Gallup Monthly Report on Eating Out,* **7(3),** 1–9.

Garg, N., Wansink, B. and Inman, J. J. (2007) The influence of incidental affect on consumers' food intake. *Journal of Marketing,* **71,** 194–206.

Gladwell, M. (2004) On spaghetti sauce. Available at http://ed.ted.com/lessons/malcolm-gladwell-on-spaghetti-sauce (accessed January 2014).

Goerlitz, C. and Delwiche, J. (2004) Impact of label information on consumer assessment of soy enhanced tomato juice. *Journal of Food Science,* **69,** 376–379.

Goldman, S. J., Herman, C. P. and Polivy, J. (1991) Is the effect of a social model on eating attenuated by hunger? *Appetite,* **17,** 129–140.

Goodwin, C. and Verhage, B. J. (1989) Role perceptions of services: A cross-cultural comparison with behavioral implications. *Journal of Economic Psychology,* **10,** 543–558.

Greifeneder, R., Bless, H. and Pham, M. T. (2011) When do people rely on affective and cognitive feelings in judgment? A review. *Personality and Social Psychology Review*, **15**, 107–141.

Gross, J. J. and Levenson, R. W. (1995) Emotion elicitation using films. *Cognition and Emotion*, **9**, 87–108.

Guéguen, N., Jacob, C. and Ardiccioni, R. (2012) Effect of watermark visual cues for guiding consumer choice: An experiment with restaurant menus. *International Journal of Hospitality Management*, **31**, 617–619.

Hammond, D. (2012) Efficacy of menu labelling: Summary of Canadian research. Available at http://cspinet.org/canada/pdf/Toronto_DavidHammond.pdf (accessed January 2014).

Harper, L. V. and Sanders, K. M. (1975) The effect of adults' eating on young children's acceptance of unfamiliar foods. *Journal of Experimental Child Psychology*, **20**, 206–214.

Hedden, J. (1997) Maximize menu merchandising power. *Restaurant USA*, May.

Heffetz, O. and Shayo, M. (2009) How large are non-budget-constraint effects of prices on demand? *American Economic Journal: Applied Economics*, **1**, 170–199.

Howard, S., Adams, J. and White, M. (2012) Nutritional content of supermarket ready meals and recipes by television chefs in the United Kingdom: Cross sectional study. *BMJ*, **345**, e7607.

Howlett, E. A., Burton, S., Bates, K. and Huggins, K. (2009) Coming to a restaurant near you? Potential consumer responses to nutrition information disclosure on menus. *Journal of Consumer Research*, **36**, 494–503.

Huber, J. and Puto, C. (1983) Market boundaries and product choice: Illustrating attraction and substitution effects. *Journal of Consumer Research*, **10**, 31–44.

Hug, R. J. and Warfel, M. C. (1991) *Menu Planning and Merchandising*. McCutchan Pub. Corp., Berkeley, CA.

Hunt, R. R. (1995) The subtlety of distinctiveness: What Von Restorff really did. *Psychonomic Bulletin and Review*, **2**, 105–112.

Huron, D. (2007) *Sweet Anticipation: Music and the Psychology of Expectation*. MIT Press, Cambridge, MA.

Iyengar, S. S. (2010) *The Art of Choosing: The Decisions We Make Everyday, What They Say About Us and How We Can Improve Them*. Little, Brown, London.

Iyengar, S. S. and Lepper, M. R. (2000) When choice is demotivating: Can one desire too much of a good thing? *Journal of Personality and Social Psychology*, **79**, 995–1006.

James, A. (2005) Identity and the global stew. In: *The Taste Culture Reader: Experiencing Food and Drink* (ed C. Korsmeyer), pp. 372–384. Oxford, Berg.

Jones, M. (2008) *Feast: Why Humans Share Food*. Oxford University Press, Oxford.

Kahkonen, P. and Tuorila, H. (1998) Effect of reduced-fat information on expected and actual hedonic and sensory ratings of sausage. *Appetite*, **30**, 13–23.

Kelson, A. H. (1994) The ten commandments for menu success. *Restaurant Hospitality*, **78(7)**, 103.

Kim, H. M. and Kachersky, L. (2006) Dimensions of price salience: A conceptual framework for perceptions of multi-dimensional prices. *Journal of Product and Brand Management*, **15**, 139–147.

Kincaid, C. S. and Cursun, D. L. (2003) Are consultants blowing smoke? An empirical test of menu layout on item sales. *International Journal of Contemporary Hospitality Management*, **15**, 226–231.

Kozup, J. C., Creyer, E. H. and Burton, S. (2003) Making healthful food choices: The influence of health claims and nutrition information on consumers' evaluations of packaged food products and restaurant menu items. *Journal of Marketing*, **67**, 19–34.

Leake, J. (2009). EU dollops out doubt on benefit of health foods. *The Sunday Times*, 19 July, 5.

Logue, A. W. (2004) *The Psychology of Eating and Drinking* (3rd edition). Brunner-Routledge, Hove, East Sussex.

Lorenzini, B. (1992) Menus that sell by design. *Restaurants and Institutions*, **102**, 106–112.

Luria, A. R. (1968) *The Mind of a Mnemonist*. Harvard University Press, Cambridge, MA.

Luttinger, N. and Dicum, G. (2006) *The Coffee Book: Anatomy of an Industry from Crop to the Last Drop*. The New Press, New York.

Macht, M. and Mueller, J. (2007) Immediate effects of chocolate on experimentally induced mood states. *Appetite*, **49**, 667–674.

Macht, M., Roth, S. and Ellgring, H. (2002) Chocolate eating in healthy men during experimentally induced sadness and joy. *Appetite*, **39**, 147–158.

Main, B. (1994) *Mining the menu. ID*, **30(11)**, 79.

Martin, D. (2007) *Evolution*. Editions Favre, Lausanne.

McDermott, K. (2013) Revealed: Pret a Manger's bizarre 'emotional labour' rules for workers who are told to 'be happy', touch each other and NEVER act moody. *Daily Mail*, available at http://www.dailymail.co.uk/news/article-2272400/Pret-A-Manger-Workers-lift-lid-firms-strict-Pret-behaviour.html (accessed January 2014).

McVety, P. J., Ware, B. J. and Ware, C. L. (2009) *Fundamentals of Menu Planning* (3rd edition). John Wiley and Sons, Hoboken, NJ.

Mennell, S., Murcott, A. and van Otterloo, A. H. (1992) *The Sociology of Food: Eating, Diet and Culture*. Sage Publications, London.

Meyer, D. (2010) *Setting the Table: Lessons and Inspirations from one of the World's Leading Entrepreneurs*. Marshall Cavendish International, London.

Miller, J. E. (1992) *Menu Pricing and Strategy* (3rd edition). Van Nostrand Reinhold, New York.

Mori, D., Chaiken, S. and Pliner, P. (1987) "Eating lightly" and the self-presentation of femininity. *Journal of Personality and Social Psychology*, **53**, 693–702.

Nowlis, S. M., Mandel, N. and McCabe, D. B. (2004) The effect of a delay between choice and consumption on consumption enjoyment. *Journal of Consumer Research*, **31**, 502–510.

Padoa-Schioppa, C. and Assad, J. A. (2008) The representation of economic value in the orbitofrontal cortex is invariant for changes of menu. *Nature Neuroscience*, **11**, 95–102.

Panitz, B. (2000) Reading between the lines: The psychology of menu design. *Restaurants USA*, August, 22–27.

Perez, S. (2005) Eating his words. *Gourmet*, January, available at http://motorestaurant.com/press/gourmet-january-2005/ (accessed January 2014).

Petrini, C. (2007) *Slow Food: The Case for Taste* (translated W. McCuaig). Columbia University Press, New York.

Piqueras-Fiszman, B. and Spence, C. (2012) The weight of the bottle as a possible extrinsic cue with which to estimate the price (and quality) of the wine? Observed correlations. *Food Quality and Preference*, **25**, 41–45.

Plassmann, H., O'Doherty, J., Shiv, B. and Rangel, A. (2008) Marketing actions can modulate neural representations of experienced pleasantness. *Proceedings of the National Academy of Sciences USA*, **105**, 1050–1054.

Platte, P., Herbert, C., Pauli, P. and Breslin, P. A. S. (2013) Oral perceptions of fat and taste stimuli are modulated by affect and mood induction. *PLoS ONE*, **8(6)**, e65006.

Polivy, J., Herman, C. P., Younger, J. C. and Erskine, B. (1979) Effect of a model on eating behaviour: The induction of a restrained eating style. *Journal of Personality*, **47**, 100–117.

Poundstone, W. (2010a) *Priceless: The Myth of Fair Value (and How to Take Advantage of it)*. Hill and Wang, New York.

Poundstone, W. (2010b) Cash and calories. *Psychology Today*, available at http://www.psychologytoday.com/blog/priceless/201003/cash-and-calories (accessed January 2014).

Reynolds, D., Merritt, E. A. and Pinckney, S. (2005) Understanding menu psychology: An empirical investigation of menu design and consumer response. *International Journal of Hospitality and Tourism Administration*, **6(1)**, 1–10.

Roberto, C. A., Agnew, H. and Brownell, K. D. (2009) An observational study of consumers' accessing nutritional information in chain restaurants. *American Journal of Public Health*, **99**, 820–821.

Roberts, G. (2010) The lowdown on restaurant markups. The inside story on why and how restaurants price their wines. Available at http://www.winemag.com/Wine-Enthusiast-Magazine/May-2010/The-Lowdown-on-Restaurant-Markups/ (accessed January 2014).

Rowley, T. (2010) What do you fancy? Pizza Express staff to be taught how to flirt with customers. *The Independent*, **14 October**, p. 24.

Schwarz, N. (2012) Feelings-as-information theory. In: *Handbook of Theories of Social Psychology* (eds P. Van Lange, A. Kruglanski and E. T. Higgins), pp. 289–308. Sage, London.

Schwarz, N. and Clore, G. L. (1983) Mood, misattribution, and judgments of well-being: Informative and directive functions of affective states. *Journal of Personality and Social Psychology*, **45**, 513–523.

Seo, H. S. and Hummel, T. (2011) Auditory–olfactory integration: Congruent or pleasant sounds amplify odor pleasantness. *Chemical Sense*, **36**, 301–309.

Sharp, R. (2009) Mind games on the menu: The psychological tricks restaurants use to part us from our money. *The Independent*, 17 December, available at http://www.independent.co.uk/life-style/food-and-drink/features/mind-games-on-the-menu-the-psychological-tricks-restaurants-use-to-part-us-from-our-money-1842872.html (accessed January 2014).

Shooter, A. (2011) The alarming truth about your 'fresh and healthy' Pret A Manger lunch. *Daily Mail*, 11 April, 23.

Sobal, J. (2000) Sociability and the meal: Facilitation, commensality, and interaction. In: *Dimensions of the Meal: The Science, Culture, Business, and Art of Eating* (ed. H. Meiselman), pp. 119–133. Gaithersburg, MD, Aspen.

Sobal, J. and Nelson, M. K. (2003) Commensal eating patterns: A community study. *Appetite*, **41**, 181–190.

Sommer, R. and Steele, J. (1997) Social effects on duration in restaurants. *Appetite*, **29**, 25–30.

Sonnefeld, A. (2003) Series editor's introduction. In: *Slow Food: The Case for Taste* (translated by W. McCuaig; ed. C. Petrini), pp. xi–xv. Columbia University Press, New York.

Spence, C. (2010) The price of everything – the value of nothing? *The World of Fine Wine*, **30**, 114–120.

Spence, C. (in press) Searching for the value of wine. *Proceedings of the AWITC*. Sydney, Australia.

Stoner, C. L. (1986) Menus: Design makes the difference. *Lodging Hospitality*, **42**, 70–72.

Swami, V. and Tovée, M. J. (2006) Does hunger influence judgments of female physical attractiveness? *British Journal of Psychology*, **97**, 353–363.

Tanner, R. J., Ferraro, R., Chartrand, T. L., Bettman, J. R. and van Baaren, R. (2008) Of chameleons and consumption: The impact of mimicry on choice and preferences. *Journal of Consumer Research*, **34**, 754–766.

Thorndike, E. L. (1920) A constant error in psychological rating. *Journal of Applied Psychology*, **4**, 25–29.

Veblen, T. (1899/1992) *The Theory of the Leisure Class*. Transaction Publishers, New Brunswick, NJ.

Verma, R., Pullman, M. E. and Goodale, J. C. (1999) Designing and positioning food services for multicultural markets. *The Cornell Hotel and Restaurant Administration Quarterly*, **40**, 76–87.

Von Keitz, B. (1988) Eye movement research: Do consumers use information they are offered? *European Research*, **16**, 217–223.

Wansink, B. and Linder, L. W. (2003) Interactions between forms of fat consumption and restaurant bread consumption. *International Journal of Obesity*, **27**, 866–868.

Wansink, B., Ittersum, K. V. and Painter, J. E. (2004) How diet and health labels influence taste and satiation. *Journal of Food Science*, **69**, S340–S346.

Wharton, R. (2008) The $175 burger is a haute handful for rarified tastes. *New York Daily News*, 20 May, available at http://www.nydailynews.com/life-style /eats/175-burger-haute-handful-rarefied-tastes-article-1.330877 (accessed January 2014)

Wurgaft, B. A. (2008) Economy, gastronomy, and the guilt of the fancy meal. *Gastronomica*, Spring, 55–59.

Yang, S. S. (2012) Eye movements on restaurant menus: A revisitation on gaze motion and consumer scanpaths. *International Journal of Hospitality Management*, **31**, 1021–1029.

Yang, S. S., Kimes, S. E. and Sessarego, M. M. (2009) $ or Dollars: Effects of menu-price formats on restaurant checks. *Cornell Hospitality Report*, **9(8)**.

Yeomans, M. R. and Coughlan, E. (2009) Mood-induced eating: Interactive effects of restraint and tendency to overeat. *Appetite*, **52**, 290–298.

3

Tastes Great, But What do We Call It? The Art and Science of Food Description

"In real life, unlike in Shakespeare, the sweetness of the rose depends upon the name it bears. Things are not only what they are. They are, in very important respects, what they seem to be." (Hubert Humphrey, cited in Robinson 1998, p. 151)

3.1 Introduction

One might think that a diner's perception of the taste, flavour, quality and ethnicity of a dish would be determined primarily by the quality and nature of the ingredients used and by how the food was prepared, not to mention how the various ingredients/elements are presented on the plate (see Chapter 4). However, as we will see later in this chapter, our response to food is driven to a much larger extent than many of us ever realize by the way in which those items are described; this includes how they are named, labelled and what sensory descriptive terms are used. These effects occur no matter whether the information happens to be written on the menu, chalked up on the daily specials board or described verbally by the waiter or waitress standing by your table.

The Perfect Meal: The Multisensory Science of Food and Dining, First Edition.
Charles Spence and Betina Piqueras-Fiszman.
© 2014 John Wiley & Sons, Ltd. Published 2014 by John Wiley & Sons, Ltd.

But what exactly is in a name or description of a dish? Why are the names of certain dishes so mystifying to many diners, while others are pretty much incomprehensible (even to the chefs themselves)? The names of others, meanwhile, continue to be written in French or Italian. Why? Furthermore, have you ever wondered why restaurateurs don't simply translate the names of the dishes into the local language? And if they did, would anything be lost apart from overcoming the challenge to the diner's powers of pronunciation (at least for anyone with a traditional British education)?

What effects do these various styles of naming a dish have on the diner who is reading, or perhaps trying to decipher, the items on the menu? 'Exotic' names tend to be chosen over more realistic (one might say down-to-Earth) descriptions for a number of reasons, such as to: (1) raise curiosity, for example in the case of the intentional use of mystifying names (Miller and Kahn 2005); (2) elicit expectations of a certain ethnic quality, for example giving a dish a foreign-sounding name might perhaps make us believe that the food has been prepared according to a traditional recipe and from ingredients of the region where that language is spoken (cf. LeClerc *et al.* 1994);[1] (3) make a dish appear more sophisticated, for example a dish with 'deconstruction' or 'aerated' in the title would very likely make us imagine the creative chef beavering away earnestly in the kitchen with tweezers, perhaps carefully pipetting the sauce onto the plate; (4) highlight the presence of expensive or unusual ingredients (when, after all, was the last time that you were surprised by the divine flavour of truffle in a dish without its presence having been made all too clear on the menu in advance?); (5) convey nostalgia (and homemade/natural qualities) by using terms such as 'old', 'Grandma's', 'natural', etc. (Wansink *et al.* 2005);[2] and last, but by no means least, (6) hide or disguise the existence of certain ingredients that the diner might not perceive to be all that appealing, at least when read on the menu by the diner (sweetbreads sound so much more appealing than pancreas or glands, do they not?).

In sum, ethnic food labels, descriptive sensory food labels, organic and natural labels and even nutritional information all play an important role in influencing the diner's behaviour towards, not to mention their perception of, food and drink (see also Asam and Bucklin 1973; Dubé and Cantin 2000). Getting the name right can therefore help a chef or restaurateur to enhance a diner's enjoyment of whatever it is that they happen to be serving. It can also help to increase the likelihood that a diner will choose one dish over another. Indeed, as Gerard Samper, the effete Englishman and ghostwriter to the stars puts it in James Hamilton-Patterson's wonderfully funny novel *Cooking with Fernet Branca*: "*I sometimes wonder if one is not more seduced by the mellifluous*

---

[1] Of course, it might also be the case that the foreign name simply cannot be translated perfectly into English. For example, *poulette* means a young hen and also a young girl or woman. According to Halligan (1990, p. 125), there simply isn't a translation that does the French culinary descriptor '*à la poulette*' justice in the whole of the English language.

[2] As Seremetakis (1994, p. 1) puts it: "*nothing tastes as good as the past*".

sound of a dish than by how it would actually taste" (Hamilton-Patterson 2004, p. 67). And while one might not necessarily agree with Gerard's claim, the latest evidence really does appear to suggest that certain food attributes/qualities are better matched with particular food names, in particular, with the speech sounds that these names contain (see Spence 2012).

"*What might appear unpalatable in print is exquisite on the plate.*" (Description of one of Pierre Gagnaire's menus; Weiss 2002, p. 119)

3.2 Snail porridge

Heston Blumenthal created quite a stir a few years back with his infamous 'Snail Porridge' (see Blumenthal 2007, 2008). Everyone was shocked: how could anyone even conceive of combining snails with a breakfast staple such as porridge? Of course, the vivid green colour of the oats in this case didn't help to allay people's concerns about this savoury dish (see Figure 3.1). However, it can be argued that this dish would never have created anything like the impact that it ultimately did had it been given a different name (see also Rayner 2004; Poole 2012). If you inspect them closely, many a popular dish already brings together seemingly bizarre, or incongruous, combinations of ingredients (e.g. foodstuffs associated with very different styles of cuisine): it is just that this fact is rarely highlighted to the diner in the name or description of the dish on the restaurant menu. Take for example tomato ketchup, which has been called an 'überfood' because of its ability to hit at least four of the five basic tastes (see Rozin 1988; Gladwell 2009; according to Vilgis 2012 the taste of bitterness is missing). This ingredient is commonly used in Pad Thai noodles, a standard dish served in any self-respecting Thai restaurant (Heldke 2005, p. 392), one that dates from the Second World War (Cowen 2012, p. 116). However, you don't need to be an experimental psychologist in order to know that describing a dish as 'noodles with ketchup' just doesn't sound the same. With 'Snail Porridge', Blumenthal used the very name of the dish to surprise and shock and to capture the attention of his diners, not to mention the adoring global media. He undoubtedly succeeded in the latter, given the huge amount of press coverage that this particular dish generated in the years since it was introduced on the menu at Heston's flagship restaurant.

Another example of the shocking approach to food naming that has emerged from the kitchens of the same chef is his 'Meat Fruit', a chicken liver parfait made to look like a mandarin, served as a starter at the Dinner restaurant in London, UK (www.dinnerbyheston.com; see Figure 3.2).[3] While it is certainly common in many types of cuisine to cook meat with some sort

---

[3] This dish references the name of the hotel, the Mandarin Oriental, in which the restaurant is located. For many diners, this dish constitutes the highlight of a meal at Dinner.

Figure 3.1 'Snail Porridge', as served at Heston Blumenthal's The Fat Duck restaurant. The name of the dish deliberately juxtaposes two seemingly incongruous ingredients. *Source*: Photograph by Ashley Palmer-Watts. Reproduced with permission of Lotus PR and The Fat Duck. *See colour plate section*

Figure 3.2 The 'meat fruit' dish from Heston Blumenthal's Dinner. In this case, the name of the dish makes it unclear whether the meat or the fruit is the dominant element. *See colour plate section*

of fruit, it can be argued that it is always clear what the dominant element in the dish is. The title 'Meat Fruit', by contrast, seems to give pretty much equal weighting to both of the constituent parts. These selected examples will hopefully have given the reader some idea of the potential impact that choosing the right name for a dish can have on the response of the diner to it.[4]

A similar conclusion regarding the importance of naming follows from the growth in sales of Patagonian toothfish (*merluza negra*) over the last decade or so. Up until a few years ago, the fishermen who brought this modern-day 'monster of the deep' up in their nets had a hard time selling their catch. The suggestion was that people simply weren't interested in buying a fish with such an ugly name! Until, that is, some bright spark came up with the idea of rebranding it as 'Chilean sea bass' (or *merluza chilena*). Suddenly, sales of this formerly obscure fish sky-rocketed. In no time at all, Chilean sea bass (as it is now widely known) started appearing on the menu at many a fine restaurant (see Dolnick 2008; Jones 2013).

A similar story concerns the renaming of pollock (or pollack), sometimes known as 'Boston blues'. In 2009, the British supermarket chain Sainsbury's renamed (or should that be rebranded?) the fish 'Colin' (the French term for it) as an eco-friendly approach to do their bit to save the dwindling cod stocks (and at the same time increase the sales of this most unfashionable of fish). The name change was launched under the banner "*Colin and chips can save British cod*" (Smithers 2009). The supermarket also considered the possibility that some of their shoppers might have been embarrassed to ask for the fish by its original name, due to its similarity to the English swear word 'bollocks' (not to mention its reputation as an inferior fish). Once again, the name given to a food transformed the way in which the consumer reacted to it.

That said, finding the right name for a food is by no means a new problem. The desire to rename dishes actually goes way back in time. For instance, the famous French chef Antonin Carême (1822) decided early on that 'Glaced nun farts' was one dish whose name really had to be changed. This was only one of a number of traditional French dishes whose title he wanted to update:

> "*It is the same with soups* à la jambe de bois [peg-leg] (*rather than saying beef marrow): if we add to that* la culotte de boeuf [beef panties], *veal* roulé en crotte d'âne [donkey droppings] *taken from* Cuisinier Gascon, culs d'artichauts [artichoke "ass"], les pets de nonne glacés [glaced nun farts], *and other entrées and desserts of that ilk; and when at table we ask the waiter the identity of an entrée: Sir, it's peg-leg soup; what an ignoble phrase! It's an entrée of pheasant socks* à la Conti; *or, again, it's beef fillets sautéed in the form of glazed boot heels, and so forth.*" (Hyman 2001, pp. 40–41)

---

[4] However, as we will see later, the art of food naming might be somewhat different in the context of the modernist restaurant than for other kinds of eating establishment (Mielby and Bom Frøst 2012).

Carême himself preferred to sign his culinary creations by referencing the most illustrious names from the French nobility or society (see Ferguson 2004, p. 74). In fact, a number of the culinary labels that Carême introduced to the dinner table are still with us today; thinkof dishes *à la reine, à la dauphine, à la royale …* Carême even named one of his finest soups *Victor Hugo* before he (the chef that is) died just short of his 50th birthday. One can certainly imagine waiters tiring of having to repeat to the curious diners the history of some of the most bizarre, not to mention long-lasting, of dish names; think here only of 'Devils on horseback' (prunes wrapped in bacon served on bread) *or* 'Angels on horseback' (fried oysters wrapped in bacon and served on fried slices of bread). It turns out that many such peculiar names date back to Victorian times.[5]

> *"Cod is a simple word associated with cheapness, so folks cock a snook at it. Rechristen it pterodactyl and some would still offer sales resistance because they could not pronounce the name easily."* (Fanny Cradock, Daily Mail, 1955, cited in Ellis 2007, p. 197)

Choosing the right name for food was also a very serious problem back in the 1940s. During the closing years of the Second World War, there was a shortage of prime cuts of meat. The only thing to offer a hungry (not to mention protein-deficient) carnivore was the ominously named 'variety' meats.[6] The question that the authorities wanted to answer was how best to encourage the populace to change their consumption habits and incorporate 'organ meat' (consisting of hearts, brains, intestines, kidneys, ears, etc.), actually a rich source of protein, into their diets. So important was this issue judged to be that the American government enlisted the help of academics in a number of fields from psychology through to sociology, including the famous North American anthropologist Margaret Mead. The evidence covered in this chapter would certainly appear to suggest that offering people 'variety' rather than 'organ' meats could have had a significant effect on people's consumption behaviours (see also Cline 1943; Time-Life 1982; Wansink 2002 for a review).[7]

---

[5] As Alan Davidson notes in his entry in the *Oxford Companion to Food*, such savoury delights seem to live in the "time warp" of London clubs and those restaurants that trade specifically on nostalgia (Damrosch 2008, p. 46).

[6] Given the recent horsemeat scandal gripping Europe (http://www.bbc.co.uk/news/world-europe-21457188; Anonymous 2013), this might be a name that we see coming back into fashion. This is by no means the only major scandal in terms of mislabelling food to have hit the headlines in recent years (see Wilson 2009). Indeed, there have been a number of reports of fish being mislabelled. According to one report, as many as 25% of fish were not the species described on the menu at take-away restaurants, fishmongers' shops and supermarkets (e.g. Anonymous 2010; Leake and Dowling 2011; Simpson 2013).

[7] Wilson (2009) highlights a similar problem with less savoury foods faced by the British Government after the end of the Second World War. In this case, the government wanted the population to embrace snook, a kind of fish. With such an unappealing name, it comes as no surprise that people didn't take to this particular source of nutrition with quite the enthusiasm that the government had hoped for; excess stocks had to be sold as cat food.

In a related vein, in her compendious *Food in History*, Reay Tannahill (1973, p. 102) noted that when times were particularly hard back in twelfth century China and some people were forced to eat human flesh in order to survive, the dish-*du-jour* was euphemistically known as 'two-legged' mutton.

More recently, there has been a similar trend toward renaming those traditional dishes whose name might offend the politically correct lobby. Take for example the classic British dish 'Spotted Dick' (no sniggering, please).[8] Several British institutions, as well as the supermarket chain Tesco, deemed the name of this steamed suet pudding with dried fruit (usually raisins) inappropriate, offensive even. The solution? Simply rename the dish 'Spotted Richard' instead (Anonymous 2009; Tozer 2009). That said, given the unexpected public backlash against the renaming of this, one of Britain's favourite Victorian puddings, the original name was thankfully soon restored. Other examples of outdated names which haven't survived so well include faggots.[9]

3.3   Can labelling enhance the taste and/or flavour of food?

The idea that the label given to a food can actually change its taste also has a long history (see Anonymous 1962 for early research on this topic). Wolfson and Oshinsky (1966) were commissioned to see what they could do about making space food more palatable for astronauts. The following quote makes quite clear what they were up against[10]:

> "*These space foods contain the necessary nutritional, vitamin, and caloric content to sustain an astronaut during long periods of space travel. Since some of the ingredients have a relatively offensive odor, taste, and texture when condensed in a small total volume of the food product, flavorings such as chocolate or vanilla may be added in an attempt to make the product more acceptable*" (Wolfson and Oshinsky 1966, p. 21)

Wolfson and Oshinsky (1966) gave the participants in their study commercially available chocolate milk and/or a chocolate-flavoured liquid space diet drink (yummy). These drinks could either be identified as 'space food' or as 'unknown'. When the participants were given the same drink twice and all that changed was the name, no changes were observed in participants' ratings on its second presentation. That is, no one was fooled by the experimental

---

[8] It has been suggested that the name of this dish, which dates back to 1850, may be a contraction of pudding or puddink (Anonymous 2009).
[9] It is our guess that anyone who prefers their Spotted Richard to their Spotted Dick probably isn't going to be a great fan of the snack called 'Nucking Futs' either.
[10] For anyone wanting to try a spot of space food at home, you need only get yourself a copy of Bourland and Vogt's (2010) fascinating book *The Astronaut's Cookbook*.

manipulation. However, when the participants received each of the drinks once, then simply changing the name of the drink really did alter people's ratings. On average, the drink labelled as 'space food' was rated 2 points higher on a 9-point preference rating scale than the drink labelled as 'unknown'. As the authors of this study went on to conclude: "*Altering the name of a food product to something related to the exotic would enhance the preference rating for the product*" (Wolfson and Oshinsky 1966, p. 23).[11]

Elsewhere, people have been shown to rate M&Ms as tasting significantly more chocolatey if they are labelled as 'dark' rather than as 'milk' chocolate (Shankar *et al.* 2009). That said, until recently it hasn't been altogether clear whether such results (e.g. Wolfson and Oshinsky 1966; Olson and Dover 1978; Shankar *et al.* 2009) really demonstrated that the taste and/or flavour of the food or drink was changed by changing the description, or whether changing the label merely modified the way in which people *responded* to (i.e. rated) it.[12] Over the last few years however, researchers have started to unravel some of the neural mechanisms underlying the effects of labelling (and expectancy) on multisensory flavour perception (Grabenhorst *et al.* 2008). In many cases, these effects have been shown to be genuinely perceptual in nature.[13] Indeed, it could be argued that this has been one of the more useful findings to have emerged from the neuroimaging research conducted in the area of flavour perception over the last decade or so.

For example, Nitschke *et al.* (2006) demonstrated that changing the verbal description given to a liquid can result in significant changes in the brain processing associated with it when a person tastes it. (Note that most of the neuroimaging research on flavour perception tends to involve the delivery of flavoured liquids rather than solid foods for purely practical reasons.) The participants in this study were given different solutions of water through a tube inserted into their mouth while lying flat on their back in a brain scanner (the participants would also have had their heads clamped still). The presented liquids varied in terms of how much quinine (a bitter-tasting substance) and glucose (which tastes sweet) had been added. The participants were sometimes informed that they were about to be given something very bitter to drink while, on other trials, they were told that they would be tasting something that wasn't bitter at all. The neural changes that were observed as a function of modifying the verbal description of the liquid took place in some of the earliest brain sites after the taste and smell signals are coded by the tongue and nose, respectively. In particular, the neural

---

[11] Reading the description of space food in this study makes the cinematic depiction of this most unusual class of foods in films such as Stanley Kubrick's *2001: A Space Odyssey* seem not so far from the truth (see Haden 2005).

[12] Note that this subtle-sounding distinction between perceptual and decisional effects is in fact one that keeps many a psychologist awake at night!

[13] Because perceptual effects can sometimes be obtained doesn't however mean that they always will be, or that the effects of food naming/description should be attributed in their entirety to effects occurring at a perceptual level.

response in the orbitofrontal cortex (OFC) to bitter-tasting compounds changed systematically as a function of whether people were expecting to taste something bitter (see also Chapter 7 on the theme of disconfirmed flavour expectations). The results of this study also demonstrated that activity in the middle and posterior insula was modulated by the verbal description regarding the intensity of the about-to-be-delivered taste stimulus.[14] Such neuroimaging results perhaps provide the beginnings of a neural explanation for Olson and Dover's (1978) early behavioural results regarding the perceived bitterness of coffee.

Verbally describing an odour (an equal mixture of isovaleric and buteric acid) as either 'parmesan cheese' or 'vomit' can change the rated pleasantness of the odour (see Herz and von Clef 2001; Manescu *et al.* 2014). What is more, changing the description also changes which part of the OFC will be activated in response to sniffing the aroma (De Araujo *et al.* 2005; Djordjevic *et al.* 2008). The OFC, a roughly walnut-sized structure lying just behind the diner's eyes, codes the reward value (how much do you like it) rather than the sensory-discriminative aspects of taste (i.e. what is it). Reading about this neuroimaging research certainly makes one wonder whether the 'Stinking Bishop' cheese (http://en.wikipedia.org/wiki/Stinking_Bishop_cheese), made popular by the Wallace and Gromit movies, would smell any less offensive should the manufacturers have decided to give it a somewhat different name.[15]

But does it matter when exactly in the course of a dinner a diner is given the name of a dish? Research from Lee *et al.* (2006) would suggest that it most certainly does. They reported, perhaps unsurprisingly, that descriptive information has the most pronounced effect when it is provided *prior* to consumption (although everyone has no doubt had the experience of not being able to remember quite what the mouth-watering names of all the various ingredients in the dish that they ordered were, suggesting that names can, on occasion, be provided too early during the course of a meal). In their research, conducted on a New England university campus, the participants (people drinking in a bar) were approached by one Professor Ariely, a famous behavioural economist currently based at Duke University in the US. He asked the drinkers to taste two beers: a regular beer and what he claimed was a new tipple called 'MIT Brew' (which was actually regular Budweiser with a few drops of balsamic vinegar added), and then to indicate which beer they preferred.

There were three conditions in Lee *et al.*'s (2006) study. In one, the participants were informed that what they were drinking was actually regular Budweiser that had been adulterated with a few drops of balsamic vinegar. A second group of participants were provided with this information only *once* they had tasted the beer, while the last group was not provided with

---

[14] However, as noted by Grabenhorst *et al.* (2008), it isn't altogether clear whether this modulation was actually taking place in the primary taste cortex or in an adjacent region.

[15] In this case, the name itself relates not to the smell of the cheese but rather to the fact that the cheese is washed in perry made from the Stinking Bishop pear.

any information about the modified beer (the blind condition). The majority (59%) of those in the latter condition preferred the MIT brew over regular Bud. By contrast, only 30% of those who were informed about the added ingredient before tasting preferred it. For those who were informed after they had already tasted the beer, the percentage was 52%. So, a bit like the Pad Thai noodles mentioned in Section 3.2, it would once again appear that sometimes we like foods and beverages more the less we know about the ingredients that actually went into making them.

The description of a food can play a particularly important role when the expectations set up by our eyes upon seeing a dish (what is sometimes referred to as the 'visual flavour'; e.g. Masurovsky 1939; Hutchings 1977) is either ambiguous or else different from the actual taste or flavour of the dish. This was beautifully demonstrated in the now-classic study by Yeomans *et al.* (2008). In this collaboration between chef Heston Blumenthal and food scientists based at the University of Sussex, three groups of participants were given a pinkish-red ice-cream to taste in the laboratory. One group was given no information about the dish. A second group was informed that they would be tasting a savoury ice-cream. Meanwhile, a third group of participants was told that they would be tasting a novel food called 'Food 398'). Those participants who hadn't been given any information about the smoked-salmon-flavoured food (and who presumably would have been expecting to taste a sweet berry-flavoured ice-cream) rated the dish as much saltier than either of the other two groups. What is more, this group of participants reported liking the dish far less, presumably due to the occurrence of a strong disconfirmation of expectation response (see also Imm *et al.* 2012).[16]

There are very often carry-over effects in this kind of experiment such that, when the participants are subsequently given another chance to eat the very same food a few weeks later, those whose eyes had led them astray the first time are still likely to report liking it less when they are subsequently re-exposed to it (at which time they consume less even though they now know what to expect). Such insights obviously have profound implications for the way in which a chef or restaurateur describes a dish on a menu. This is especially likely to be the case when the visual appearance of the food may be expected to mislead the diner, or when catering to an international clientèle who may arrive at the dinner table with a head full of very different colour–flavour associations (see also Chapters 6 and 7).

Our very own experience of disconfirmed expectation came when on a visit to Japan some years ago. The street vendors outside Kyoto's rightly famous temples lying along 'The Philosophers Way' were selling a light-green

---

[16] Okamoto *et al.* (2009) report that participants who tried various solutions made to stimulate the basic tastes (sour, bitter, sweet and salty) liked them significantly more when they were given food names such as: 'lemon', 'coffee jelly', 'caramel candy' and 'consommé soup', respectively, than when these solutions were identified with numerical labels instead (to codify the samples, rather than to create an air of mystery as in Yeoman *et al.*'s 2008 study).

ice-cream one hot spring afternoon. "*What refreshing pistachio-flavoured ice-cream*", we thought. You can imagine our shock when, on taking a mouthful, the flavour of the creamy concoction turned out to be not at all as we had expected. It was not sweet as pistachio ice-cream ought to be; the taste was even a little bitter. It turned out that green tea, which is drunk in vast quantities in Japan, is also one of the most popular flavours for their ices. Now, whenever visiting Japan and given a helping of green tea ice-cream in a formal banquet setting, we can't help but think back to that first exposure to this unexpectedly flavoured ice. The research of Yeomans *et al.* (2008) therefore suggests that if the ice-cream had been clearly labelled (or better said, clearly labelled in English) the first time that it had been tasted, we would have enjoyed the dish far more all these years later than we currently do. The main point to remember here is that the effects of naming not only have immediate effects, but also much longer-term consequences.

3.4   Interim summary

Taken together, the results of the research reported so far in this chapter demonstrate that the way in which a food is described or labelled can exert a significant influence on what it tastes like, and even on how much we like it. In theory, then, it ought to be possible to use food names in order to modify (either to enhance or to reduce) the intensity of a particular taste or flavour (e.g. Olson and Dover 1978; Shankar *et al.* 2009). Obviously, changing the name of a dish cannot literally change its chemical composition or physical make-up, but what it likely does change is what the diner chooses to focus his/her attention on and hence how the diner experiences the food that is before him/her. Changing the name can also have an impact on a diner's hedonic response to a food. The name and description of a dish may also provide a label on which the diner can hang an otherwise ambiguous flavour experience. Anyone wanting to bring out a particular note in a specific dish should therefore ensure they name that taste, flavour or ingredient on the menu and/or verbally when the waiter delivers the dish to the table (thus ensuring that the diner hasn't forgotten what it is that they had ordered some time ago). Furthermore, and as we will see in Section 3.7, the description can also be used to enhance the perceived ethnicity of a meal.

3.5   On the neuroscience of naming food

It is important to note that attention plays a crucial role in determining what it is that we perceive in a dish (especially in a dish where there are multiple competing elements all vying for a diner's attention). Although all of us have the impression that we can see everything that comes before our eyes, and presumably that we can taste everything that enters our mouth, the evidence from over half a century of cognitive psychology tells a very different story. In fact, the available research now unequivocally demonstrates that we are creatures with a very limited capacity to process incoming sensory information (Gallace *et al.* 2012); taste/flavour is no exception here. Contrary to what most of us intuitively believe, we are aware of only a very small proportion of the information that our various senses are picking up at any given time (e.g. Driver 2001 Chabris and Simons 2011). Selective attention is the mechanism by which our brain prioritizes its limited neural resources on the processing of certain more relevant information over other putatively less-important information (Spence 2010b).

Descriptive food labels may work, at least in part, by helping to direct a diner's attention to one element or flavour in a dish. In doing so, that element then becomes more salient (or becomes the foreground) against the background competition of a variety of other tastes, flavours, aromas, textures and mouth-feels all competing for the diner's limited pool of attentional resources. (Remember that the brain is the body's most blood-thirsty organ; Allen 2012). Naming a particular flavour or oral-somatosensory texture can help to make it stand out in the diner's mind against all of the other sensations that they may be experiencing at the same time. Stimuli that we attend to also tend to appear more salient than those stimuli (think tastes or flavours) we are not concentrating on (Ashkenazi and Marks 2004; Stevenson 2012); what is more, they may even be perceived slightly earlier in time (von Békésy 1964; Spence and Parise 2010) relative to other unattended, or less well attended, stimuli. This may be part of the reason why providing a name *before* tasting a food or drink has more of an effect than providing the same name *after* tasting (see Lee *et al.* 2006). What is more, people typically rate those flavours/aromas that they can put a name to as being more intense than those elements in a dish that they find difficult to name. They also tend to rate them as being more familiar and hence pleasant.

Furthermore, simply reading the name of a food or ingredient that has a distinctive aroma (such as, for example, reading the word 'cinnamon') can also lead to the increased activation of the olfactory parts of the brain, despite the fact that a diner might not be smelling anything at the time (see González *et al.* 2006). It is almost as if the smell is being imagined; indeed, when most people try to do just that, they will end up breathing in through their noses as if there really was a smell there (Bensafi *et al.* 2003; Royet *et al.* 2013). Barrós-Loscertales *et al.* (2012) recently reported that simply reading taste-related words can result in increased activation of the primary and secondary gustatory areas (namely in the anterior insula, the frontal operculum and the OFC). What is more, it has been suggested that even just reading a food name can be enough to make some of us salivate (Winer *et al.* 1965; see also Spence 2011 for a review), not to mention affect how easy we find it to swallow certain foods (Nakamura and Imaizumi 2013).

In another study, Grabenhorst *et al.* (2008) described an umami (e.g. savoury) solution with one of several different labels designed to vary people's expectations concerning the pleasantness, rather than the intensity, of that which they were about to taste. The descriptors included terms such as 'rich and delicious taste' versus 'monosodium glutamate' for a water solution to which umami had been added, and 'rich and delicious flavour' versus 'boiled vegetable water' versus 'monosodium glutamate' when a vegetable odour had been added to the solution. Once again, the results demonstrated that the words used to describe the solutions resulted in a change in participants' brain activity.[17] Grabenhorst *et al.* 2008 (p. 1549) went on to conclude that "*top-down language-level cognitive effects reach far down into the earliest cortical areas that represent the appetite value of taste and flavour*".

Arana *et al.* (2003) used a neuroimaging technique called positron emission tomography (PET) to compare the patterns of neural activity that were associated with the processing of the incentive value in a range of food items, as presented in the form of a restaurant menu. This involved a sequence of food choices that participants had to make. The menu items were tailored to each participant's individual food preferences: one menu had been designed to contain their very favourite dishes (these were described as high-incentive items) and another included those that they liked but were clearly not their favourite (these were the low-incentive items). In half of the trials, the participants simply had to read the menu items as if they were sitting in a restaurant; in the other half, they had to make a food choice. Three major findings emerged from this study. First, both the amygdala (a region known to be related to appetitive motivation) and the medial OFC were activated when individuals considered the appetitive incentive value of foods. Second, the medial OFC also lit up when the incentive value informed participants' goal selection, even showing that activity varied as a function of the difficulty of the choice that they were being asked to make. Finally, the selection process also activated the right-lateral OFC, specifically when the participants had to choose between the high-incentive alternatives (which proved to be a difficult task for them).

While all of these neuroimaging results are undoubtedly most impressive, it is worth remembering that the change in the label is the only thing that is going on for the poor participant in one of these studies lying prone with their head clamped still in the noisy brain scanner. This situation is obviously completely different from those of diners distractedly looking at the menu in a real restaurant setting while chatting amiably with their friends. Hence, we would argue that it is going to be hard (if not impossible) to infer the consequences of the change in the way a food is described documented in one situation (in the brain scanner, for example) for people's experience in the

---

[17] In particular, activity was modulated in brain regions such as the pregenual cingulate cortex and the ventral striatum that are connected to the OFC (an area whose activity was also affected). By contrast, no such changes were observed in the insular taste cortex. This latter result makes sense given that people's expectations regarding the intensity of the taste were not manipulated in this study.

other (e.g. while dining out at a restaurant). The best we can hope for here is to acquire convergent findings from the two settings.

There is also an important link to be made to the literature on mental imagery. People differ markedly in terms of how vividly they can imagine (either imagine visually or imagine the taste/flavour of something given a verbal description; see Kemps and Tiggemann 2013). Perhaps, then, those individuals who are lucky enough to have particularly vivid mental imagery abilities may be better able to picture a dish in their mind's eye, almost to taste it (on the mind's tongue), based on nothing more than the description written on the menu or the verbal recounting of the specials by the waiter or waitress (Djordjevic *et al.* 2004; Kobayashi *et al.* 2004, 2011; Lacey and Lawson 2013). Basically, all this means is that dish naming may be more evocative for some diners than for others.

3.6   Naming names

One other obvious reason for name-checking certain ingredients in the description of a dish is to highlight the expense. "Truffle this", "Caviar that" ... perhaps there isn't anything fundamentally wrong with highlighting ingredients that are only present in very small quantities, as long as they contribute substantially to the overall flavour of the dish. Consider truffles: while the amount of this ingredient in a dish might well fall below one part in 100, that doesn't mean that its addition doesn't have a major impact on the overall flavour of the food. Of course, it does (and if it doesn't, you should send the dish straight back to the kitchen!). However, it is not always the case that expensive ingredients present in small concentrations actually do contribute significantly to the flavour of a dish.

When it comes to naming ingredients, one sometimes has the luxury of choosing between alternative descriptors. Think *maracuyá* versus passion fruit, or tree tomato versus *tamarillo*. Even in this case, it may be worth carefully considering what the best name for a given ingredient is. What is the effect that the cook is attempting to achieve in the diner's mind? As Heldke (2005, pp. 391–392) puts it, would galangal be rated as tasting more like ginger were it to be given its alternate name of Thai ginger? Heldke reminisces about the memorable occasion on which she first ate *kha gai*, a Thai chicken soup made with galangal. When she first came across galangal on a restaurant menu, it was an exotic ingredient to her. She wondered:

> *"But would it have tasted the same? If that first menu had offered "chicken soup with ginger and coconut milk," what would I have tasted? ... What would I have tasted had the menu simply said "ginger," but presented me with Thai ginger? ... Surely I would remember, no? After all, my experience of the flavor was not wholly shaped by the unfamiliarity of the word used to describe it – was it?"*

Another reason for choosing one name over another is that of pronounce-ability. Take for example the Hawaiian fish that goes by what has to be the most prize-winningly unpronounceable name of humuhumunukunukuapua'a. Or the wedge-tail triggerfish, whose Latin name is *rhinecanthus rectangulus* (both from Anonymous 2011a). The marketers tell us that consumers like names which they find easy to process or pronounce (Labroo *et al.* 2008). As a chef or restaurateur, one therefore might be well advised to maximize the processing fluency associated with the names of one's dishes. After all, this is likely a large part of the success of critter brands when it comes to the sale of New World wines.

3.7   Does food labelling influence the perceived ethnicity of a dish?

Why do we, in the English-speaking world, still see dish names (or parts of them) in French or Italian (those being the most common foreign languages for foods)? Could it perhaps be because the way in which a food is described on a menu can help to enhance the perceived ethnicity of the dish? The answer would appear to be 'yes'. Take for instance one of the classic studies in this area conducted by Meiselman and Bell (1993). These researchers investigated the effect of adding an Italian name to a range of different pastas. The British par-ticipants who took part in this study were given pasta dishes with four different sauces to evaluate, either named or unnamed. The names chosen for the study were 'pasta *Italiana*', 'pasta *Napolitana*', 'pasta *Bolognesa*' and 'pasta *Parmigiana*'. Those who ate a dish with an Italian name rated it as being more Italian, while lowering their ratings of their perceived Britishness (whatever that is ... see also Bell and Paniesin 1992).

One limitation with Meiselman and Bell's (1993) study is that their results were obtained in the laboratory; it is unclear to what extent similar results would have been found under more realistic testing conditions (e.g. had the participants actually been ordering and consuming the dishes in a restaurant context, say).

Bell *et al.* (1994) were able to deal with this concern in a follow-up study by replicating the experiment in a British university campus restaurant; they obtained essentially the same pattern of results. However, while the impact of food naming was no less striking, it is still not so easy to unambiguously inter-pret the results of the latter study. Apart from having Italian (versus English) dish names on the menu, in the Italian ethnicity condition the restaurant in which the study was conducted was also decorated with Italian flags, posters and wine bottles on the tables (with the classic red-and-white checked table-cloths, of course). It is therefore not possible to know for sure how much of the change in participants' ratings in the latter study should be attributed to the change in the name of the dishes versus the change in the atmosphere in the restaurant where those dishes happened to be served (see Chapter 9 on the profound role played by the atmosphere of the places where we eat on our perception of what we eat).

In other instances, using the French name for a dish instead of its English version could give the dish a touch of sophistication or nobility, as Carême might have said (e.g. *quiche* versus tart and *au chocolat* versus with chocolate). Halligan (1990, p. 98) suggests that describing a food as a *casserole* will make it sound a little more exotic than referring to it as a stew (although French cuisine is perhaps no longer held in quite as high esteem as it once was; Gopnik 1997; Lubow 2003; Steinberger 2010). In fact, people the world over have traditionally thought of French *cuisine* as being delicate and exquisite (cf. LeClerc *et al.* 1994). In any case, the inclusion of foreign names on the menu is always going to transmit the impression that the ingredients or, at the very least, the cooking methods are (or were) originally from that region. The only thing left for the chef or restaurateur to do is to make sure that the expectations of the foodies sampling their culinary offerings are met and, by so doing, avoid any disconfirmation of expectation related specifically to the perceived ethnicity of the food that they happen to be serving.

> *"'Sir', said Dr. Johnson, more than 200 years ago, 'my brain is obfuscated after the perusal of this heterogeneous conglomeration of bastard English ill-spelt and a foreign tongue'."* (Halligan 1990, p. 98)

It is also worth mentioning one of the other reasons for the use of foreign terms: to bamboozle the customer. That, at least according to Aradna Krishna (2013, p. 120), is part of the reason why Starbucks uses terms such as *grande* and *venti* on their drinks menu. Apparently, it also makes their coffee sound foreign, exotic and more expensive than if normal names had been used instead. This strategy doesn't work for everyone, however. Once upon a time Dunkin' Donuts introduced a hot sandwich and called it a *'panini'*. However, they soon had to change that to a 'stuffed melt' when their regulars complained that the Italian name was just too fancy (see Adamy 2008; Cowen 2012, p. 82).

> *"'Heirloom', 'sustainable', 'organic', and 'local' are labels that have become just another form of branding in a label-obsessed society, complete with a low-end line at Wal-Mart. As evidence of this trend, the Per se menu reads like a fashion magazine, only instead of the cut, color, and designer, we had the cut, color, and farmer. Potatoes weren't potatoes, they were pureed purple marble potatoes from Mr. McGregor's garden. Cavendish Farms quail, Snake River Farms beef, Four Stry Hills veal, Hallow Farms rabbit, Thumbelina carrots, Pink Lady apples, and wild arugula were similarly gussied."* (Damrosch 2008, p. 157)

3.8 Natural and organic labels

In recent years, especially in many developed countries, there has been a dramatic growth of interest in (not to mention desire for) all things natural/organic, especially when it comes to food (e.g. Rozin *et al.* 2004).[18] The explosion in organic foods started at the end of the 1990s in the UK (see Luttinger and Dicum 2006, p. 197) and can be linked to the emergence of the slow food movement (Petrini 2007, p. 88), fears about biotechnology in food and a response to the widespread belief that many fruits and vegetable have lost much of their taste as they have been increasing bred for their looks, size, colour and uniformity (see Engel 2001; Haden 2005; Wilson 2009). To put this change into perspective, organic food sales have increased tenfold from £105 million to £1.2 billion over the last decade. Given such a trend, one might ask what effect labelling a food as 'natural' or 'organic' has on people's perception of the taste/flavour of a dish in a restaurant or, for that matter, anywhere else.[19] The cognitive neuroscientists have shown that labelling food as organic can lead to increased activity in the ventral striatum, a part of the brain involved in controlling our motivation to eat and acquire food (Linder *et al.* 2010).

To date however, only a few studies have attempted to investigate whether organically grown products do actually taste better than commercially grown fruit and vegetables. For instance, Schutz and Lorenz (1976) reported that people did not rate organically grown products as being tastier than their commercially grown counterparts (note that in this study the organic food was not given an organic or natural label). That said, in their comprehensive review of the literature Bourn and Prescott (2002) did not find any clear evidence to show that consumers would rate organic foods as tasting better when tried blind. Meanwhile, Zhao *et al.* (2007) reported that people couldn't detect any difference between organic and conventionally grown vegetables, despite the fact that 28% of the consumers who were tested in this study thought that the organic veg would taste better. This is not to say that people might not rate a food as tasting better once it has been labelled as 'organic' or 'natural', of course (cf. Spence 2010a). In fact, Schuldt and Hannahan (2013) reported recently that while organic foods are generally perceived as being healthier than conventional foods, consumers also expect them to taste worse (their study was conducted in the US). This was especially true among those participants who weren't so concerned about the environment.

---

[18] Although MacClancy (1992, chapter 23) sees the growth of 'natural' associated with foods as nothing more than a cynical marketing ploy, given the vagueness surrounding what exactly the term itself means. MacClancy points to the fact that the Oxford English Dictionary lists 26 different definitions of the term (see also Tannahill 1973, p. 368).

[19] See Frewer *et al.* (1995) for tips on how to change the attitude of consumers towards, and their acceptance of, genetically engineered foods.

Separate from the question of how good organic food (or foods labelled as such) may appear to taste, Schuldt and Schwarz (2010) observed that cookies were rated as having fewer calories when they were described as organic. This would seem to represent an example of a halo effect.[20] A similar effect was also observed when participants actually tasted an identical cookie that had been labelled as 'organic' (they were also rated as being higher in fibre; Lee *et al.* 2013). Meanwhile, Wansink *et al.* (2000) reported that nutrition bars described as containing 'soy protein' were rated as more grainy and less flavourful than when the word 'soy' was not included in the description of the product (see also Wansink and Park 2002; Wansink *et al.* 2004). Given the widespread growth of interest in this area, more research is clearly going to be needed before any definitive conclusions can be reached regarding the influence of natural and/or organic food labels in a meal setting (whether in a restaurant or in the context of home dining). That said, what can safely be concluded at the present time from the research reported above is that changing the description of a food will likely alter a diner's response to it. What is harder to predict, certainly in the case of natural/organic labelling, is the direction of that change.

3.9   Health/ingredient labels

As we saw in Chapter 2, the information given to a diner about the nutritional content (mainly the number of calories, and the fat and salt contents) can have important consequences in the food choices that they make (Burton *et al.* 2006; see also Ford 1994). In those cases where one has the option of serving oneself (and repeating) as many times as one wants (this is known as eating *ad libitum*), is providing this information beneficial for the diners/consumers? What effect does this information have on the diner's/consumer's perception of the food?

Let's start by looking at some case studies and trying to answer these questions. It has been shown that people like strawberry yoghurt and cheese spreads more when they are labelled as 'full-fat' than when they are labelled 'low-fat' (see Wardle and Solomons 1994). That said, this may depend on the specific level of fat in the spread that a consumer normally happens to have with his/her toast (see Daillant-Spinnler and Issanchou 1995), as well as on the consumer's attitudes/beliefs towards one type of spread versus the other (Levin and Gaeth 1988; Aaron *et al.* 1994). Apparently, people eat more vanilla ice-cream if it has been accurately labelled as 'high-fat' than when (inaccurately) labelled as 'low fat' (Bowen *et al.* 1992). At least,

---

[20] In this context, a 'halo effect' is said to occur when the rating of one attribute (in this case the label 'organic') is transferred to another (in this case, the calorie count/fat content) in such a way that it affects the consumer's judgment about the product ("it says organic, so this food must be healthy and low in fat!").

that was the case 20 years ago when this study was originally conducted. The story nowadays would appear to be rather more complex than this simple reading of the older literature would suggest. Indeed, several recent studies have documented an apparently contradictory pattern of results (e.g. Piqueras-Fiszman *et al.* 2011).

For instance, Provencher *et al.* (2009) reported that people (regardless of whether they happened to control their diet or not) ate significantly more cookies (35%) when they were described as having been made with 'healthy' ingredients as compared to those purportedly made with 'unhealthy' ingredients. There are, however, some interesting cross-cultural differences to be noted here: Werle *et al.* (2013) found that the French implicitly associate 'healthy' with 'tasty' and equate 'unhealthy' with 'untasty' when shown images of foods that are obviously healthy and unhealthy (such as salad and pizza, respectively), as well as when given a neutral food (a mango milkshake) labelled as either 'healthy' or 'unhealthy'. North Americans, by contrast, tend to associate 'unhealthy' with 'tasty' instead (see Raghunathan *et al.* 2006).

Cavanagh and Forestell (2013) recently extended this study to look at healthy and unhealthy snack brands. The participants in their study rated cookies given a healthy brand label as being more satisfying and as having a better taste and flavour.[21] Furthermore, restrained eaters consumed more of the healthy brand than of a brand that had been described as less healthy. By contrast, the intake of unrestrained eaters did not differ as a function of the label given to the food. Perhaps helping to explain the contradictory results that have now been documented across a variety of foods, Wansink *et al.* (2005) reported that a meal description that includes 'healthy' or 'diet' has a stronger positive effect (that is, it results in higher liking ratings) for hedonic foods (e.g. desserts) than for side dishes or entrées (the idea here being that people may already expect the latter to be at least somewhat healthy).

Elsewhere, Irmak *et al.* (2011) presented participants with a pasta salad consisting of a mixture of chopped tomatoes, onions, red peppers, pasta shells, salami and mozzarella, served on romaine lettuce and then dressed with a vinaigrette. This dish could legitimately be identified either as a salad or as a pasta item. However, when the dish was identified as a pasta, dieters rated it as being less healthy and less tasty than the non-dieters. Conversely, when described as a salad, the dieting tendencies of the participants had no effect on their overall evaluation of the dish. This result was explained by the non-dieters' insensitivity to food cues as well as their reliance on cues indicating a lack of healthiness and their tendency to use a heuristic information processing strategy when they evaluated the foods. In a similar follow-up study, people were shown to eat twice as much fruit candy when it was labelled as 'fruit chews' rather than as 'candy chews'!

---

[21] Once again, the cookies in this study were actually identical in both conditions.

The potential, if unrecognized, societal importance of food labelling has also been highlighted by Wansink and Chandon (2006). These influential researchers have put forward the controversial claim that the introduction of low-fat nutrition labels might actually, not to mention somewhat counter-intuitively, contribute in some small way to the growing obesity crisis. Their suggestion is that, at the end of the day, people might eventually consume more because they think that a food is light (but see Miller *et al.* 1998). Note also that this effect extends to the names of restaurants (or better said, to the connotations that those restaurant names give rise to in the mind of the consumer). As an example of this, people would likely underestimate the calories in a meal from Subway (long marketed as a healthy alternative to other fast-food chains)[22] when compared to a meal from McDonald's (see Chandon and Wansink 2007). Furthermore, people tend to select less-caloric items (an average calorie content of M = 618 kcal) in fast-food eateries that are considered healthy when the calories of all of the items are displayed compared to when they are not (M = 714 kcal; Wei and Miao 2013). By contrast, for fast-food outlets that are generally considered as being unhealthy, the opposite pattern of results has been observed. That is, when the calories contained in a food are shown, people tend to select those items with a higher caloric content (M = 668 kcal) than when such information is not made explicit (M = 612 kcal).

3.10   Local labels

Why is it no longer possible to buy a straightforward cheese and onion sandwich? Why do supermarkets always seem to be offering a 'sharp Cheshire cheddar with Wiltshire red onion' sandwich, for example, or succulent 'Mr Jones' turkey and first-growth cranberry sandwich' rather than just a standard old-fashioned turkey and cranberry roll? Part of the reason for this change likely comes down to the fact that people appreciate food more, and will pay more for it, if it has been given a label that is more descriptive (note that Walkers in the UK has also adopted this strategy for its main range of crisps; Gyekye 2013). Certainly, no one expects that the person eating one of these sandwiches would ever be able to distinguish between Mr Jones' and Mr Smith's turkey breasts (for those who remain doubtful, see Makens 1965).

---

[22] It is however a little hard to see how anyone who advertises one of their products as a "comforting, splodgilicious Sub with tasty meatballs" can be all that healthy (see Wallop 2012). Here, one might also want to consider the recent explosion of burger chains with superlative-filled names – think of 'Gourmet Burger', 'Ultimate Burger' or Giles Coren's (2012, p. 50) suggestion of 'Wicked Blinding Mental Burger with Knobs On'! Coren continues " … *they all boast grass-fed, arse-licked, pan-killed, dry-aged, stone-seared fillet of a half-cow/half-mermaid Sports Illustrated cover girl, ground by pixies and char-tickled over baby maple sapling smoke … and you get there, and it's just another minced mammal in a bap.*"

"When was the last time a good piece of British cheese wrapped itself around your taste buds? Have you ever made the acquaintance of the spritely Cheshire made by the Appelby family; the majestic mature Lancashire of Mrs Kirkham; the Irish Cashel Blue; the beguiling Spenwood or the infinitely beguiling Wigmore; or the imperial Stilton from Colston Bassett?" (Matthew Fort, The Guardian, 9 October 1993; cited in James 2005, p. 381)

Such descriptive food naming should make the diner all the more suspicious given the results of a study by (Local Government Regulation) inspectors who tested hundreds of items in restaurants, shops, markets, etc. They came across a worryingly large number of examples of misleading food labels: *"including 'Welsh lamb' which actually came from New Zealand, 'Somerset butter' from Scotland, and 'Devon ham' from Denmark"* (all examples taken from Anonymous 2011b). In terms of mislabelling food as local when it was in fact not, the restaurants were the worst offenders with nearly 20% of claims being proved false (see also Wilson 2009).[23]

3.11   Descriptive food labelling

Even something as simple as describing a dish as 'tender chicken' rather than simply as just plain old 'chicken' will likely improve a diner's reaction to it (see Ariely 2008, pp. 164–165). This could be the reason why, on a recent flight to Miami, one of your authors was asked by the stewardess: *"Madam, what do you prefer, delicious pasta or delicious chicken?"* She's still uncertain as to whether the stewardess was joking or whether she was really asked to describe the menu like this (given the poor reputation of in-flight menus, the airlines presumably need all the linguistic help possible to make their offerings taste a little bit more palatable. "Heston, come quick!" Whittle 2013).

"Red rich marbled lovingly aged thoughtfully slaughtered premium rump steak from the contented cattle of the legendary Darling Downs farcied with a mousse of lobster that only this morning swam in the sea served on a coulis of succulent king prawns and garnished with our very own garden-young vegetables adorned with the chef's special mango and raspberry vinaigrette dressing." (Halligan 1990, pp. 198–199)

In what has become the classic study of descriptive food naming, Wansink *et al.* (2005) served people a range of savoury main courses that had been given either a basic name or else a more descriptive food label (e.g. 'Seafood Filet' versus 'Succulent Italian Seafood Filet'; 'Chicken Parmesan' versus 'Home-style Chicken Parmesan'; or 'Chocolate Pudding' versus 'Satin Chocolate Pudding'). In this study, the use of descriptive food labels actually led to a doubling

---

[23] Worryingly, the report continues *"There is currently no legal definition of the term 'local' in food labelling legislation."*

in the number of positive comments that the dishes attracted, as compared to when the more basic food labels were used instead (though one should not abuse either; Wansink *et al.* 2001).[24] More recently, Brian Wansink *et al.* (2012) have gone on to show very similar effects in children. The 8–11 year olds tested in their study ate twice as many 'X-ray Vision Carrots' than simple anonymous 'carrots'. This effect is apparently at least moderately long-lasting given that the same results were observed in another month-long study. Such results should obviously be of interest to those restaurants/chains offering a children's menu (but see Robinson *et al.* 2007 on the more worrying effects of food labelling/branding in children).

3.12   Labelling culinary techniques

Mielby and Bom Frøst (2012) demonstrated that the story regarding food naming may be rather more complex than was suggested by the results of the classic study by Wansink *et al.* (2005). In particular, these researchers worked with one of the sous chefs at Noma, currently one of the world's top restaurants (http://noma.dk/). An eleven-course tasting menu was specially created and served to the diners/participants taking part in this study (the kind of study that everyone would like to take part in, no?). Each of the dishes was given one of four different names. The results revealed, once again, that the diner's experience of the dishes was affected by the information that was provided about the dish. That said, their results also highlighted the presence of a complex interaction with the dish being served. This means that there is no simple answer to the question of how changing the name will influence a diner's response to a dish. For example, a Brie parfait rolled in rye bread crumble and a rhubarb sherbet could either be described to the diners in a hedonically evocative manner such as: 'Cheese and rhubarb: a delicious and creamy parfait is united with a refreshingly cooling ice-cream of rhubarb which assembles in the mouth in pure enjoyment' or, in order to emphasize the culinary process, as: 'Cheese and rhubarb: this dish was frozen at a very low temperature (−22°F/−30°C) and the ice crystals were comminuted using a Pacojet' or, to highlight the experience, as: 'Cheese and rhubarb: parfait of cheese and sorbet of rhubarb in another texture'.

Somewhat surprisingly, given what we have seen so far in this chapter, those diners who received the hedonically evocative food description liked the dish

---

[24] Here, we are reminded of what Coren (2012, p. 228) calls the M&S marketing principle. *"You take the same old product and identify that words like 'organic' and 'free range' are playing well in the marketplace, but realize that legally you can't say that about your food so you just slide other, subjective and essentially meaningless words into the template and hope that nobody will notice. You know the sort of thing: 'These are not just tomatoes, these are red, round, bulbous, tomatoey tomatoes, with skins and a little green stalky bit on.'"*

least. This would imply that if you are going to serve a particularly complex dish at your next dinner party, one that (heaven-forbid) makes use of the latest modernist tools and techniques, then you would be well-advised to give it a name that highlights the technique used rather than necessarily describing the specific sensory experiences that your dinner guests might be expected to have as a result of eating it (assuming that you have managed to follow all of the steps in the recipe; Youssef 2013). While these recent results prevent the drawing of any simple conclusions about what kind of name is best for a given kind of food, the main point still remains that very often people's perception of, and response to, a dish can be dramatically altered simply by changing the name/description that it is given. In other words, the naming of a dish is far too important a decision to be left to whim or chance.[25]

One can see a combination of technical description with an inimitable hint of Câreme's style of naming dishes after famous scientists in the following example from what Hervé This describes as the first note-by-note art and science menu: "*Kientzheim butter on a fillet of sole veiled with white corn flour ... Polyphenol sauce seasoning a blue lobster* fricasée *poached in a* noisette *butter*" (Note by Note N° 1, This 2012). Although it is of course important not to sound too pretentious and to avoid Giles Coren's imaginary "*Lobster Pierre Choderlos de Laclos, sauce Antoine de Saint-Exupéry, sur son lit de pommes de terre façon, merde, je ne me rappelled plus le nom, tu sais, le mec qui etait le chef de Tallyrand, je voulais dire Antonin Worrall-Thompson, mais ce n'est pas lui ...* " (Coren 2012, p. 176).

That said, in certain cases (and here we are mostly talking about top-end restaurants) the descriptions are simply incomprehensible. Some examples of this kind of food naming include "Spherical egg of white asparagus with false truffle", the name of one of the dishes served at the world-famous elBulli restaurant in Spain before it sadly closed its doors for the last time (see Figure 3.3). Note here that even when one has seen the dish, it is still unclear what exactly is being served. Perhaps the only thing one does know is what to expect in terms of the taste/flavour? Another example is "Lamb ....?????............!!!!!!!!!!!!!", served at Grand Achatz's flagship Chicago restaurant Alinea (www.alinearestaurant.com).

Other chefs prefer to describe their food offerings in a 'short and sweet' style. This is undoubtedly the case of Fergus Henderson's approach in his acclaimed London restaurant, St John. There, the ingredients speak for themselves. For example, you might find: 'Veal, chicory and anchovy'; 'Mussels, cucumber and dill'; or 'Brill, fennel and green sauce' on the menu. These descriptions certainly sound as though they have come straight from a shopping list rather than from one of London's premier dining venues.

---

[25] In the early days, breakfast cereals were described as "pre-digested" in line with their nutritionist movement background (Belasco 2006, p. 171). One can't imagine them having been so successful had that particular descriptor not been removed.

Figure 3.3 'Spherical egg of white asparagus with false truffle', a dish served at elBulli before it closed. Just by reading the name of the dish on the menu, one knows that most probably it will be useless to search for the egg or the truffle (or even the asparagus). *Source*: Reproduced with permission of Blake Jones. *See colour plate section*

The exact-same naked style of description can be found on the menu of any of the Polpo restaurants in London. Russell Norman (founder of the group of restaurants; http://www.polpo.co.uk/) finds the bombastic style of dish naming very off-putting. As he put it: "*I'm particularly unsympathetic to florid descriptions*" (Fleming 2013). These minimalist or intriguing examples of naming/describing undoubtedly draw the diner's attention to the waiter bringing the mysteriously titled dish to the table rather than letting the diner be distracted by the conversation around the table. That said, as Giles Coren (2012; food critic for the *The Times* newpaper for a number of years) has noted, the very absence of florid adjectives can itself become problematic: "*Brevity and the low vernacular: 'beef in crust', 'burnt cream', 'cock in wine'. It's a weird, strangulated form of pretentious unpretentiousness. And it feels to me like the most pretentious sort of naming there could be.*" (Coren 2012, p. 177)

Gay Bilson's minimalist menu for the meal served at a Symposium of Australian Gastronomy a few years ago (taken from Halligan 1990, p. 198), read:

> The menu
> *LOAVES*
> *OYSTERS*
> *PRAWNS*
> *RICE*
> *SALAD*
> *HARE*
> *SWEETS*
> *PIES*

The absence of any sensory descriptors in this pared-down menu sends its own message.

3.13 Surprise!

In contrast, the names of various other dishes don't even mention a single ingredient, such as those from the menu of the Unsicht-Bar (a dine-in-the-dark restaurant in Berlin). Take for instance the main course of the 'River and Sea' menu, described as: 'Like Zeus this nobleman dreams in his golden bed about guests from all over the world that get drunk on fine wine'. What's ironic at this restaurant is that there's also a 'Surprise menu', which has no description at all (see Chapter 8). These are examples of the mysterious/intriguing style of dish naming mentioned at the start of the chapter.

Just as intriguing are the names given to dishes that have little to do with the contents of the plates themselves. Take for example the 'Bombay duck' sometimes found on the menu in curry houses in the UK (and, for all we know, elsewhere). The dish is actually not made from duck but from the lizardfish, a native of Mumbai, India. In this case, the origin of the dish's name isn't altogether clear (see https://en.wikipedia.org/wiki/Bombay_duck). Just imagine the face of those naïve tourists ordering sweetbreads in some English-speaking country, obviously expecting what it says on the label. One wonders whether this is another of those names that may have been chosen to obscure the actual nature of what is on offer (reminding one of the two-legged mutton mentioned in Section 3.2). Meanwhile, prairie oysters, animelles or external kidneys are just some of the euphemistic names given to plated testicles (Halligan 1990, p. 99).[26] Another form of labelling that doesn't seem

---

[26] For anyone who is still hungry, things can get much, much worse; just take a look at Harding (2006) for the perfect accompaniment to a side of breaded animelles!

altogether accurate occurs when the menu describes the seafood as 'resting' on a bed of vegetables.[27]

3.14   Expectations and reactions

As we have seen throughout this chapter, there's much more to a name than one might initially have thought. The name and/or description that is provided before a dish is presented to a diner will necessarily create expectations about its likely qualities: what ingredients it will have been made from; perhaps which country the ingredients originate from; the possible texture it will have; how it will be presented; which techniques may have been used to prepare it; or simply: 'What will it be?' The chef/restaurateur undoubtedly has the power to play around with those expectations. For instance, they might want to highlight the presence of the most expensive ingredients on the plate (e.g. 'with roses from the Himalayas'), enhance the ethnicity of the food (e.g. '*confit de canard*'), avoid confusion, set up expectations regarding specific sensory qualities (e.g. a 'rich and creamy soup with crunchy croutons'), increase the perceived value (and hence price) and quality by identifying the particular farm from which the produce came (e.g. 'with fresh organic vegetables grown in our own fields'), elicit surprise (e.g. 'Space food', 'Food 386' or 'Surprise Omelette', sometimes called '*Omelette à la Norvégienne*' or, in other words 'Baked Alaska'; see Halligan 1990, p. 83) and/or capture the diner's (or media's) attention (think 'Snail Porridge'). Importantly, all these techniques are possible and all constitute relevant naming strategies in different contexts. It is also worth pointing out here that even those who try not to set expectations inevitably do so; remember that there are still expectations around a dish with no name or the courses that are served on a surprise menu. As the philosopher Jonathan Cohen notes when it comes to the blind tasting of wine, we always create our own expectations and beliefs around food/drink even when we are served blind (e.g. in black glasses in the case of a double-blind wine tasting; Cohen, in press) or presumably while sampling the surprise menu at the dark dining restaurant (as we'll see in Chapter 8).

One day I said to the barman, 'Do you think you could do me thin chips for a change?'

'How do you mean?'

'You know, like in France – the thin ones.'

'No, we don't do them.'

'But it says on the menu your chips are hand-cut.'

---

[27] There is a wonderful line in Steve Coogan and Rob Bryden's TV series *The Trip* where, after a dish is introduced in this manner by the waiter, Steve Brydon points out "*Rather optimistic to say they're resting. Their days of resting have been and gone. They are dead*" (see Poole 2012).

'Yes.'

'Well, can't you cut them thinner?'

The barman's normal affableness took a pause. He looked at me as if he wasn't sure whether I was a pedant or an idiot, or quite possibly both.

'Hand-cut chips mean fat chips.'

'But if you hand-cut chips, couldn't you cut them thinner?'

'We don't cut them. That's how they arrive.'

'You don't cut them on the premises?'

'That's what I said.'

'So what you call "hand-cut" chips are actually cut elsewhere, and quite possibly by a machine?'

'Are you from the council or something?'

'Not in the least. I'm just puzzled. I never realised that "hand-cut" meant "fat" rather than "necessarily cut by hand".'

'Well, you do now.'" (Barnes 2011, pp. 145–146)

Different names will elicit different expectations (and consequently reactions) to a dish from a diner. For instance, if a brownie is named as a 'Brownie to die for', the chef had better make sure that it is suitably spectacular if he or she doesn't want the one being murdered! The same holds true for the 'crunchy pork crackling' or the 'crispy fries' that are nowadays a perennial feature in gastropub menus in the UK.[28] Many of those writing these menus feel the need to fill them with sensory descriptors without necessarily having to skill to deliver the said properties on a consistent basis.[29] Hence, the name of the dish (what it conveys) should be consistent with what the diner is going to experience. Note that it doesn't need to be congruent (as when intending to elicit a surprise reaction), but if the name/description creates very high expectations the food obviously needs to meet them (or at least set up an expectation that is pretty close to the reality). Ensuring that the expectations set by the name are just a little higher than what the diner really gets is most probably okay (it may even be a desirable naming strategy; Spence 2012). After all, the difference might not be noticed by the diner and, what is more, the expectations set up by the name/description could actually end up enhancing the perceived or remembered flavour of the food itself.

It seems probable that the diner in a modernist restaurant likely questions his/her taste buds more than someone eating in a regular dining establishment.

---

[28] Anyone wanting to avoid the wrath of Elizabeth David would also be well advised to delete the 'y' in 'crispy'. As she once said to the fine British chef Simon Hopkinson: *"I do not understand this obsession with crispy; crisp is finite"* (Whittle 2013). Elsewhere, one finds Hopkinson himself asking why menus always have to 'drizzle' their balsamic. Doesn't pouring or splashing the vinegar sound just as appealing? One other complaint that is currently levelled at the gastropub menu is why exactly that reassuring adjective 'proper' has started popping up everywhere, as in "proper sausage and mash" or "proper pork pie" (Poole 2012).

[29] For anyone wanting tips on how to get the perfect crispy-crunchy finish, see Young and Myhrvold (2012)

We would therefore argue that any discrepancy between label and food is more likely to be noticed in a modernist restaurant than perhaps elsewhere. In many cases, the aim of the chef (or whoever it is who takes up the challenge of deciding on the names of the dishes) is to elicit surprise or mystery (for instance, by using names that do not give any clue as to what is actually on the plate). This will most probably make the diner wonder and become impatient to see the dish and immediately take a bite. However, if the food itself is not consistent with the surprise element that the name promises (say, it turns out to be just a plain old steak), then the diner will likely experience a negative disconfirmation of expectations (a bad thing; see Chapters 6 and 7) not necessarily because the steak is bad, but because the exciting surprise factor disappears at the very same time as the dish appears.

> "'The naming of a cat is a difficult matter' wrote the Old Possum T. S. Elliot. So is the naming of anything. Names become their objects, objects their names; they reveal and conceal. The naming of food is full of surprises. Tell me what a man eats and I will tell you what he is, says Brillat Savarin. Telling you what he calls it is even more enlightening." (Halligan 1990, p. 94)

3.15   Conclusions

The research that has been reviewed in this chapter clearly highlights how our perception of and response to food and drink can be dramatically affected by the names and labels that they are given, whether it is the name of the dish on the restaurant menu or the description provided (on the menu or by the waiter). The name of a food may influence our perception of its taste/flavour either by setting up an expectation (or end anchor; Stewart 2009) that tends to attract a diner's attention and hence focus/bias their perception (Cardello 2007). Getting the name and description right is especially important in those cases in which people are likely to assume the wrong taste or flavour, that is, when they may be misled by the visual attributes of the dish or where there are big cross-cultural differences involved (Shankar *et al.* 2010). The importance of choosing the right name therefore makes one wonder about all those dishes whose name seems to have been chosen on the basis of nothing more than whim, such as 'Egg in the Hole' (an egg fried inside a hole carved out of a slice of bread) or 'Bubble and Squeak' (another traditional English dish made of the leftover vegetables from a roast, normally potatoes and cabbage, which are mashed up and then fried together in a pan). In this chapter, we have seen a number of examples documenting how the semantic (meaning) aspects of the words we use to describe the foods we consume influence our perception of and responses to a variety of different food items (see Table 3.1 for a summary of some of the key strategies and goals of food naming and description).

Table 3.1 Summary of studies that have looked at the effect of modifying the name on the consumer/diner's perception of the food being served (and, on occasion, their liking of the dish)

Effect of the dish name	Study
Enhance the perceived naturalness/ healthiness of the food	Bowen et al. 1992; Wardle and Solomons 1994; Wansink et al. 2000; Wansink and Chandon 2006; Chandon and Wansink 2007; Provencher et al. 2009; Schuldt and Schwarz 2010; Irmak et al. 2011; Cavanagh and Forestell 2013; Schuldt and Hannahan 2013; Lee et al. 2013; Wei and Miao 2013
Playing with the disconfirmation of expectations	Cardello and Sawyer 1992; Yeomans et al. 2008
Highlight the culinary technique/ skills of the culinary team	Mielby and Bom Frøst 2012; This 2012
Enhance the perceived ethnicity of the dish	Bell and Paniesin 1992; Meiselman and Bell 1993; Bell et al. 1994; LeClerc et al. 1994
Increase uncertainty/mystery concerning the identity of the dish	Wolfson and Oshinsky 1966; Robinson et al. 2007; Yeomans et al. 2008; Wansink et al. 2012
Enhance the sensory qualities (e.g. flavour/taste/aroma) of a food or drink	Olson and Dover 1978; Daillant-Spinnler and Issanchou 1995; Herz and von Clef 2001; Wansink et al. 2005; Nitschke et al. 2006; Okamoto et al. 2009; Shankar et al. 2009

The appropriate labelling of a dish can be particularly important in those situations in which the other (e.g. visual) attributes of the food (or atmosphere) are likely to lead the diner astray in terms of correctly identifying the food on their plate. It is also important in those areas where various brands and varieties of a given food may be difficult to distinguish on the basis of their flavour alone. While hedonic and descriptive sensory labels for a dish often enhance the response of diners, it is important to remember that the appropriateness of a particular kind of description may well depend on the dish that is being served, or even the kind/style of venue in which it is going to be consumed (see Mielby and Bom Frøst 2012). For anyone tempted to try out their own modernist cooking techniques on their friends at their next dinner party, remember that describing the techniques used may be more appropriate than the use of one of the descriptive sensory labels made famous by the work of Brian Wansink (Wansink et al. 2005). Finally, when thinking about the naming of a dish, it is always worth bearing in mind that it is sometimes better not to tell the diner about all of the ingredients that may have gone into its creation (Heldke 2005; Lee et al. 2006).

To sum up, while the ultimate aim when thinking up the most appropriate name and description for a dish may vary, what the results reviewed in this chapter all make absolutely clear is that the name you give to a dish nearly always has an effect on a diner's expectations and hence on their experience (both sensory-discriminative and hedonic). So, having dealt with the menu and the basics of food naming, we can now move on through the course of the perfect meal. In the next chapter we will look at the next elements that are likely to contribute to one's dining experience, namely the plateware and the plating of the food.

References

Aaron, J. I., Mela, D. J. and Evans, R. E. (1994) The influence of attitudes, beliefs and label information on perception of reduced-fat spread. *Appetite*, **22**, 25–37.

Adamy, J. (2008) Dunkin' Donuts tries to go upscale, but not too far. *The Wall Street Journal*, April 8–9, A1, A7.

Allen, J. S. (2012) *The Omnivorous Mind: Our Evolving Relationship with Food*. Harvard University Press, London.

Anonymous (1962) Does the label 'change' the taste? *Printers Ink*, **278(1)**, 55–57.

Anonymous (2009) To prevent sniggering, pudding is renamed Spotted Richard. *The Daily Telegraph*, 9 September, 14.

Anonymous (2010) Caught: Inferior fish sold as cod and haddock. *The Guardian*, 24 April, 5.

Anonymous (2011a) Fish name cods staff. *The Times*, 2 February.

Anonymous (2011b) Local labels on many foods are false, study suggests. Available at http://www.bbc.co.uk/news/uk-12582056 (accessed January 2014).

Anonymous (2013) Can we prevent another horsemeat scandal? *New Food*, **16(2)**, 19–23.

Arana, F. S., Parkinson, J. A., Hinton, E., Holland, A. J., Owen, A. M. and Roberts, A. C. (2003) Dissociable contributions of the human amygdala and orbitofrontal cortex to incentive motivation and goal selection. *Journal of Neuroscience*, **23**, 9632–9638.

Ariely, D. (2008) *Predictably Irrational: The Hidden Forces that Shape our Decisions*. Harper Collins, London.

Asam, E. H. and Bucklin, L. P. (1973) Nutrition labeling for canned goods: A study of consumer response. *Journal of Marketing*, **37** (April), 32–37.

Ashkenazi, A. and Marks, L. E. (2004) Effect of endogenous attention on detection of weak gustatory and olfactory flavors. *Perception and Psychophysics*, **66**, 596–608.

Barnes, J. (2011) *The Sense of an Ending*. Jonathan Cape, London.

Barrós-Loscertales, A., González, J., Pulvermüller, F., Ventura-Campos, N., Bustamante, J. C. *et al.* (2012) Reading salt activates gustatory brain regions: fMRI evidence for semantic grounding in a novel sensory modality. *Cerebral Cortex*, **22**, 2554–2563.

Belasco, W. J. (2006) *Meals to Come: A History of the Future of Food*. University of California Press, Berkeley, CA.

Bell, R. and Paniesin, R. (1992) The influence of sauce, spice, and name on the perceived ethnic origin of selected culture-specific foods. In: *Product Testing with*

Consumers for Research Guidance: Special Consumers Groups, Vol. **2** (eds L. S. Wu and A. D. Gelinas), pp. 22–36. American Society for Testing and Materials, Philadelphia, PA, STP 1155.

Bell, R., Meiselman, H. L., Pierson, B. J. and Reeve, W. G. (1994) Effects of adding an Italian theme to a restaurant on the perceived ethnicity, acceptability, and selection of foods. *Appetite*, **22**, 11–24.

Bensafi, M., Porter, J., Pouliot, S., Mainland, J., Johnson, B. *et al.* (2003) Olfactory activity during imagery mimics that during perception. *Nature Neuroscience*, **6**, 1142–1144.

Blumenthal, H. (2007) *Further Adventures in Search of Perfection: Reinventing Kitchen Classics*. Bloomsbury Publishing, London.

Blumenthal, H. (2008) *The Big Fat Duck Cookbook*. Bloomsbury, London.

Bourland, C. T. and Vogt, G. L. (2010) *The Astronaut's Cookbook: Tales, Recipes, and More*. Springer, New York.

Bourn, D. and Prescott, J. (2002) A comparison of the nutritional value, sensory qualities, and food safety of organically and conventionally produced foods. *Critical Reviews in Food Science and Nutrition*, **42**, 1–34.

Bowen, D. J., Tomoyasu, N., Anderson, M., Carney, M. and Kristal, A. (1992) Effects of expectancies and personalized feedback on fat consumption, taste, and preference. *Journal of Applied Social Psychology*, **22**, 1061–1079.

Burton, S., Creyer, E. H., Kees, J. and Huggins, K. (2006) Attacking the obesity epidemic: The potential health benefits of providing nutrition information in restaurants. *American Journal of Public Health*, **96**, 1669–1675.

Cardello, A. V. (2007) Measuring consumer expectations to improve food product development. In: *Consumer-Led Food Product Development* (ed H. J. H. MacFie), pp. 223–261. Woodhead Publishing, Cambridge.

Cardello, A. and Sawyer, F. (1992) Effects of disconfirmed consumer expectations of food acceptability. *Journal of Sensory Studies*, **7**, 253–277.

Carême, M.-A. (1822) *Le Maitre d'Hôtel Francais V1 (The French Headwaiter)*. Kessinger Publishing, Whitefish, MT.

Cavanagh, K. V. and Forestell, C. A. (2013) The effect of brand names on flavor perception and consumption in restrained and unrestrained eaters. *Food Quality and Preference*, **28**, 505–509.

Chabris, C. and Simons, D. (2011) *The Invisible Gorilla and Other Ways Our Intuition Deceives Us*. HarperCollins, London.

Chandon, P. and Wansink, B. (2007) The biasing health halos of fast-food restaurant health claims. Lower calorie estimates and higher side-dish consumption intentions. *Journal of Consumer Research*, **34**, 301–314.

Cline, J. A. (1943) The variety meats. *Practical Home Economics*, **21**, 57–58.

Cohen, J. (in press) Wine tasting, blind and otherwise: Blindness as a perceptual limitation? In *Wine Expertise* (ed. O. Deroy). Oxford University Press, Oxford.

Coren, G. (2012) *How to Eat Out: Lessons from a Life Lived Mostly in Restaurants*. Hodder and Stoughton, London.

Cowen, T. (2012) *An Economist Gets Lunch: New Rules for Everyday Foodies*. Plume, New York.

Daillant-Spinnler, B. and Issanchou, S. (1995) Influence of label and location of testing on acceptability of cream cheese varying in fat content. *Appetite*, **24**, 101–105.

Damrosch, P. (2008) *Service Included: Four-Star Secrets of an Eavesdropping Waiter*. William Morrow, New York.

De Araujo, I. E., Rolls, E. T., Velazco, M. I., Margot, C. and Cayeux, I. (2005) Cognitive modulation of olfactory processing. *Neuron*, **46**, 671–679.

Djordjevic, J., Zatorre, R. J., Petrides, M. and Jones-Gotman, M. (2004) The mind's nose: Effects of odor and visual imagery on odor detection. *Psychological Science*, **15**, 143–148.

Djordjevic, J., Lundstrom, J. N., Clément, F., Boyle, J. A., Pouliot, S. and Jones-Gotman, M. (2008) A rose by any other name: Would it smell as sweet? *Journal of Neurophysiology*, **99**, 386–393.

Dolnick, E. (2008) *Fish or foul?* Available at http://www.nytimes.com/2008/09/02 /opinion/02dolnick.html?_r=1&scp=1&sq=chocolate%20strawberry%20yogurt &st=cse (accessed January 2014).

Driver, J. (2001) A selective review of selective attention research from the past century. *British Journal of Psychology*, **92**, 53–78.

Dubé, L. and Cantin, I. (2000) Promoting health or promoting pleasure? A contingency approach to the effect of informational and emotional appeals on food liking and consumption. *Appetite*, **35**, 251–262.

Ellis, C. (2007) *Fabulous Fanny Cradock*. Sutton Publishing Limited, Stroud.

Engel, M. (2001) That green vegetable: Who needs it? *Sydney Morning Herald*, 25 February.

Ferguson, P. P. (2004) *Accounting for Taste: The Triumph of French Cuisine*. University of Chicago Press, Chicago.

Fleming, A. (2013) Restaurant menu psychology: Tricks to make us order more. Available at http://www.guardian.co.uk/lifeandstyle/wordofmouth/2013/may/08 /restaurant-menu-psychology-tricks-order-more (accessed January 2014).

Ford, G. T. (1994) The effects of the new food labels on consumer decision making. *Advances in Consumer Research*, **21**, 530.

Frewer, L. J., Howard, C. and Shepherd, R. (1995) Genetic engineering and food: What determines consumer acceptance. *British Food Journal*, **97(8)**, 31–36.

Gallace, A., Ngo, M. K., Sulaitis, J. and Spence, C. (2012) Multisensory presence in virtual reality: Possibilities and limitations. In: *Multiple Sensorial Media Advances and Applications: New Developments in MulSeMedia* (eds G. Ghinea, F. Andres and S. Gulliver), pp. 1–40. IGI Global, Hershey, PA.

Gladwell, M. (2009) The ketchup conundrum: Mustard now comes in dozens of varieties. Why has ketchup stayed the same? In: *What the Dog Saw and Other Adventures*, pp. 32–50. Little, Brown and Company, New York.

González, J., Barros-Loscertales, A., Pulvermüller, F. *et al.* (2006) Reading *cinnamon* activates olfactory brain regions. *NeuroImage*, **32**, 906–912.

Gopnik, A. (1997) Is there a crisis in French cooking? *New Yorker*, 28 April–15 May, 150.

Grabenhorst, F., Rolls, E. T. and Bilderbeck, A. (2008) How cognition modulates affective responses to taste and flavor: Top-down influences on the orbitofrontal and pregenul cortices. *Cerebral Cortex*, **18**, 1549–1559.

Gyekye, L. (2013) Walkers crisps unveils new packs. Available at http://www .packagingnews.co.uk/design/new-packs/walkers-crisp-unveils-new-packs/ (accessed January 2014).

Haden, R. (2005) Taste in an age of convenience. In: *The Taste Culture Reader: Experiencing Food and Drink* (ed C. Korsmeyer), pp. 344–358. Berg, Oxford.

Halligan, M. (1990) *Eat my Words*. Angus and Robertson, Sydney.

Hamilton-Paterson, J. (2004) *Cooking with Fernet Branca*. Faber and Faber, London.

Harding, A. (2006) Beijing's penis emporium. Available at http://news.bbc.co.uk /1/hi/5371500.stm (accessed January 2014).

Heldke, A. (2005) But is it authentic? Culinary travel and the search for the "genuine article". In: *The Taste Culture Reader: Experiencing Food and Drink* (ed C. Korsmeyer), pp. 385–394. Berg, Oxford.

Herz, R. S. and von Clef, J. (2001) The influence of verbal labelling on the perception of odors: Evidence for olfactory illusions? *Perception*, **30**, 381–391.

Hutchings, J. B. (1977) The importance of visual appearance of foods to the food processor and the consumer. *Journal of Food Quality*, **1**, 267–278.

Hyman, P. (2001). Culina Mutata: Carême and l'ancienne cuisine. In: *French Food: On the Table, on the Page, and in French Culture* (eds L. R. Schehr and A. S. Weiss), pp. 71–82. Routledge, New York.

Imm, B., Lee, J. H. and Lee, S. H. (2012) Effects of sensory labels on taste acceptance of commercial food products. *Food Quality and Preference*, **25**, 135–139.

Irmak, C., Vallen, B. and Robinson, S. R. (2011) The impact of product name on dieters' and nondieters' food evaluations and consumption. *Journal of Consumer Research*, **38**, 390–405.

James, A. (2005) Identity and the global stew. In: *The Taste Culture Reader: Experiencing Food and Drink* (ed C. Korsmeyer), pp. 372–384. Berg, Oxford.

Jones, G. (2013) Seafood savvy chefs love sinking their teeth into this monster of the deep. *The Telegraph (Australia)*, July 13, 9.

Kemps, E. and Tiggeman, M. (2013) Imagery and cravings. In: *Multisensory Imagery* (eds S. Lacey and R. Lawson), pp. 385–396. Springer, New York.

Kobayashi, M., Takeda, M., Hattori, N., Fukunaga, M., Sasabe, T., Inoue, N. *et al.* (2004) Functional imaging of gustatory perception and imagery: "Top-down" processing of gustatory signals. *NeuroImage*, **23**, 1271–1278.

Kobayashi, M., Sasabe, T., Shigihara, Y., Tanaka, M. and Watanabe, Y. (2011) Gustatory imagery reveals functional connectivity from the prefrontal to insular cortices traced with magnetoencephalography. *PLoS ONE*, **6**, e21736.

Krishna, A. (2013) *Customer Sense: How the 5 Senses Influence Buying Behaviour*. Palgrave Macmillan, New York.

Labroo, A. A., Dhar, R. and Schwartz, N. (2008) Of frog wines and frowning watches: Semantic priming, perceptual fluency, and brand evaluation. *Journal of Consumer Research*, **34**, 819–831.

Lacey, S. and Lawson, R. (eds) (2013) *Multisensory Imagery*. Springer, New York.

Leake, J. and Dowling, K. (2011) Fishy labels: What's really in that pack of haddock. The Sunday Times, 24 April (News), 8–9.

LeClerc, F., Schmitt, B. H. and Dubé, L. (1994) Foreign branding and its effects on product perceptions and attitudes. *Journal of Marketing Research*, **31**, 263–270.

Lee, L., Frederick, S. and Ariely, D. (2006) Try it, you'll like it: The influence of expectation, consumption, and revelation on preferences for beer. *Psychological Science*, **17**, 1054–1058.

Lee, W.-C., Shimizu, M., Kniffin, K. M. and Wansink, B. (2013) You taste what you see: Do organic labels bias taste perceptions? *Food Quality and Preference*, **29**, 33–39.

Levin, I. P. and Gaeth, G. J. (1988) How consumers are affected by the framing of attribute information before and after consuming the product. *Journal of Consumer Research*, **15**, 374–378.

Linder, N. S., Uhl, G., Fliessbach, K., Trautner, P., Elger, C. E. *et al.* (2010) Organic labeling influences food valuation and choice. *NeuroImage*, **53**, 215–220.

Lubow, A. (2003) A laboratory of taste. *The New York Times*, 10 August. Available at http://www.nytimes.com/2003/08/10/magazine/a-laboratory-of-taste.html?pagewanted=all&src=pm (accessed January 2014).

Luttinger, N. and Dicum, G. (2006) *The Coffee Book: Anatomy of an Industry from Crop to the Last Drop*. The New Press, New York.

MacClancy, J. (1992) *Consuming Culture: Why You Eat What You Eat*. Henry Holt, New York.

Makens, J. C. (1965) Effect of brand preference upon consumers' perceived taste of turkey. *Journal of Applied Psychology*, **49**, 261–263.

Manescu, S., Frasnelli, J., Lepore, F. and Djordjevic, J. (2014) Now you like me, now you don't: impact of labels on odor perception. *Chemical Senses*, **39**(2), 167–175.

Masurovsky, B. I. (1939) How to obtain the right food color. *Food Engineering*, **11**(13), 55–56.

Meiselman, H. L. and Bell, R. (1993) The effects of name and recipe on the perceived ethnicity and acceptability of selected Italian foods by British subjects. *Food Quality and Preference*, **3**, 209–214.

Mielby, L. H. and Bom Frøst, M. (2012) Eating is believing. In: *The Kitchen as Laboratory: Reflections on the Science of Food and Cooking* (eds C. Vega, J. Ubbink and E. van der Linden), pp. 233–241. Columbia University Press, New York.

Miller, D. L., Castellanos, V. H., Shide, D. J., Peters, J. C. and Rolls, B. (1998) Effect of fat-free potato chips with and without nutritional labels on fat and energy intakes. *American Journal of Clinical Nutrition*, **68**, 282–290.

Miller, E. G. and Kahn, B. E. (2005) Shades of meaning: The effect of color and flavor names on consumer choice. *Journal of Consumer Research*, **32**, 86–92.

Nakamura, A. and Imaizumi, S. (2013) Auditory verbal cues alter the perceived flavour of beverages and ease of swallowing: A psychometric and electrophysiological analysis. *BioMed Research International*, Article ID 892030, doi:10.1155/2013/892030.

Nitschke, J. B., Dixon, G. E., Sarinopoulos, I. *et al.* (2006) Altering expectancy dampens neural response to aversive taste in primary taste cortex. *Nature Neuroscience*, **9**, 435–442.

Okamoto, M., Wada, Y., Yamaguchi, Y. *et al.* (2009) Influences of food-name labels on perceived taste. *Chemical Senses*, **34**, 187–194.

Olson, J. C. and Dover, P. A. (1978) Cognitive effects of deceptive advertising. *Journal of Marketing Research*, **15**, 29–38.

Petrini, C. (2007) *Slow Food: The Case for Taste* (translated by W. McCuaig). Columbia University Press New York.

Piqueras-Fiszman, B., Ares, G. and Varela, P. (2011) Semiotics and perception: Do labels convey the same messages to older and younger consumers? *Journal of Sensory Studies*, **26**, 197–208.

Poole, S. (2012) *You Aren't What You Eat: Fed up with Gastroculture*. Union Books, London.

Provencher, V., Polivy, J. and Herman, C. P. (2009) Perceived healthiness of food. If it's healthy, you can eat more! *Appetite*, **52**, 340–344.

Raghunathan, R., Naylor, R. W. and Hoyer, W. D. (2006) The unhealthy = tasty intuition and its effects on taste inferences, enjoyment, and choice of food products. *Journal of Marketing*, **70**(4), 170–184.

Rayner, J. (2004) The man who mistook his kitchen for a lab. *The Observer*, 15 February. Available at http://www.guardian.co.uk/lifeandstyle/2004/feb/15/foodanddrink.restaurants (accessed January 2013).

Robinson, J. (1998) *The Manipulators: A Conspiracy to Make us Buy*. Simon and Schuster, London.

Robinson, T. N., Borzekowski, D. L., Matheson, D. M. and Kraemer, H. C. (2007) Effects of fast food branding on young children's taste preferences. *Archives of Pediatric and Adolescent Medicine*, **161**, 792–797.

Royet, J.-P., Delon-Martin, C. and Plailly, J. (2013) Odor mental imagery in non-experts in odors: A paradox? *Frontiers in Human Neuroscience*, **7**, 87.

Rozin, E. (1988) Ketchup and the collective unconscious. *Journal of Gastonomy*, **4**(2), 45–56.

Rozin, P., Spranca, M., Krieger, Z. *et al.* (2004) Preference for natural: Instrumental and ideational/moral motivations, and the contrast between foods and medicines. *Appetite*, **43**, 147–154.

Schuldt, J. P. and Schwarz, N. (2010) The 'organic' path to obesity? Organic claims influence calorie judgments and consumption recommendations. *Judgment and Decision Making*, **5**, 144–150.

Schuldt, J. P. and Hannahan, M. (2013) When good deeds leave a bad taste. Negative inferences from ethical food claims. *Appetite*, **62**, 76–83.

Schutz, H. G. and Lorenz, O. A. (1976) Consumer preferences for vegetables grown under 'commercial' and 'organic' conditions. *Journal of Food Science*, **41**, 70–73.

Seremetakis, C. N. (ed) (1994) *The Senses Still*. The University of Chicago Press, Chicago.

Shankar, M. U., Levitan, C. A., Prescott, J. and Spence, C. (2009) The influence of color and label information on flavor perception. *Chemosensory Perception*, **2**, 53–58.

Shankar, M. U., Levitan, C. and Spence, C. (2010) Grape expectations: The role of cognitive influences in color-flavor interactions *Consciousness and Cognition*, **19**, 380–390.

Simpson, J. (2013) Fishy business leaves royal caviar supplier on the hook. *The Times*, 20 May, 1, 8.

Smithers, R. (2009). *A colin and chips? Sainsbury's gives unfashionable pollack a makeover*. Available at http://www.guardian.co.uk/business/2009/apr/06/sainsburys-pollack-colin-fish-stocks (accessed January 2014).

Spence, C. (2010a) The price of everything – the value of nothing? *The World of Fine Wine*, **30**, 114–120.

Spence, C. (2010b) Crossmodal attention. *Scholarpedia*, **5**(5), 6309.

Spence, C. (2011) Mouth-watering: The influence of environmental and cognitive factors on salivation and gustatory/flavour perception. *Journal of Texture Studies*, **42**, 157–171.

Spence, C. (2012) Managing sensory expectations concerning products and brands: Capitalizing on the potential of sound and shape symbolism. *Journal of Consumer Psychology*, **22**, 37–54.

Spence, C. and Parise, C. (2010) Prior entry. *Consciousness and Cognition*, **19**, 364–379.

Steinberger, M. (2010) *Au Revoir to All That: The Rise and Fall of French Cuisine*. Bloomsbury, London.

Stevenson, R. J. (2012) The role of attention in flavour perception. *Flavour*, **1**, 2.

Stewart, N. (2009) The cost of anchoring on credit-card minimum repayments. *Psychological Science*, **20**, 39–41.

Tannahill, R. (1973) *Food in History*. Stein and Day, New York.

This, H. (2012) Molecular gastronomy is a scientific activity. In: *The Kitchen as Laboratory: Reflections on the Science of Food and Cooking* (eds C. Vega, J. Ubbink and E. van der Linden), pp. 242–253. Columbia University Press, New York.

Time-Life (1982) *Variety Meats*. Time-Life Books, Chicago.

Tozer, J. (2009) Spotted Richard! Pudding gets a new name as staff tire of sniggering. *Daily Mail*, 9 September, 28.

Vilgis, T. (2012) Ketchup as tasty soft matter: The case of xantham gum. In: *The Kitchen as Laboratory: Reflections on the Science of Food and Cooking* (eds C. Vega, J. Ubbink and E. van der Linden), pp. 143–147. Columbia University Press, New York.

von Békésy, G. (1963) Interaction of paired sensory stimuli and conduction in peripheral nerves. *Journal of Applied Physiology*, **18**, 1276–1284.

Wallop, H. (2012) How hot is that sarnie? Subway in new VAT row. *The Sunday Times*, 24 June, B1.

Wansink, B. (2002) Changing eating habits on the home front: Lost lessons from World War II research. *Journal of Public Policy and Marketing*, **21**(Spring), 90–99.

Wansink, B. and Park, S.-B. (2002) Sensory suggestiveness and labeling: Do soy labels bias taste? *Journal of Sensory Studies*, **17**, 483–491.

Wansink, B. and Chandon, P. (2006) Can 'low-fat' nutrition labels lead to obesity? *Journal of Marketing Research*, **43**, 605–617.

Wansink, B., Park, S., Sonka, S. and Morganosky, M. (2000) How soy labeling influences preference and taste. *International Food and Agribusiness Management Review*, **3**, 85–94.

Wansink, B., Painter, J. and van Ittersum, K. (2001) Descriptive menu labels' effect on sales. *Cornell Hotel and Restaurant Administrative Quarterly*, **42**(December), 68–72.

Wansink, B., van Ittersum, K. and Painter, J. E. (2004) How diet and health labels influence taste and satiation. *Journal of Food Science*, **69**, S340–S346.

Wansink, B., van Ittersum, K. and Painter, J. E. (2005) How descriptive food names bias sensory perceptions in restaurants. *Food Quality and Preference*, **16**, 393–400.

Wansink, B., Just, D. R., Payne, C. R. and Klinger, M. Z. (2012) Attractive names sustain increased vegetable intake in schools. *Preventive Medicine*, **55**, 330–332.

Wardle, J. and Solomons, W. (1994) Naughty but nice: A laboratory study of health information and food preferences in a community sample. *Health Psychology*, **13**, 180–183.

Wei, W. and Miao, L. (2013) Effects of calorie information disclosure on consumers' food choices at restaurants. *International Journal of Hospitality Management*, **33**, 106–117.

Weiss, A. S. (2002) *Feast And Folly: Cuisine, Intoxication and the Poetics of the Sublime*. State University of New York Press, Albany, NY.

Werle, C. O. C., Trendel, O. and Ardito, G. (2013) Unhealthy food is not tastier for everybody: The "healthy = tasty" French intuition. *Food Quality and Preference*, **28**, 116–121.

Whittle, N. (2013) Chef talk: Simon Hopkinson. *FT Magazine*, May 25/26, 45.

Wilson, B. (2009) *Swindled: From Poison Sweets to Counterfeit Coffee: The Dark History of the Food Cheats*. John Murray, London.

Winer, R. A., Chauncey, H. H. and Barber, T. X. (1965) The influence of verbal or symbolic stimuli on salivary gland secretion. *Annals of the New York Academy of Sciences*, **131**, 874–883.

Wolfson, J. and Oshinsky, N. S. (1966) Food names and acceptability. *Journal of Advertising Research*, **6(1)**, 21–23.

Yeomans, M., Chambers, L., Blumenthal, H. and Blake, A. (2008) The role of expectancy in sensory and hedonic evaluation: The case of smoked salmon ice-cream. *Food Quality and Preference*, **19**, 565–573.

Young, C. and Myhrvold, N. (2012) On superb crackling duck skin. In: *The Kitchen as Laboratory: Reflections on the Science of Food and Cooking* (eds C. Vega, J. Ubbink and E. van der Linden), pp. 176–185. Columbia University Press, New York.

Youssef, J. (2013) *Molecular Cooking at Home: Taking Culinary Physics out of the Lab and into your Kitchen*. Quarto, London.

Zhao, X., Chambers, E., Matta, Z., Loughin, T. M. and Careym E. E. (2007) Consumer sensory analysis of organically and conventionally grown vegetables. *Journal of Food Science*, **72**, S87–S91.

4
Plating and Plateware: On the Multisensory Presentation of Food

"The first taste is always with the eyes." (Apicius, first century)
"The visual sensation of a dish is as important as its flavour." (Yang 2011)

4.1 Introduction

The food is ready, the garnishes set. Now, how should the dish be presented to the diner waiting expectantly at the table? If, as the popular expression goes, we "eat with our eyes", then the visual presentation of the food (e.g. Imram 1999), the aesthetics on (or even of) the plateware, may turn out to be almost as important as the sensory qualities of the food itself in determining how a diner will respond to that special dish. The truth is, there's a whole 'science' (or art) of food presentation designed to help ensure that the dish passes from 'appetizing' to 'mouth-watering' and 'alluring' in the mind of the diner before they have even picked up their cutlery (assuming they have any; see Chapter 5). But how is the chef or restaurateur to achieve this goal? Are there clear rules of presentation that should be followed, or display guidelines that can be used by any of us to help prepare a more delicious meal?

At the outset here, it is worth noting that chefs sometimes spend months in planning and perfecting the meals that they intend to serve, thinking about the ingredients that will be in season at the time and the visual composition of the elements on the plate, all to make sure that the food is balanced and

The Perfect Meal: The Multisensory Science of Food and Dining, First Edition.
Charles Spence and Betina Piqueras-Fiszman.
© 2014 John Wiley & Sons, Ltd. Published 2014 by John Wiley & Sons, Ltd.

harmonious. Take the following planning timeline for the 2008 menu from elBulli (Adrià *et al.* 2007):

"*OCTOBER 2007*
Day 2: elBulli closes
Day 10: The creative team commences work

NOVEMBER 2007–FEBRUARY 2008
Attending specialised courses; searching for new foods and new equipment; visits to USA, Japan, other countries; visits to designers in other industries

MARCH 2008
Day 26: elBulli opens

APRIL 2008
First week: Start with the 2007 menu and scattered new recipes; gradual incorporation of new recipes to the menu

MAY 2008–AUGUST 2008
2008 menu finalised"

As one can imagine, the selection of the plateware on which the food is going to be presented often enters into the planning process at a very early stage, since everything from the proportions to the combination of colours present on the plate has to be thought out and perfectly balanced (on some occasions, literally; see Section 4.6.1). Of course, one doesn't have to be working at a top-end restaurant in order to be able to prepare and enjoy such delightfully presented creations. Any amateur cook with at least a dash of aesthetic sensitivity likely has the relevant skills, if not the instrumental paraphernalia that 'molecular' chefs use, in order to impress their guests. Indeed, books, courses and step-by-step tutorials on this art are now easier to come by than ever before (e.g. Styler and Lazarus 2006). The purpose of all this information is to go beyond simply serving the food in a way that just looks nice; it's to express creativity and originality, breaking previous moulds, conventions and rules and, in the process, challenging the properties of the ingredients, which then become elements of construction used to create that unique dish. As we will see in the pages that follow, there is also an emerging science here.

When it comes to the selection of the plateware, a growing number of chefs are no longer content with the extensive commercial ranges of plateware found in the hospitality and catering catalogues. Instead, in order to fulfil their desires the plateware may even be designed (that is, tailor-made) specifically to fit in with the food itself (of course, here we're talking about the star chefs). In a number of other cases, natural or easily available materials are used instead. The crucial point is that regardless of whether it's a standard issue round white plate (apparently half of all plates that are sold are white; Hultén *et al.* 2009) or a reclaimed brick that the food is going to be served on, all of its (multisensory) attributes – its size, shape, colour and even its material properties – likely matter (Butter 2013). Indeed, a number of recent scientific studies have demonstrated the profound influence that each and every one of these factors can exert on a diner's perception of the food, how much they serve themselves and even how much they will eventually end up consuming. What is more, the plateware can impact on how much a diner likes whatever it is that (s)he happens to have been served. Most people believe that they are simply enjoying the food that is put in front of them when in fact they are, in a very real sense, tasting the plate as well! In this chapter, we take a closer look at the various factors concerning the presentation of food that affect our eating behaviours, our multisensory dining experiences and ultimately our enjoyment of food.

4.2   A potted history of food presentation

Nowadays there's more to effective plating (the term 'plating' being part of the vocabulary of *nouvelle cuisine* according to Halligan 1990, p. 241) than merely garnishing a dish with a sprig of fresh parsley. However, it is a little harder to explain quite how we got from tasting that elegant garnish sitting on top of our mashed potatoes to eating the centrepiece on the table at Noma (currently one of the world's top restaurants), which constitute the first two courses on the menu! Nevertheless, let's give it a try. The focus here will primarily be on the trends in food presentation that have emerged from behind the doors of some of the most innovative restaurants of each historical period which, sooner or later, will percolate down to more down-to-Earth restaurants and even to the context of home dining.

In the Middle Ages, meals eaten by the peasantry basically consisted of the ladling of stews or porridge into hollowed-out, improvised 'plates' that had been cut from loaves of old bread. In this case of course, the staler the better (Tannahill 1973). As one might expect, royalty enjoyed much more elaborate, heavily meat-based, feasts with all sorts of garnishes, sauces and fruits piled onto large plates or trays.

It was in the seventeenth and eighteenth centuries that French cuisine really began to resemble what so many of us know and love today. Louis XIV placed *cuisine* as an integral part of French culture, both for its flavour and for its aesthetic appeal. It was however Antonin Carême, the celebrated French pastry chef or pâtissier (Gopnik 2011) born in 1783 (see previous chapter) who brought plating into the modern world. Carême himself was an avid amateur student of architecture and was the chef to distinguished figures all the way up to Napoleon Bonaparte (see Frascari 2004). He became famous partly

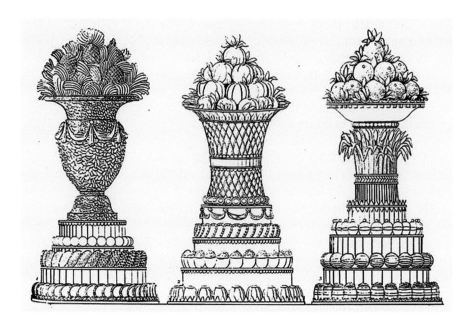

Figure 4.1   Engraving from *Le Patissier Royal Parisien*. *Source*: Carême 1828, p. 158

because he would often present his culinary creations in the shape of famous monuments, waterfalls and pyramids (see Figure 4.1). Carême is thought to have invented the famous *croquembouche* (French for 'crunch in mouth' made with custard-filled cream puffs coated with caramel and stacked to make a towering pyramid).[1]

Up until the late nineteenth century however, it was still only society's elite who got to enjoy Carême's innovations in the art of cooking and presentation. Interestingly, the French artists of the period (such as Manet) opposed this pomposity in food presentation and, in response, tended to paint simple and natural foods instead (Bendiner 2004; see also Wilson 1991). As Desbuissons (2012, p. 53) put it: *"Manet developed alimentary motifs which valorised a diet representing a negation of the complicated and expensive preparations the imperial France would boast about."* Elsewhere, Reay Tannahill suggests that painters *"prefer the raw material of dining, the tantalizing promise of wonderful food–tomorrow"* (see Halligan 1990, p. 64). From the early twentieth century onwards another French chef, Auguste Escoffier (born two years after the death of Carême), decided to reorganize the classic methods of cuisine, further refining the recipes and establishing a new way for French restaurants to

---

[1] Carême didn't just revolutionize the world of pastry with his culinary constructions; he also came up with the idea of 'mother sauces', has been credited with reducing the size of the portions (particularly appropriate given that he was mostly working on large banquets with numerous courses) and also tried to emphasize complementary flavours and food pairings in the dishes that he presented.

be run. One of Escoffier's most revolutionary ideas was that individual chefs should specialize in certain elements on a menu. This quickly became standard practice in restaurants the world over, hence reaching a much broader cross-section of the population. Now, smaller and more individualized plates could be served without the food getting cold. It was presumably precisely this approach that enabled Escoffier to prepare 500 plates in an hour when he applied this system at the Hotel Ritz).[2]

"No doubt the pressures of earning awards and positive reviews have pushed French chefs to take a more artistic approach to plating." (Yang 2011)

Early in the twentieth century, Fernand Point, considered by many as yet another of the fathers of modern French cuisine, introduced elements that would later become the signature of *nouvelle cuisine*: seasonal ingredients with a focus on natural flavours and, above all, simplicity and elegance. He even made the now-ubiquitous baby vegetables a regular addition to the plate. Point's style was consolidated by his most famous *protégé*, Paul Bocuse, whose neat and detailed presentation of food soon provided the iconic images of the increasingly popular *nouvelle cuisine* movement in the 1960s. Indeed, it was around this time that chefs started adding their monograms to the plateware (Halligan 1990). Up until the 1960s, French cuisine had tended to be rich, heavy and lacking in any real visual appeal. According to one account: *"The most common kind of French plating was to show all the ingredients, as they are side by side, placing the starches next to the vegetables, next to the meat or fish. The food would quite often also be stacked, by placing the main item on a bed of vegetables or potatoes"* (Yang 2011).

"20 years ago French presentation was virtually non-existent. If you ordered a coq au vin *at a restaurant, it would be served just as if you had made it at home. The dishes were what they were. Presentation was very basic."* (Sebastian Lepinoy, Executive Chef at L'Atelier de Joel Robuchon, quoted in Yang 2011)

Around this time, Japanese cuisine started to influence the plating style of Western cuisine. The Japanese chef Shizuo Tsuji opened the first French culinary school in Japan in 1960, leading to a much greater cultural exchange between the Japanese and leading French chefs, including Paul Bocuse. Soon, seasonality and simplicity, so characteristic of Japanese cuisine (Yang 2011), started to become key elements on the plate in French cuisine (Youssef 2013). The next generation of international chefs, led by the likes of Ferran Adrià and Grant Achatz among others, took minimalism in cooking and presentation even further (sometimes to the extreme). Now that we have in mind the evolution from the elaborate French feasts of Carême (not to mention the

---

[2] One might consider Escoffier to be the Henry Ford of the kitchen.

eccentric meals of the Italian Futurists; Berghaus 2001) all the way through to the foamy impossibilities constructed (until it closed in 2011) by the culinary team at elBulli, one might wonder what exactly the future holds in terms of food styling.[3]

> *"Today, the goal is to feature the ingredient as close as possible to the source. If we have very fresh microgreens, or a fresh fish, we put it right on the plate. When we have a fresh item, we don't need to do much to it. The freshness of the ingredients guides the presentation."* (Sergio Remolina, chef and professor at the Culinary Institute of America; Park 2013)

And *voilà*, from medieval banquets to the minimalistic and natural presentations of fresh ingredients is how, today, we arrive at Noma's Moss and Cep (crispy deer lichen dusted with cep mushrooms resting on a bed of brilliant green moss; see Figure 4.2), as if straight from one of Denmark's forests!

In the sections that follow, we'll examine the influence of each of the visual elements of the food and the plateware (assuming that there *is* a plate) on our perception of the food, our eating behaviour and our appraisal of the overall multisensory dining experience. We will see a number of examples in which the inseparable combination of these two elements (food and plateware) together delivers the truly memorable dining experiences.

Figure 4.2 Fried deer moss from the Noma restaurant in Copenhagen. *Source:* Reproduced with permission of Sonali Fenner. *See colour plate section*

---

[3] Here we are mostly talking about the challenges associated with designing each of the 20+ course marathon tasting menus now being served at the world's top-end modernist restaurants.

4.3   The plate: the essential element of our everyday meal

What is obvious, but has received surprisingly little attention up until now from researchers, is the fact that in our everyday lives food is never presented (or served) in isolation. That is, the food is always placed in a container (a bowl, a plate, etc. even if it's dispensable, as in the case where we eat food direct from the packaging; see Spence and Piqueras-Fiszman 2012). Wherever we are, even at home, the food we eat is served from or in some kind of receptacle (we'll use the term plate in most cases for the sake of simplicity).[4] Let's start by talking about the basics: the plateware (the sort that we all have at home in our kitchen cupboards). This commonly neglected element of the meal can affect everything from how we perceive the taste of food to how many calories we consume.

As mentioned already, when plating food in the most attractive manner possible it is important to remember that the actual plate is critical to the final presentation; the plate constitutes the frame for the food. This is something that the Japanese have known for many years. In Japan, chefs carefully select their vessels, utensils and other paraphernalia in order to make sure that each dish complements the food that is going to be served from it, in tune with the season. They also try to enhance the visual appearance of the food as much as possible. Today, there are endless sizes, shapes, colours and patterns available (although we won't mention those gaudy plates, picked on occasion by some hosts who seemingly lack any semblance of good taste). Choosing the correct size for the plateware is also important (Sobal and Wansink 2007). While food mustn't appear too crowded on the plate, it should instead convey the notion that the portion is adequate and not too measly either. Increasingly, one also sees textures being embossed onto the surface of plates in order, presumably, to add some visual interest (not to mention differentiation). Furthermore, specialty plates may be used for very specific dishes, but we will cover this emerging trend in Section 4.5.5.

4.3.1   On the colour of the plate

Nowadays, most chefs use white plates to showcase their culinary offerings (though black is sometimes used too, and there are undoubtedly many other colours that are available for anyone who wants to spend their time perusing the pages of the catering catalogues or the shelves of the kitchen stores), but what effect does the colour of the 'canvas' really have on our perception of food? Some years ago, in his book on the psychology of food Lyman (1989) suggested that purple grapes don't look quite right when served from

---

[4] Apparently, we eat as much as 30% of our food direct from the packaging (Wansink 1996).

a blue plate. It is, however, only over the last couple of years or so that such anecdotal claims (specifically that the colour of the plateware might exert a significant impact on the taste/flavour of whatever happens to be served from it) have been assessed empirically.

> *"Colour is important, so instead of having boring white plates, we have a chef that comes in every morning to paint the plates to match the dishes on the menu. The dishes must catch the attention of the guest at first sight. It must entice the appetite."* (Sebastian Lepinoy, Executive Chef at L'Atelier de Joel Robuchon, quoted in Yang 2011)

In one recent laboratory study, Harrar *et al.* (2011) had their participants sample sweet or salty popcorn from four differently coloured bowls: white, blue, green and red. The salty popcorn was rated as tasting significantly sweeter when taken from a blue or red bowl, while the sweet popcorn was rated as tasting saltier when lifted from the blue bowl. Although the magnitude of these effects were pretty small (averaging only a 4% change in participants' responses for the food taken from the coloured bowl as compared to from the white bowl), they were nevertheless statistically reliable. In another study, which this time took place in one of the rooms at Ferran Adrià's Alícia Foundation (a culinary school and research institute in Catalonia, Spain), Piqueras-Fiszman *et al.* (2012) compared the taste of a strawberry-flavoured mousse (of a homogenous texture and colour) that was served on either a black or a white plate (though it should be noted that, strictly speaking, these are not colours). Amazingly, the dessert served from the white plate was perceived as being 15% more intense, 10% sweeter and was liked 10% more than exactly the same mousse when served from a black (otherwise identical) plate instead. Funnily enough, several of the participants in this study were actually students or interns at the Alícia Foundation. One might therefore have expected that they would have some expert background knowledge on taste perception of the flavours, etc. That said, whenever we have tested those working in the culinary sector we have found that they are normally just as susceptible to these psychological influences as anyone else!

> *"Simultaneous color contrast suggests that foods can be arranged in combinations so that their colors are subtly enhanced, subdued, or otherwise modified. Yellow scrambled eggs on a yellow plate will look paler because of contrast. Purple grapes will look less purple on a purple plate and will look redder on a blue plate. A green salad will look less green on a green plate than on a plate that has no green in it. Red food on a blue plate will look more orange. Broccoli served with red fish will make the fish look redder, and slices of lime surrounding a grape mousse will enhance the color of both."* (Lyman 1989, p. 112)

In this case, the colour of the plate may have affected the perceived colour of the food by means of the well-known colour contrast illusion. In the phenomenon of simultaneous colour contrast (Ekroll *et al.* 2004), a foreground object appears to have a different colour (or contrast) depending on the colour of the background (Leibowitz *et al.* 1955; Hutchings 1994). According to such a perceptually based interpretation of their findings, the colour of the mousse served in the study of Piqueras-Fiszman *et al.* (2012) may have appeared more salient when set against the background of the white plate than when served from the black plate. The rated intensity of the food's taste/flavour will presumably have been influenced by its perceived colour saturation which, in turn, will have been influenced by the colour saturation of the plate itself.[5]

More recently, Piqueras-Fiszman *et al.* (2013) have gone on to test the extent to which the colour of the plateware (again, black and white plates were used) influences the gustatory and hedonic experiences of a complex food (desserts with layers and decorations having different colours, textures, tastes and flavours). Importantly, this study was performed in an entirely naturalistic setting (one of the restaurants at the Institut Paul Bocuse, in Lyon, France), under conditions that were as ecologically valid as possible. A between-participants experimental design was used, in which the participants/diners were able to interact and consume the three-course meal at their own pace and, what is more, had the privilege of paying for the pleasure. Over the course of the two weeks during which the experiment was conducted, three different desserts were served. The results showed that the colour of the plateware exerted a significant influence on people's perception of the food, but that this effect varied as a function of the type of dessert that was being served.

Interestingly, the perceptual pattern for each dessert was constant for each plate used; that is, for all of the attributes rated (including how appetizing the food was, its appearance, colour intensity, the flavour intensity and the participants' overall liking), higher scores were obtained with the same plate for each of the desserts. However, these results could not be accounted for solely in terms of contrast effects, since it was the dessert that had a darker brownish hue that participants rated more highly when served off of the black plate. The other two desserts, which were red and creamy in colour, were rated as looking more delicious when served from the white plate instead: "*[Each dish was] photographed on pure white crockery ... In this way, we have ensured that the food looks both as delicious and natural as possible*" (Anonymous 1997, p. 11). It could be that the desserts simply looked better (that is, more

---

[5] Most of us have had the experience of choosing a paint colour from one of those colour cards at the hardware store, only to find that the colour of the painted living room wall looks different. This may be the same phenomenon that impacted on people's ratings of the taste of the food in the study of Piqueras-Fiszman *et al.* (2012).

visually appealing) on a plate of a certain colour, and that this visual appraisal of the overall product offering was what produced a halo effect on the other scores that the diners made (Asch 1946; Churchill *et al.* 2009). The key message here then concerns the realization of just how important the colour of the plate (i.e. the background colour) can be to people's perception of food, even in realistic and hence less-controlled conditions such as that of a restaurant.

That said, colour contrast cannot be so easily used to explain the effects of the coloured bowls on the perception of popcorn reported by Harrar *et al.* (2011). One of the reasons for this is that the popcorn in Harrar's study was eaten by hand and would therefore likely always have been seen against a constant colour background (the participant's hand) just before being popped into the participant's mouth. However, an alternative possibility here is that the results of Harrar *et al.* demonstrate another example of sensation transference, given that red is typically associated with sweetness whereas blue is more often associated with saltiness (and green with sourness) instead (Maga 1974; O'Mahony 1983; Spence *et al.* 2010; Harrar and Spence 2013). But where exactly did such colour–taste associations originate in the first place? We believe that the most likely explanation here is that consumers are simply attuned to the natural statistics of the environment (Maga 1974) and/or, more importantly, to the regularities in packaging and product colouring that are typically used in the supermarket (Piqueras-Fiszman and Spence 2011; Spence *et al.* 2010; Spence and Piqueras-Fiszman 2012).

In his restaurant in Vevey (http://www.denismartin.ch/), Denis Martin uses so-called charger plates (a large flat dish placed under the normal plate, used primarily for decorative purposes) in the primary colours (blue, red and green; a different-coloured plate for each diner) as the base on which to place the different white dishes on which the food will actually be served to his guests. We guess this is for decorative purposes or to add a touch of colour and fun to the overall dining experience (in addition to the mooing cow we mentioned back in Chapter 2). However, as the results of the research outlined in this chapter make all too clear, there is no reason why the colour of the plateware should not also be chosen for more functional reasons.[6] For instance, Crumpacker (2006, p. 143) claims that " ... *the term blue plate special became popular during the Great Depression because restaurant owners found that diners were satisfied with smaller portions of food if it was served on blue plates.*"

Interestingly, a recent study conducted at Salisbury District Hospital (UK) documented an increase in food consumption among dementia patients of

---

[6] Potentially worrying here is the observation that those patients who require special nutritional attention in UK hospitals are normally served their meals from a red tray, so that they can easily be identified by health professionals (see AgeUK 2010; Bradley and Rees 2003). We are unaware of whether anyone has actually studied the effect that being served food from a red tray may have on patients' eating behaviours (but see Genschrow *et al.* 2012; Bruno *et al.* 2013 for the negative consequences of red on eating behaviours). Certainly, choosing the colour of a patient's dish/tray is a decision that really ought not to be left to chance. That said, as best-selling US economist Tyler Cowen (2012) wryly notes in his book *An Economist Gets Lunch*, hospitals are not famed for their fine cuisine.

nearly one-third (see Adams 2013). Hultén *et al.* (2009, p. 119) have also reported that older individuals ate more white fish when it was served from a blue plate. Why should blue sometimes increase consumption whereas at other times (in history) it reduced it? One idea here is that the majority of the food served in hospitals tends to be both bland in taste and in colour, and hence may simply fail to stand out against white crockery. This simple intervention has proved so successful it is apparently being rolled out across the whole hospital trust, according to head of catering Ian Robinson.

In conclusion, although explanations for the fact that the colour of the plate impacts taste/flavour perception are not yet fully worked out, the results of a growing number of studies now show that the colour of the plate really does matter. From the results that have been reported to date, desserts really do taste sweeter when served off of a white plate while black plates may be more appropriate for savoury foods. We hope that such results will hopefully make a few more innovative chefs, and those working in the catering industry, think a little more carefully about the colour of their plateware and its potential effects on their customers' perception of taste and flavour.

4.3.2   The shape of the plate

At the Tsunami restaurant in West Palm Beach, Florida (now closed), owners Shamin Abas and Cilione created dramatic table settings that were designed to be compatible with Asian fusion cuisine: *"Fusion can become confusion, but our flavours come out distinctively in each dish. We wanted a clean, modern design, and we use a different plate for each entrée. Each menu item has its own dish and people love that."* The stylish dinnerware at this unfortunately named restaurant came from Fortessa (http://www.fortessa.com) and included triangular, rectangular and even eye-shaped plates. While apparently differently shaped dishes may seem too fancy for the tastes of many diners, there is as yet surprisingly little research on the topic of the shape of the plateware on people's perception of the food that is served from it.

In one of the first studies to have been conducted in this area, Gal *et al.* (2007) reported that people (a group of more than 200 university students) rated a piece of cheese as tasting significantly sharper (by around 10%) after having been made to stare at jagged shapes (as opposed to after staring at rounded shapes). While it is important to bear in mind that such results come from a highly contrived laboratory study, they nevertheless hint at the potential impact that the angularity of the plateware might have on taste/flavour perception (see also Liang *et al.* 2013). A few years later, Piqueras-Fiszman *et al.* (2012) went on to investigate whether the shape of the plate would influence people's taste/flavour perception. However, these researchers failed to find any effect of the shape of the plateware on the taste of a strawberry-flavoured mousse. The plates used in this study were square, round and triangular. By contrast, others have reported that eating food from a round versus star-shaped plate can exert a small but nonetheless significant effect on the perceived sharpness of the food tasted from them (see Piqueras-Fiszman *et al.* 2012 for a review).

What might explain the inconsistency in the results reported between these latter two studies? The most likely explanation relates to the fact that previous research used a star-shaped plate (with 5 points), whereas Piqueras-Fiszman used triangular and square plates (3- and 4-points) with fairly rounded corners/points. The angularity of the plates therefore differed somewhat between the studies conducted in this area. It may well be that you need to use really angular plateware in order to emphasize the sharpness of a dish. Note also that the mousse in the study of Piqueras-Fiszman *et al.* was formed into an angular pyramid on the plate; it is therefore possible that the angularity of the food itself (seen in the foreground) may have overridden the angularity/roundness of the plate (which may have faded into the background).[7]

It should also be borne in mind that certain attributes of the food (such as its perceived sharpness, in the case of cheese) may be more susceptible to being modified by the shape (or sharpness/angularity) of the plateware than others (Spence 2012; Deroy and Spence 2013). Remember that 'sharpness' is fundamentally a tactile property; however, the term sharp is now used synaesthetically (or crossmodally; Williams 1976; Deroy and Spence 2013) in order to describe certain taste/flavour attributes (Spence *et al.* 2013a). It is therefore possible that this term may be more likely to exhibit sensation transference effects when used to describe specific foodstuffs. According to this argument, taste terms that have primarily only ever been used to describe gustatory qualities (e.g. saltiness) may be less susceptible to the effects of sensation transference from the shape/haptic qualities of the plateware.

While the studies reviewed in the last couple of sections have tended to vary only a single attribute at a time (i.e. just the colour of the plate or, separately, just the shape of the plate), in the future it will become more important to vary several attributes of the plateware at the same time and see how they interact. A step in the right direction in this regard comes from a recent study by Stewart and Goss (2013). These researchers conducted a study in which they had their participants eating from black or white plates that could be either round or square. In this case, the researchers found that the dessert (a cheesecake) was rated as 20% sweeter when tasted from the round white plate than any of the other three plates. The effects on flavour intensity were even more dramatic: 30% higher when served from the round white plate than when served from any of the other plates. Taken together, the results of a number of studies now suggest that the best way to keep the calorie count down when eating dessert is to serve the food from a round white plate.

---

[7] Note that such a contrast between the shape of the food and the shape of the plateware is a common feature of Japanese cuisine. According to Yang (2011), "*When selecting tableware, a law of opposites is employed. If the food is round in shape, then a square or long, narrow flat dish is used.*"

4.3.3   The size of the plate

In addition to any effect that colour and shape of the plateware has on the perception of food diners are also influenced by the size of the plateware, mostly when it comes to calculating the volume (amount) of food to serve and/or to consume (Levitsky and Youn 2004). This is particularly relevant considering the fact that since early last century the average size of the plate has increased by around 20% (Wansink 2010), as has the average waist size of humans (Nielsen and Popkin 2003; see also Young and Nestle 2002). For example, in one influential study Wansink *et al.* (2006) investigated the effect of varying the size of the bowls on food consumption at a social event. When the participants in their study were given a larger bowl, they served themselves in excess of 50% more ice-cream than those given a much smaller bowl. Furthermore, since the participants nearly always finished their food (as is apparently generally the case under self-serve conditions; see Diliberti *et al.* 2004; Wansink and Cheney 2005), those eating from a larger bowl ended up consuming far more ice-cream overall. Here, one might wonder what happens when the diner doesn't manage to finish their food: is the amount of waste also correlated positively with the size of the plate?

Van Kleef *et al.* (2012) obtained similar results when the bowls that differed in size were the everyday containers from which a pasta was served in a canteen setting; the difference in capacity in this case was nearly double (3.8 versus 6.8 L). Despite the fact that the diners' individual plates were all of the same size (approximately 23 cm in diameter), those who served themselves from the larger communal bowl ended up with 77% more pasta and felt more satiated than the others who found themselves in the smaller common bowl condition (see also Marchiori *et al.* 2012).

Wansink and his colleagues attempted to account for these results in terms of the Ebbinghaus–Titchener size-contrast illusion and/or the Delboeuf illusion (Titchener 1908; Lyman 1989). In perhaps the most well-known version of the former illusion, we 'see' a circle as larger when surrounded by smaller circles than another identical circle that happens to be surrounded by bigger circles (see Figure 4.3a). The latter illusion makes us think that, of two identical circles, the one that is placed within a larger ring looks smaller than the other one without a ring or else placed within a smaller ring (see Figure 4.3b). Wansink and his colleagues have suggested that such visual illusions may result in a given amount of food being perceived as much smaller against the background of a larger bowl, and as larger when presented in a smaller bowl instead (Wansink and Cheney 2005).[8]

It is however important to note here that the effects of the size of the plateware on a diner's consumption behaviour are somewhat inconsistent. For instance, Rolls *et al.* (2007) were unable to find a significant difference between the diameter of the plate (17, 22 or 26 cm) and the amount of food

---

[8] More generally, it would appear that there is a rich vein to be tapped here in terms of creating edible versions of a whole host of visual illusions.

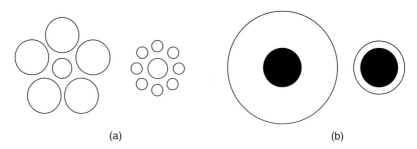

(a)   (b)

Figure 4.3 (a) Representations of the Ebbinghaus–Titchener size-contrast illusion. *Source*: http://commons.wikimedia.org/wiki/File:Mond-vergleich.svg and (b) the Delboeuf illusion. *Source*: http://en.wikipedia.org/wiki/File:Delboeuf.jpg

that was consumed at a meal in three separate laboratory-based experiments. In this case, the discrepancy between the significant results reported by Wansink *et al.* (2006) and the null results reported by Barbara Rolls and her colleagues may point to the existence of important differences between people's food consumption behaviours in the laboratory versus those seen under more realistic dining conditions. Remember here that Rolls *et al.* tested people's eating in the laboratory whereas Wansink *et al.* had people fill out questionnaires at a company picnic (a real-world event), which may also help to explain the inconsistencies. Such differences should certainly always be kept in mind when trying to extrapolate from the results of laboratory-based psychological research to predicting people's real-world eating behaviours (e.g. de Graaf *et al.* 2005).

Why is it that the French continue to be slimmer than the North Americans, not to mention having a lower mortality rate from heart disease, despite eating meals that contain a greater amount of fat, more saturated fat and less reduced-fat foods than Americans? One explanation for this phenomenon, known as the 'French Paradox', is in terms of cross-cultural differences in portion sizes (Rozin *et al.* 2003). The suggestion here is that the French serve themselves smaller portions of food, whether in full-service restaurants, buffets or supermarkets or even as indicated in cookbooks.

To summarize: the colour, size and shape of the plateware can all influence people's perception of the food placed on it (as well as how much they end up eating; see Wansink *et al.* 2009). It is now time to go beyond the visual aspects of the plateware. From this point on, we'll take a look at the additional contribution made by the tactile attributes of the plateware (i.e. by the feel of the plateware or crockery), any sound that it makes and even any fragrance/aroma that it might release upon a diner's multisensory dining experiences.

4.3.4 On the haptic aspects of the plateware

The weight On many occasions, such as at catering events, take-away meals or just while sitting comfortably on the sofa at home, one sometimes holds the plate or bowl in one hand while eating with the other.[9] What role does the weight of the dish have on our perception of how much we eat (or even how full we feel)? We'll address these questions in the following section. Despite its potential importance, only one study (as far as we are aware) has examined the impact of the haptic/tactile attributes of the plateware on the perception of food to date. Specifically, Piqueras-Fiszman *et al*. (2011) explored whether the weight of the bowl from which people tasted a yoghurt exerted any influence on their multisensory flavour perception.

Three bowls, identical except for their weights, were filled with exactly the same amount of yoghurt. Participants were instructed to hold each of the bowls in their hand while rating the taste and flavour of the yoghurt on four scales. The food sampled from the heaviest bowl was rated as being 13% more intense in flavour, 25% denser, 25% more expensive and was liked 13% more than when sampled from the lightest bowl. Since weight properties are often used to describe the density of a food (e.g. as when we describe a food or meal as being 'heavy'), the attributes that we associate with the heavier bowl may well have been transferred (subconsciously or otherwise) onto the participant's perception of the qualities of the food in the bowl itself (cf. Ackerman *et al*. 2010; Spence and Gallace 2011). As such, these results can perhaps be best explained in terms of psycholinguistic and metaphoric transfer effects. That is, the terms that we use to describe the plateware are transferred metaphorically onto how we feel about the food itself (Schneider *et al*. 2011).

In a follow-up study, Piqueras-Fiszman and Spence (2012a) went on to demonstrate similar effects on people's ratings of yoghurt when served from light (20 g) versus heavy (95 g) plastic bowls. This time, though, it should be noted that the absolute variation in the weight of the plateware was much smaller; the yoghurt tasted from the heavier bowl was nevertheless still estimated as being significantly more satiating (prior to consumption), denser (once again) and as likely to be more filling.[10] All of this research might make the curious chef think a little more carefully about whether or not to serve the soup with any cutlery next time.

The feel of the plateware   As 'Object' (1936) by Meret Oppenheim (a fur-covered cup, saucer and spoon; see http://www.moma.org/collection/object .php?object_id=80997) illustrates, the feel (either real or expected) of a particular piece of plateware against our skin (especially against sensitive regions of the body such as the lips; Weinstein 1968) can generate unpleasant sensations. Over the last couple of years or so, researchers have started to investigate just what effect varying the weight and texture of the plateware may have on people's perception of the food that is served from it.

---

[9] In several restaurants nowadays, such as Chicago's Alinea, the diner is instructed to hold the bowl/plate in one hand.

[10] Note that these were within-participants studies. As compared to a between-participants design, such an experimental design likely emphasizes any relative differences in weight between conditions. It would therefore be ideal in future research if these findings could be reproduced in an authentic restaurant setting when using a between-participants experimental design.

While the visual pun of Oppenheim's fur-covered cup was very much about something that people would likely not want to put their lips to, psychologists and designers, together with modernist chefs and a growing number of food and beverage producers, are currently trying to make plateware and/or packaging (for the many foods that we eat direct from the packaging; Wansink 1996) that by more effectively stimulating the diner's (or consumer's) sense of touch can enhance their experience of food and drink. One contemporary example of plateware that may successfully achieve this goal is Nao Tamura's silicone leaf plates. Their texture resembles that of a leaf, possibly making the eating experience more natural (http://naotamura.com /projects/seasons-milano-salone-covo). However, apart from stimulating the sense of touch, does the texture of the plateware (especially when it is held in the hand and hence felt by the diner) also affect the rated texture of the food in the diner's mouth? Suggestive evidence in this regard comes from a recent study by Piqueras-Fiszman and Spence (2012b). We found that people rated pieces of stale or fresh digestive biscuit served from a small plastic yoghurt pot as tasting both significantly crunchier and significantly harder when the container itself had been given a rough sandpaper finish, as compared to when exactly the same food was served from a container with the usual smooth plastic feel of a yoghurt pot.[11] We have also heard anecdotally that if you were to hold a handful of gravel in one hand then you would end up rating a sample of ice-cream as tasting grittier than if you were to grasp a handful of cotton wool instead, once again echoing one of the ideas first put forward by the Italian Futurists (see Spence *et al.* 2013b). Obviously, what you hold in the non-eating hand shouldn't influence your rating of food sampled with the other; apparently, though, it does. This then might be yet another example of sensation transference, what Spence and Gallace (2011) refer to as affective ventriloquism.

While we don't yet know to what extent these laboratory-based haptic findings would have an effect in a real-world setting such as that of a restaurant, we are increasingly seeing chefs selecting plates with textures in order to play with the contrast between the food and the background (which is achieved at a visual level). At restaurant Per Se in New York (http://www.perseny.com/) for example, chef Thomas Keller plates one of his signature dishes 'Oysters and Pearls' on a set of four (yes, four, and on occasions even five) evenly stacked custom-designed plates (see Figure 4.4). Each of the plates is designed to nestle within the plate below. The aim here is to create the illusion of one large rim (with interior waves that draw the diner's attention to the centre). In this case, the effect of touching these textured plates on the dining experience is unclear.

---

[11] Although the feel of the container influenced people's perception of a dry food product, it had no effect on their ratings of yoghurt. Further research will therefore be needed in order to understand the limiting conditions on this particular effect (i.e. the effect of what people hold on what they taste/experience).

Figure 4.4 Thomas Keller plates his star classic dish 'Oysters and Pearls' on this set of four evenly stacked, textured plates. *Source*: Reproduced with permission of Jennifer Che, Tiny Urban Kitchen, www.tinyurbankitchen.com

An example of plates (and bowls) that have been created especially to stimulate the diner's sense of touch have been designed by Jinhyun Jeon (http://jjhyun.com). These items of crockery are made of soft materials such as silicon, which have different levels of transparency and make use of embossed textures. Jeon says: *"I propose that tableware should be designed to suit our intuitive and mostly subconscious sensorial abilities rather than forcing the people to adapt to the tools … it should not just be a tool for placing food in our mouth, but it should become a sensorial appetizer, teasing our senses in the moment when the food is still on its way to being consumed."*

Moving away from the domain of the restaurant for a moment, but nevertheless still relevant to this section, is the set of bowls that were designed by the London-based team of Bompas & Parr for the Heinz Beanz Flavour Experiences packs launched in 2013 (and only available from London's very own Fortnum & Mason department store). The textures and colours of the bowls here were all designed to enhance the flavour associations in the mind of the person trying the food from them. The idea was that this would enhance the diner's experience of that which they were eating. For instance, the Heinz Beanz Garlic and Herbs Flavour Experience bowl had the organic shape of a garlic bulb, made from 96 layers of 750-micron-thick card, giving it a soft, natural appearance (see Figure 4.5, left). Meanwhile, the Heinz Beanz Cheese Flavour Experience included a circular bowl made of cast yellow

Figure 4.5   Heinz released a limited number of these Beanz Flavour Experience packs which included a matching bowl with the texture and shape of its main ingredient for each flavour. Left: the Beanz Garlic and Herbs Flavour Experience bowl; right: the Beanz Cheese Flavour Experience bowl. Pictures taken by Nathan Pask. *Source*: Photograph by Nathan Pask Photography. Reproduced with permission of Ann Charlott Ommedal, Bompas and Parr, London, and Nathan Pask. *See colour plate section*

wax, based on the form, texture and colour of a traditional round of cheese (see Figure 4.5, right).

However, unless we happen to be seated in one of the tactile dinner parties organized by the Italian Futurists, in which courses were sometimes served from small bowls covered with different tactile materials (Marinetti 1932, p. 125), or we somehow manage to get hold of one of those Heinz Beanz Flavour Experience bowls (something of a collector's item by now), plateware (and we remind the reader that we're talking about the most traditional type of plateware) is usually made from porcelain or ceramic. Plastic and paper versions are obviously available for home parties or takeaways and of course for children. The material of the plateware, then, depends on the occasion and who happens to be eating from it.

Certainly no one expects to eat from a porcelain plate in a fast-food joint or in an aircraft (unless you're a regular traveller at the front of the plane, or else you happen to be old enough to remember what it was like in the very early days of dining in the sky; see Hudson and Pettifer 1979), just as no one would

Figure 4.6   Ceramic paper plates, conveying the feeling of a picnic while sitting at one of the best restaurants in New York (Eleven Madison Park)

expect to be served from a plastic or paper plate in a smart restaurant (see Chapter 5 when this question is addressed with regards to the use of stainless steel versus plastic cutlery). In fact, at Eleven Madison Park in New York, the cheese course nowadays comes in a picnic basket which includes the tableware and the food. The diners have to open the basket and then set the table. The plates for this course are made of ceramic, but imitate paper plates (they are in fact moulded based on a real paper plate). This is an example of how the feeling of a picnic can be conveyed while maintaining the quality standard of the restaurant (see Figure 4.6).

And what about a silver plate? Although it is not a common material for plateware today, silverware has always been associated with high quality, an association that we have from our ancestors. In his book *Cookery and Dining in Imperial Rome* back in the first century, Apicius suggested that *"an expensive silver platter would enhance the appearance of this dish materially"* when referring to the plating of his Apician Dish (number 141; Apicius 1936, p. 103). Today, at least for elegant or special dinner occasions, a silver or gold-finish charger plate might be used as part of the tableware and convey that touch of luxury and sophistication to the guests. In restaurants however, unless they are of a particularly traditional and elegant kind, one doesn't tend to see them anymore.[12] But just what effect does serving the food on

---

[12] Your first author thought this practice was extinct until he got to know his 75-year-old Colombian father-in-law.

one plate versus the other have on a diner's perception? Although this is still a matter awaiting some serious empirical research, one can find various examples where a sense of luxury has been conveyed by means of the use of shiny stainless steel sheets that have been thoughtfully shaped (designed by Luki Huber; http://www.lukihuber.com/) for elBulli. On such silver-finish trays, any morsel of food is likely going to look good. We would not be surprised to find that the same food served on plastic plates was perceived as being of a lesser quality when served from porcelain/metallic plates. These things perhaps shouldn't matter, but they do.

4.4   Interim summary

We have now seen that any of the parameters of the conventional (traditional) plates, such as their size, shape, colour, material and weight, have an important role in shaping our perception of the food and how much we serve ourselves, not to mention how much we end up eating. There's a growing body of laboratory-based research demonstrating just what a profound effect the plateware has, not only on how much people enjoy what they are eating but also on the sensory-discriminative attributes of the food. There is also a growing number of experimental chefs pushing the boundaries of what constitutes plateware (see Butter 2013). We might also point out here that, to date, only a few studies have ventured out into naturalistic settings (i.e. outside of the science lab). It would therefore be wonderful to see more research being conducted in experimental restaurants, in arty pop-up venues and at the ever more popular science/gastronomy events in order to know just how generalizable and robust some of these findings really are outside of the laboratory setting. Nevertheless, there is hopefully already enough evidence to convince the chef or adventurous home cook that there is much more to flavour and to multisensory dining experiences than simply what should happen to enter our mouth.

4.5   The plate that is not a plate

Let's talk about those 'plates' that are not the conventional dishes we're used to have in the cupboard or that we see in the local corner café. Here, we are referring to those elements that serve to support and present the food in one or another manner: standing on a base or hanging from something or somewhere, designed specifically for the purpose or else brought back from the forest, the owner's backyard or reclaimed from the nearest building site (such an approach to sourcing the plateware gives a whole new meaning to the idea of foraging for one's dinner). Sometimes, and this is something that one is never really sure of, even the plate itself may turn out to be edible. Seated at the table, one is left wondering just what on the table to eat and how to begin. Here, we'll review some of the most innovative (not to say bizarre) examples that can be found at a number of top-end modernist restaurants (Butter 2013). Even the simplest of everyday materials, everything from roofing slates to reclaimed red bricks, are now starting to appear in front of the diners on the tables of some of the world's most fashionable restaurants.

4.5.1   Reaching new heights

As mentioned earlier in the chapter some chefs have taken the notion of a food canvas to new heights, perhaps so that the diner can enjoy every point of view of the food in front of them. Such food/dish designs may even require special supporting elements. To illustrate, let's take some of the dishes created by one of North America's most innovative chefs, Grant Achatz. How to serve a dehydrated translucent piece of bacon wrapped in butterscotch and apple leather? Hanging from a bow, of course! That's how Achatz designed this dish, seeking the best possible way in which to accentuate the lightness and translucency of the components (see Figure 4.7a). Indeed, Achatz's on-going collaboration with multidisciplinary design group Crucial Detail (http://www.crucialdetail.com) has enabled him to present his creations in ways that are both highly unique and visually impressive. (The term 'culinary architecture', first applied to Carême's famous pastry creations, certainly seems appropriate for some of these modernist dishes.) At the same time Achatz involves the diner, who has to pay more attention to how to handle each of the morsels of food with which they have been presented. Caroline Hobkinson is also fond of serving the food hanging from the ceiling (suspended from balloons; see Figure 4.7b). As one can imagine, the diner doesn't need any cutlery to enjoy this kind of eating experience.

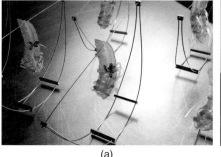

(a)   (b)

Figure 4.7   (a) Bacon hung in a steel bow designed by Crucial Detail for the Alinea restaurant in Chicago. *Source*: Photograph by Lara Kastner. Reproduced with permission of Martin Kastner, Crucial Detail LLC, Chicago. (b) Morsels of food tied to virtually invisible fishing wire and suspended from helium balloons that nestled under the ceiling. *Source*: Reproduced with permission of Caroline Hobkinson. *See colour plate section*

4.5.2   On the smell and sound of the plateware

As we'll see in Chapter 6, while eye appeal might be half the meal (as the popular expression goes), the smell and the sound of the food also play a significant role in the diner's experience of a dish. In this section, we will come across a number of examples of dishes in which the plate (or whatever the supporting element for the food happens to be) is used to deliver an aroma and thus to enhance the diner's multisensory dining experience. Sometimes it is the smell of the food itself that is being delivered, while in other cases distinct aromas are released that, together with the food's own unique smell, create a more memorable dish.

In one much more practically oriented study, Wansink and Cardello tested whether impregnating plastic cereal bowls with an aroma would induce US Army troops to eat more oatmeal for breakfast (see Wansink 2006, pp. 111–112). When the smell of cinnamon-and-raisin was impregnated into the plateware, the troops did indeed eat significantly more. However, by contrast, releasing the smell of macaroni-and-cheese wasn't judged all that great a success in terms of enhancing the troops' desire for a bowlful of porridge in the morning (presumably due to the likely ensuing sensory incongruity; see Chapter 7).

Focusing now on the upper end of the marketplace for food, Grant Achatz also likes to play with delivering aromas in the dishes that he serves. The Scallop dish that currently appears on the menu at his Chicago restaurant Alinea is served in a caldron filled with seaweed bubbling and smoking from dry ice. When the waiter opens the caldron, the diner will find the scallop (inside its shell) sitting atop the seaweed, and a wave of citrus aroma fills the diner's nostrils.[13] Another distinctive means by which to deliver scent, often used in this restaurant, involves placing an air pillow underneath the food itself. The waiter first comes to the table with a pillow filled with air and then places the dish on top of it. As the diner interacts with their food (which is normally served on an oversized bowl rather than a flat plate to avoid any accidental spills, one imagines), the pillow starts (unexpectedly) to release a stream of lavender-scented air. Here, while the food itself is undoubtedly worth noting in its own right, it is the integration of the dish with the aromatics of the 'plateware' (and the ensuing spectacle that this creates) that makes all the difference to the lucky diners' overall multisensory experience. The question however remains of whether the scented air from the pillow enhances the aroma of the food or of the immediate environment in which the dish is served (Stuckey 2012, pp. 65–67).

Indeed, this is one of several dishes that integrates aroma as a distinct element in the dish, a technique that immediately opens up so many other

---

[13] Of course, we have been serving scallops from their shells in the dish Coques Saint-Jacques for many years. What is this if not another example of the congruent plateware enhancing the diner's experience of the meal?

possibilities for experimentation. Meanwhile, another popular means of delivering the fragrance of flowers while the diner is eating is to place the food plate/bowl on top of a larger dish full of artfully arranged hyacinths, and pour some boiling water over the dish. Once again, this aromatic idea comes from the creative kitchen of Grant Achatz. For those who are familiar with the style of Japanese food presentation, this dish can perhaps be thought of as a kind of olfactorially supercharged Kaiseki meal.[14]

Sometimes, what really attracts us to the barbeque or to the steak house is that smell of smoky wood and the sound of the food sizzling over the grill. Design studio Blanch & Shock (http://blanchandshock.com/) serve a course of duck with Jerusalem artichokes and melilot on a 'plate' of charred chestnut wood. In this case, it is the charred plateware itself that gives such a fantastic aroma to the dish (see Figure 4.8a). At Eleven Madison Park meanwhile, clam chowder is served on some sizzling hot rocks that make the seaweed pop (based, apparently, on an ancient New England tradition). The sound increases as the waiter pours some water into the rocks (see Figure 4.8b). The latter example, in which the food container is also warm, helps to enhance the sound of the food as it keeps warming at the table. This is obviously reminiscent of all those Mexican restaurants where the fajitas come to the table sizzling on the hotplate (Stuckey 2012, p. 127). In these cases, the sound

(a)   (b)

Figure 4.8 Two examples of olfactorily enhanced plateware: (a) a Blanch & Shock dish served on burnt wood. *Source*: Reproduced with permission of Josh Pollen, Blanch and Shock, London and (b) clam chowder served on top of sizzling hot rocks designed to give the diner something of the impression of being at a beach, where this ancient tradition would have taken place, at Eleven Madison Park. *See colour plate section*

---

[14] In this very ornate and traditional style of Japanese cuisine, the decorations would not themselves release any fragrance (see Oshima and Cwiertka 2006).

of the sizzle can contribute to the overall multisensory dining experience (in part by building up sensory expectations).[15]

> "*As in painting and sculpture, the visual aspects of a dish can enhance the appreciation of it. As in theatre, the way something is presented – by people in costumes, with music playing, with precise timing – is as much part of the experience as the thing itself.*" (Dornenburg and Page 1996, p. 2)

4.5.3   Camouflage

"*Your first two courses are already on the table.*" Yes, that's what the diners are told by the waiter at Noma after being seated. This all seems very strange however, as there is nothing on the table. Well, nothing that is other than a collection of plant pots in the centre (the clue being the fact that there are exactly as many flowerpots as there are diners at the table). The diner is expected to bring the pot closer and then to start digging with nothing else but their hands into the 'soil', eating everything they find, including the roots! This starter, camouflaged as it is in a terracotta flower pot, consists of carrots and radishes planted in a dark brown, crunchy edible mixture of malt and hazelnut flour (which gives a pretty good visual impression of soil) and a greenish yoghurt-based sauce lying at the very bottom (see Figure 4.9a). As a friend who was lucky enough to try this dish put it: "*The crunchy malt crumble hit the palate followed by the creamy dip and the spicy and fresh crunchy radish. It was a good harvest.*" This dish, in the way that it is presented and in the given context (of a top-end designed experience), provides a very nice example of hidden conceptual incongruity as we'll discuss in Chapter 7.

On other occasions, the plateware itself may be conspicuous by its very absence, that is, the food is sometimes served directly onto the table itself. How? At Alinea, a number of the desserts require a performance that can last for several minutes. For one dish, the waiters first lay a waterproof tablecloth over the table. They then bring all sorts of small ramekins and bowls with sauces and other ingredients having different textures. Next, one of the chefs emerges from the kitchen and, in front of the intrigued diners, starts to 'plate the table' by breaking solid elements, painting with the sauces (both drop-by-drop and by 'Jackson Pollocking' the table-top) and spreading powders with the aesthetic skill, delicacy and control that only the best artists have (see Figure 4.9b). No doubt this dessert has been designed to change the diner's perception of how food can be plated. The question, then, is how to eat that 'art'. Just grab your spoon and help yourself! (Although, after reading the next chapter, some diners might perhaps be emboldened to ask for a paintbrush instead.) Indeed, it could be argued that the *nouvelle cuisine*

---

[15] This idea was captured by Elmer Wheeler's (1938) legendary marketing strapline: "*Sell the sizzle not the steak*".

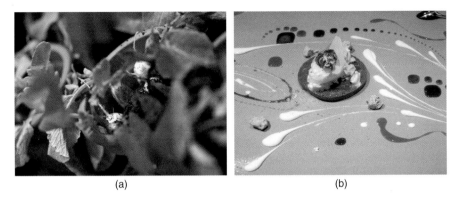

(a)   (b)

Figure 4.9 (a) One of Noma's camouflaged starters: radish and carrots in edible soil. Basically, the empty pot is the only thing that the diner should leave on the table. *Source*: Reproduced with permission of Jannie Vestergaard. (b) Alinea's most impressive dessert, served using the table as a canvas. *See colour plate section*

notion of 'plating' the food no longer seems to capture what is currently going on at the tables of at least certain modernist restaurants.

4.5.4 Improvised plateware

If dining off the table or from a plant-pot should strike you as bizarre, one need only mention the chicken liver parfait that was served from a caramel-glazed red house brick in one of London's hottest new restaurants which opened in 2012, John Salt (http://john-salt.com).[16] Forget the black slate (which diners are nowadays used to finding everywhere; see also Nolan 2013). Some of the materials used in the good old days to serve certain foods (think of sheets of old newspaper, in which fish and chips were traditionally wrapped in the UK) were chosen for their availability and price (or lack thereof). However, the experience of eating this staple dish in that particular format while strolling along the beach was so enjoyable that it became a tradition for the Brits, who say with nostalgia that this classic food combination tastes better when the fish and chips are wrapped in newspaper (Crumpacker 2006).[17] Here, once again, one is reminded of Seremetakis's (1994, p. 1) line that "*nothing tastes as good as the past*". Nowadays, though, the tables have turned and innovative chefs, in striving to present their dishes in the most uncommon and memorable way possible (while also telling a story), have started to return to this solution despite being able to count on an incredibly wide range of plateware (bricks, black slate, disks of charred rough-hewn wood, hay, banana leaves and packaging).

---

[16] Unfortunately, the chef and creator of the 'Chicken on a brick' dish, Ben Spalding, was only at the restaurant for a couple of months before leaving (Masters 2012).

[17] This is an experiment waiting to be performed …

Returning to Chicago's Alinea, perhaps one of the most conventional utensils that are used as plates is a spoon. In a dish that goes by the name of 'Squab', no plates or similar receptacles are to be seen. The waiters simply appear at the table bearing a set of nine forks and spoons, each containing morsels of different foods. The waiters then arrange the cutlery in an apparently random manner (a different arrangement for each diner). This dish pays homage to Miró's *Still Life with Old Shoe* (1937). Although the actual painting isn't shown, upon looking it up afterwards (for those who aren't familiar with it) one can clearly see the dish's similarity to the painting.

Other more natural options such as a thin slice of tree trunk or a handful of hay or leaves can also be seen in restaurants. The involvement of natural materials in the presentation of the dish may not only be designed to deliver an aroma; at the same time, one imagines, it is also meant to convey the freshness of the ingredients and to transmit the idea of naturalness, as if the diners were in a forest/island-like habitat.

"I think we're going to need designers that think about food and design in ways that we've never thought about before." (top chef Homaro Cantu, quoted in Hardy 2007)

4.5.5   Purpose-made plateware

In this chapter, we have seen an array of utensils being used to present the dishes. All of them have been carefully designed in order to display the ingredients (or the cooking procedure) in the best light possible, while at the same time delivering a memorable dining experience (and possibly also enhance the flavour). As mentioned in Section 4.1, the choice of the elements that are meant to support and/or present the food is certainly not left to chance; in fact, sometimes the meal and the means of presentation are designed at the same stage of planning a dish. On occasion however, the whimsical chef isn't content with using the existing plateware to present his/her latest culinary creation (which, in itself, is already difficult to imagine) and requires a new design, a unique design that can only be used for that course and *vice versa* (signature plateware if you will). As you can imagine, as the course is eventually retired from the menu after a successful period, so may the plateware be quietly retired from service. We have already seen several examples of purpose-designed plateware in this chapter, but an example taken to the extreme is found as the base of Heston's Mad Hatter Tea Party served at The Fat Duck restaurant (see Figure 4.10). The higgledy-piggledy tiered sandwich stand, designed specifically for the dish by Reiko Kaneko (www.reikokaneko.co.uk), was made in response to a design brief from the culinary team at The Fat Duck. It is hard to imagine this plateware being used for any other dish. Note that the design process in these cases often requires

Figure 4.10   Heston Blumenthal's Mad Hatter's Tea Party plateware, designed specially for this course served at The Fat Duck by Reiko Kaneko. *Source*: Reproduced with permission of Reiko Kaneko Ltd, Stoke-on-Trent, UK

a number of interactions before the culinary team are finally happy with the resulting design (see also Bernard 2007).

> "*I have a clear idea of how the plates should look and taste. You have a starting point and then you develop it further and constantly optimize it in terms of the visual presentation so you can reveal the secret of each ingredient in its purest form with excellent flavour and in harmony with the others.*" (Rasmus Kofoed, The World's Finest Chef 2011)

4.6   On the multiple contributions of the visual appearance of a dish

It is worth highlighting that there are several components to the visual make-up of a dish (what one might want to call the visual aesthetic). Below, we focus on the layout and composition of the various edible elements that go into making up the dish. This is a topic that has seemingly been neglected by sensory scientists over the last few decades (e.g. there is no mention of this topic in the more than 350 pages of Meiselman's 2000, book *Dimensions of the Meal*). These include, for example, the colour of the food (often broken down by researchers into studies of hue and colour intensity) and/or elements of the dish, the texture of the components and their combination (also important but little studied), the shape of the individual components and then the higher-order spatial arrangement of the various elements on the plate. Even the number of items on the plate (odd versus even) may potentially make a difference. On many occasions, the dishes are not only considered as edible art as the reader will soon see, but are also sometimes designed based on some of the very same art/aesthetic rules that have been dictated by the famous visual artists/critics (Deroy *et al.* in press).

The majority of the research (at least in recent years) has tended to focus on flat 2D representations of the visual arrangement of the food. It is however worth remembering the early work on the aesthetics of plating, and here we are thinking of the work of Carême and his ilk (cf. Figure 4.1, Carême 1828; Horowitz and Singley 2004), was very much about the vertical arrangement of the food on the plate (the 'architecture' of cuisine if you will). In the following, we will look at these elements and the psychological principles that underlie them and see what the chefs are doing to utilize these principles in the design of their dishes.

4.6.1   On the importance of harmony on the plate

We often see dishes whose construction would appear to defy the laws of gravity. Others are displayed in such a harmonious manner that the diner very likely feels that it is something of a shame to have to eat the food (they might well prefer to frame it instead; many diners are increasingly doing just that as we'll see in Chapter 10). Take for instance a number of the dishes served at Eleven Madison Park (http://elevenmadisonpark.com); they have been carefully designed to balance the colours, shades, textures and proportions. In order to enhance the beauty of the elements, the plates are often white, flat and nearly rimless (see Figure 4.11a). Chef Michel Bras (http://www.bras.fr) may well be rightly considered as the inventor of infinity plating, where there is usually no rim in sight. In addition, he uses what's called 'negative space' – that is, he plays with the contrasting background elements – in order to accentuate the many colourful ingredients that he works with. These seem to have been placed spontaneously, but actually require around 100 separate actions (see Figure 4.11b).

> "*I love serving our dishes to guests and watching their reactions. Usually, they will look at it first. Then discuss it with their friend, before pulling out a camera and taking a picture. It is great fun watching different responses.*" (Sebastian Lepinoy, Executive Chef at L'Atelier de Joel Robuchon, quoted in Yang 2011)

(a)   (b)

Figure 4.11   (a) One of the dishes served at Eleven Madison Park in New York: beef, roasted onion, cherries and ginger, aesthetically simple and asymmetric. (b) A typical multi-colourful Michel Bras dish. *Source*: (b) Reproduced with permission of Andy Hayler. *See colour plate section*

Furthermore, a number of chefs have been tempted to make edible copies of famous paintings. For example, self-confessed lover of contemporary art Michel Troisgros talked of seeing one of Mondrian's classic paintings in an art exhibition in 2010 prior to the planning of one of his own culinary creations (Csergo 2012, p. 21). He went on to name this dish "Laquered mullet *façon* Mondrian" (this title may remind the reader of the names of Carême's dish mentioned in Chapter 3).

> "*I try to interpret the artist's message and to make it mine, to translate it in life and in the dishes.*" (Massimo Bottura, Chef at Osteria Francescana, voted the World's 3rd Best Restaurant in 2013 (according to the San Pellegrino list); Barba 2013)

These dishes are obviously impressive, and we can all imagine the likely response of the diner who happens to be sitting in front of such a piece of art[18]: amazement. If treating it as such, it is likely that diners will respond more positively and hence appreciate the dish more (this could also be a kind of halo effect/sensation transference; see Sections 3.8 and 4.3.1) if the composition happens to follow the main principles of art. These include balance, unity, harmony, variety, rhythm, emphasis, contrast, proportion, pattern and movement. Balance is commonly considered to be the overriding principle in art, and it certainly helps if a well-planned meal is balanced in terms of its colours,

---

[18] Csergo (2012) refers to a survey in which 30 people were asked whether they believed that cuisine could be considered art just as paintings are. Seven replied that it wasn't art because it is meant to be "good to eat", even if we seek to make it more appetizing by making it look nicer. The others all affirmed that cuisine was one of the arts and that the visual aesthetic of a dish played a role that was judged to be just as important as taste (see also Dornenberg and Page 1996; Neill and Ridley 2002; Monroe 2007; Achatz 2009).

texture, shapes, sizes and the flavour of the various elements that have gone into making it (although we are only talking about the visual attributes, we also know that the other sensory inputs should be harmoniously incorporated).

As a basic principle of art, balance refers to the ways in which the various elements of a piece are arranged. The arrangement can be symmetrical (formal), where the elements are given equal weight from an imaginary middle point of the piece, or asymmetrical (informal or dynamic). Asymmetrical is certainly the way that many of the dishes struck us when dining at Eleven Madison Park recently (e.g. see Figure 4.11a). The latter occurs when the elements are arranged unevenly in a piece, but work together to produce an overall harmony. Art theorists have long considered balance, or notions of visual harmony, as central features of images that are judged to be aesthetically pleasing (e.g. Bouleau 1980; Arnheim 1988). Moreover, people tend to prefer visual images that are dynamically balanced over those that are not (Wilson and Chatterjee 2005).

Those working in the food sector can take actions based on theories within the field of psychology of aesthetics that might be applied to make their food products more visually appealing and thereby presumably positively affect people's food perception (and on occasion, influence the food choices that they make). According to Berlyne's (1970) collative motivation model,[19] a model based on an analysis of people's responses to visual illustrations, there is a bell-shaped relationship between hedonic appreciation and the arousal potential of a given work of art (or visual pattern). Collative properties such as complexity, novelty and variability contribute most to the arousal property of a work of art. Visual patterns with a low arousal potential (and therefore, low levels of perceived collative properties) simply aren't stimulating enough. As a consequence, they may leave the observer (or diner) feeling indifferent. On the other hand, patterns with a very high arousal potential (and thus high levels of perceived collative properties) may be too difficult to grasp and hence be considered unpleasant. The preferred patterns are considered to be those with an arousal potential that lies at a medium (or optimum) level, leading to the inverted U-shaped relationship between hedonic appreciation and arousal potential (see also Berlyne *et al.* 1968; Hekkert and Leder 2008 on the dimensionality of visual complexity, interestingness and pleasantness). Given this knowledge, how can a chef know when the composition of the food on the plate has become too complex or balanced to appeal to the diner's eyes?

Up until just a few years ago, no one had explored this topic. However, a number of researchers have conducted some intriguing studies that have finally started to address the impact of the layout of the elements in a dish on people's responses to it (Reisfelt*et al.* 2009; Zellner *et al.* 2010, 2011).

---

[19] The term 'collative properties' was introduced to describe the effects of comparisons among elements which are presented either simultaneously or in succession. These collative properties map onto subjective dimensions such as novel–familiar, simple–complex and ambiguous–clear (Cupchik and Berlyne 1979).

For instance, Reisfelt *et al.* investigated people's preference for variations of a visually presented convenience meal among a large group of Danish consumers in a study carried out in Danish malls. They used a discrete choice of experimental design in which the consumers were shown different series of eight photos consisting of combinations of variations of a meal, defined in the following way: dish (modern or traditional); vegetable mix (root vegetable mix or wok mix), meat (slices or whole pieces of tenderloin pork); sauce (either present or absent); and herbs (dishes either with or without a parsley garnish). The participants in this study had to select the meal that they preferred, the dish they would choose second and their least favourite dish. Perhaps the most relevant variable for us here is the former. The modern dish was selected significantly more often than the traditional one, with women liking the modern dish more than the men. Additionally, a preference for the traditional dish was positively correlated with age, with the elderly participants preferring the traditional option more than younger participants.

Debra Zellner and her colleagues (2010) studied whether balance and complexity[20] of the food on a plate would affect the attractiveness of the presentation. In addition, the willingness of the participants (68 under-graduates) to try the food, and their liking for it, was also assessed in four different visual conditions (monochrome–balanced, coloured–balanced, monochrome–unbalanced and coloured–unbalanced). The food consisted of four slices of water chestnut and four lines and a dot of tahini (the colour manipulation in the coloured conditions was achieved by artificially colouring the tahini). While the addition of colour increased the participants' rating of the attractiveness of the balanced food presentation, it had no effect on their rating of the unbalanced presentation.[21] The participants were more willing to try the monochrome (plain light brown) than the colourful plates of food, but there was no effect of colour or balance on people's overall liking of the flavour of the food itself. So, while manipulating the colour and balance in a dish may affect how attractive it looks, these results suggest that it doesn't necessarily alter how much people like the flavour of the food. This research was extended (Zellner *et al.* 2011) with dishes composed of lettuce, carrots, tomatoes and hummus. Even though participants did not report different attractiveness ratings for the two presentations (again balanced and unbalanced) as for their previous study, the hummus from the more balanced presentation was liked more than that from the less balanced (once tasted). However, in a second experiment during the same study, they found that chicken salad placed in a messy way in the centre of a lettuce salad was liked more than when the chicken salad was moulded into a half sphere in the centre of the salad. Consequently, Zellner and her colleagues concluded that

---

[20] Defined here as an increase in the use of colour.
[21] One shouldn't necessarily expect the latest of Grant Achatz's creations as stimuli in such laboratory-based studies.

whether their participants 'liked' the dish probably had more to do with the degree of neatness than with the degree of balance (King 1980).

It could of course be argued that complexity depends on much more than the variety of colour that happens to be present in a dish. It is also affected by the size of the elements, the variety of colours, the number of products, the familiarity of the products and the mix of products/elements that are involved. Mielby and her colleagues examined the relationship between people's visual preferences and perceived complexity for vegetables, fruits and combined mixtures of fruits and vegetables (see Mielby *et al.* 2012). As had been reported previously (see Berlyne 1970), these authors observed strong correlations between designed collative properties and rated complexity. Inverted U-shaped relationships between visual preference and complexity were documented for both the vegetable and fruit mixes. However, no such effect was observed for the combined fruit and vegetable plates (perhaps because they were already deemed complex). The take-home message here is not to allow the composition to become too simple but, if you want a dish to be successful (or at least to be visually appealing), not to overdo it either! (Where would the world be without psychologists, eh?).

In a recent collaboration between young Franco-Colombian chef Charles Michel and the Crossmodal Research Laboratory here in Oxford, we have been investigating the impact of even more complex arrangements of food than those studied in the above-mentioned studies. In particular, we were interested in knowing whether dishes in which the food is displayed in order to mimic (or has been inspired by) an artist's work (such as Grantz Achatz's dish mentioned in Section 4.5.4, served entirely on a display of cutlery) would be appreciated more (overall, and in terms of the taste/ flavour) than when the elements were arranged in a more standard manner (see also Barba 2013).

Inspired by the work of Wassily Kandinsky, Charles Michel developed an edible copy of one of Kandinsky's paintings (namely 'Painting number 201', also called 'Panel for Edwin R. Campbell No. 4; see Moma Museum website http://www.moma.org/collection/provenance/provenance_object.php?object _id=79452). Fresh ingredients, sauces and purées were arranged as in a common tossed salad (see Figure 4.12a) in one condition. In another condition, the very same 17 ingredients were served on a small canvas (Figure 4.12b). In a final condition, the elements were arranged side-by-side in a neat but non-artistic manner. The participants filled out pencil-and-paper questionnaires both before eating the food and after having finished the dish. The items in the questionnaire were designed to evaluate people's expectations and actual experience of the dish.

The Kandinsky-inspired artistic presentation resulted in the food being considered more artistic (as expected) but, more importantly, significantly more liked as well (a difference of around 16%). While the participants' perception of the basic tastes of bitterness, sweetness, saltiness and sourness were unaffected by the plating condition, the food arranged in the art-inspired

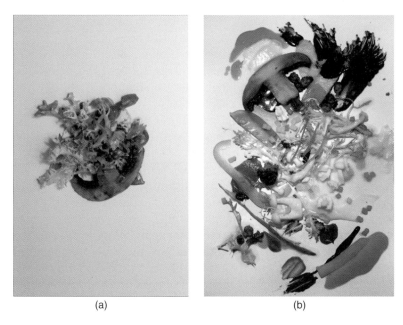

(a)   (b)

Figure 4.12   (a) 17 ingredients disposed as in a tossed salad and (b) 'Kandinsky on a plate' created by Charles Michel: an experimental dish with the same 17 ingredients arranged to look like one of Kandinsky's paintings. *See colour plate section*

condition was rated as 21% tastier overall.[22] These results therefore suggest that, assuming that there is nothing too unpalatable on the plate when they eventually come to taste it, the diner's hedonic and sensory perception will largely be determined by the expectations of the diner (Michel *et al.* 2013).

4.7   Individual diner responses to the visual presentation of food

When it comes to the visual appeal of various arrangements of the elements of a dish, there may well be some important individual differences to bear in mind here. Two obvious candidates are cross-cultural differences and the diner's general sensitivity to visual aesthetics. We will look at research on each of these themes in turn. There would seem to be no *a priori* reason why one should expect that what looks nice to a North American diner would appeal equally to an Italian or Japanese diner? Zampollo *et al.* (2012) recently addressed this question. Their results showed that across these three countries, there were consistent preferences in terms of the number

---

[22] Sometimes when this dish is served, people are given a scented paintbrush with which to move the food from canvas to mouth.

of colours on the plate (three), the number of components on the plate (three to four) and how full the plate should appear (nearly empty versus crowded, as they put it). Where there were cross-cultural differences was in terms of how diners in the different countries thought the items should be organized on the plate, where the main item should appear and the degree of casualness/organization of the presentation.

Of course, it could be that a diner's response to a plate of food will depend on how sensitive they are to visual aesthetics. If these participants had been tested for this trait (e.g. using the visual aesthetic sensitivity test or VAST; Goetz et al. 1979) and the results for visual aesthetics sensitivity and flavour appreciation for artistically presented meals found to be highly correlated, one might hypothesize that the visual results could extend to the culinary sphere (i.e. those who score more highly would likely appreciate the dish more when it is presented in a balanced manner; Becker et al. 2011).

Should such a correlation be found, much of the preliminary research on the design of new dishes could presumably be conducted simply over the internet (perhaps using digitally modified images in a kind of virtual proto-typing, but this time applied to the world of gastronomy; Kildegaard et al. 2011; Wei et al. 2012). But let's not forget here that while visual appeal can undoubtedly help to increase a diners' appreciation of the dish, the balance of textures and flavours should also be appropriate. Nowadays, managing to get the perfectly balanced mix of all of these attributes isn't quite as difficult as it once was, thanks to all the technology that is available to the professional chef (and the molecular gastronomy kits that are increasingly being made available to the amateur; see Youssef 2013) together with some knowledge of physics and chemistry.

There are certainly plenty of books and blogs from where one can learn the basics of how to present dishes beautifully (e.g. Siple and Sax 1982; Mitchell 1999). From chef Josef Youssef's book *Molecular Cooking at Home* (2013), we can extract the following useful recommendations when it comes to dressing the plate: (1) use large plates; (2) use small portions of food; (3) use different food elements proportionally balanced (in terms of their colour and texture); (4) focus on vibrant colours; (5) use seasonal ingredients; (6) elevation: do not just pile the food up, use alternative elements such as parmesan or bacon crisps to build the volume of the dish upward; (7) use cookie cutters and moulds to help make layer-able shapes; (8) use repetition in odd numbers, for example by using 3 pieces of food for small dishes or 5 for main courses; (9) be artistic: use paintbrushes and squeezy bottles to distribute the sauces; and (10) bear in mind that the presentation should not compromise the taste, temperature or practicality of the dish.[23]

---

[23] As pointed out by Dornenburg and Page (1996), it is important not to forget about the tastes and textures when contemplating how a dish will look. One needs to make sure that what is served is more than merely 'food porn' as Anthony Bourdain calls it. In other words, pleasures that are merely voyeuristic are not at all what we are after here; let's leave that for the pages of the cookery books (Damrosch 2008, p. 14).

4.8 Conclusions

In conclusion, the results of the research reviewed in this chapter clearly demonstrate that the plateware (as has been broadly defined here) can contribute to the diners' overall multisensory eating experience. Furthermore, we have also seen how a growing number of chefs and designers are now increasingly trying to enhance the diner's experience of food by serving it from plateware that has been specifically developed in order to enhance the sound, the aroma and even the diner's perception of texture. When it comes to delivering the perfect meal, it would appear as if there really are no limits to plating and plateware other than a chef's imagination (see Butter 2013).

The contribution of the plateware to the multisensory experience of a meal obviously starts with its visual input (or perhaps the sound of the sizzle if you happen to have ordered a hot plate dish). In this chapter, we have also seen that it can be used to deliver and enhance the sound of the food as well (think only of the sizzle of the steak or the fajitas on the hotplate). Of course, food is normally silent until we manipulate it (think of the sound of an airy mousse when you tuck your spoon into it); amplifying those sounds is practically impossible. However, as we will discover in Chapter 10, the latest in digital technology is now capable of doing a lot more to help facilitate the multisensory experience of diners, and sound is no exception. We have also seen various innovative examples where the plate (or its support; remember Alinea's lavender-scented pillow) has been used to diffuse scent. Many chefs are also becoming increasingly sensitive to the texture or feel of the 'plate' and the role that it plays in enhancing the experience (both by visual and haptic cues). Getting the plateware right will clearly come to constitute an increasingly important element in the perfect meal. The perfect meal, after all, undoubtedly deserves a quality frame.

We now move on to discover a bit more about the utensils used (in case we decide not to use our own hands) to taste these appetizing and extremely elaborate dishes.

References

Achatz, G. (2009) Food tasting or art installation? Available at http://food.theatlantic .com/back-of-the-house/food-tasting-or-art-installation.php (accessed January 2014).

Ackerman, J. M., Nocera, C. C. and Bargh, J. A. (2010) Incidental haptic sensations influence social judgments and decisions. *Science*, **328**, 1712–1715.

Adams, S. (2013). How to rescue NHS food? Put it on a blue plate: Simple switch has helped elderly and weak patients eat nearly a third more. DailyMail Online. Available at http://www.dailymail.co.uk/news/article-2520058/How-rescue-NHS-food-Put-blue-plate-Simple-switch-helped-elderly-weak-patients-eat-nearly-more .html (accessed January 2014).

Adrià, F., Soler, A. and Adrià, J. (2007) *Un Día en elBullli (One Day at elBulli)*. RBA, Barcelona.

AgeUK (2010) Still hungry to be heard. Available at http://www.ageuk.org.uk /BrandPartnerGlobal/londonVPP/Documents/Still_Hungry_To_Be_Heard _Report.pdf (accessed January 2014).

Anonymous (1997) The Silver Spoon. Originally published as *Il Cucchiaio D'argento* (1950). Phaidon Press, London.

Apicius (1936) *Cooking and Dining in Imperial Rome* (c. first century; translated by J. D. Vehling). University of Chicago Press, Chicago.

Arnheim, R. (1988) *The Power of the Center: A Study of Composition in the Visual Arts*. University of California Press, Berkeley, CA.

Asch, S. E. (1946) Forming impressions of personality. *Journal of Abnormal and Social Psychology*, **41**, 258–290.

Barba, E. D. (2013) My cuisine is tradition in evolution. Available at http://www.swide .com/food-travel/chef-interview/michelin-starred-chef-an-interview-with-massimo-bottura/2013/4/23 (accessed January 2014).

Becker, L., Van Rompay, T. J. L., Schifferstein, H. N. J. and Galetzka, M. (2011) Tough package, strong taste: The influence of packaging design on taste impressions and product evaluations. *Food Quality and Preference*, **22**, 17–23.

Bendiner, K. (2004) *Food in Painting: From the Renaissance to the Present*. Reaction Books, London.

Berghaus, G. (2001) The futurist banquet: Nouvelle Cuisine or performance art? *New Theatre Quarterly*, **17(1)**, 3–17.

Berlyne, D. E. (1970) Novelty, complexity and hedonic value. *Perception and Psychophysics*, **8**, 279–286.

Berlyne, D. E., Ogilve, J. C. and Parham, L. C. C. (1968) The dimensionality of visual complexity, interestingness, and pleasingness. *Canadian Journal of Psychology*, **22**, 376–387.

Bernard, S. (2007) Coming to a table near you. *New York Magazine*, 28 May. Available at http://nymag.com/shopping/features/28505/ (accessed January 2014).

Bouleau, C. (1980) *The Painter's Secret Geometry*. Hacker Books, New York.

Bradley, L. and Rees C. (2003) Reducing nutritional risk in hospital: The red tray. *Nursing Standard*, **17**, 33–37.

Bruno, N., Martani, M., Corsini, C. and Oleari, C. (2013) The effect of the color red on consuming food does not depend on achromatic (Michelson) contrast and extends to rubbing cream on the skin. *Appetite*, **71**, 307–313.

Butter, S. (2013) Why dinner in London now comes in a bag, bath and flowerpot. *London Evening Standard*, 31 October. Available at http://www.standard.co.uk /goingout/restaurants/why-dinner-in-london-now-comes-in-a-bag-bath-and-flowerpot-8915106.html?origin=internalSearch (accessed January 2014).

Carême, M.-A. (1828) *Le Pâtissier Pittoresque, Précédé d'un Traité des Cinq Ordres d'Architecture (The Picturesque Pastry Chef, Preceded by a Treatise of the Five Orders of Architecture)*. Reanouard, Paris.

Carême, M.-A. (1854) *Le Pâtissier Royal Parisien (The royal Parisian pastrycook)*. Reanouard, Paris.

Churchill, A., Meyners, M., Griffiths, L. and Bailey, P. (2009) The cross-modal effect of fragrance in shampoo: Modifying the perceived feel of both product and hair during and after washing. *Food Quality and Preference*, **20**, 320–328.

Cowen, T. (2012) *An Economist Gets Lunch: New Rules for Everyday Foodies*. Plume, New York.

Crumpacker, B. (2006) *The Sex Life of Food: When Body and Soul Meet to Eat*. Thomas Dunne Books, New York.

Csergo, J. (2012) L'Art culinaire ou l'insaisissable beauté d'un art qui se dérobe. quelques jalons, xviiie-xxie siècle (The art of cooking as an imperceptible beauty of an art which slips away: some milestones, 18th–19th century). In: *L'Artification du Culinaire* (eds E. Cohen and J. Csergo), pp. 13–36. Publications de la Sorbonne, Paris.

Cupchik, G. C. and Berlyne, D. E. (1979) The perception of collative properties in visual stimuli. *Scandinavian Journal of Psychology*, **20**, 93–104.

Damrosch, P. (2008) *Service Included: Four-Star Secrets of an Eavesdropping Waiter*. William Morrow, New York.

de Graaf, C., Cardello, A. V., Kramer, F. M., Lesher, L. L., Meiselman, H. L. and Schutz, H. G. (2005) A comparison between liking ratings obtained under laboratory and field conditions: The role of choice. *Appetite*, **44**, 15–22.

Deroy, O. and Spence, C. (2013) Quand les goûts & les formes se répondent. *Cerveau and Psycho*, **55**(Janvier–Février), 74–79.

Deroy, O., Michel, C., Piqueras-Fiszman, B. and Spence, C. (in press) The plating manifesto (I): From decoration to creation. *Flavour*.

Desbuissons, F. (2012) Yeux ouverts et bouche affamée: le paradigme culinaire de l'art modern (1850–1880) (Eyes open and mouth closed: the culinary paradigm of modern art). In: *L'Artification du Culinaire* (eds E. Cohen and J. Csergo), pp. 49–71. Publications de la Sorbonne, Paris.

Diliberti, N., Bordi, P. L., Conklin, M. T., Roe, L. S. and Rolls, B. J. (2004) Increased portion size leads to increased energy intake in a restaurant meal. *Obesity Research*, **12**, 562–568.

Dornenburg, A. and Page, K. (1996) *Culinary Artistry*. John Wiley and Sons, New York.

Ekroll, V., Faul, F. and Niederée, R. (2004) The peculiar nature of simultaneous colour contrast in uniform surrounds. *Vision Research*, **44**, 1765–1786.

Frascari, M. (2004) Semiotica ab edendo (Taste in architecture). In: *Eating Architecture* (eds J. Horowitz and P. Singley), pp. 191–203. MIT Press, Cambridge, MA.

Gal, D., Wheeler, S. C. and Shiv, B. (2007) *Cross-modal influences on gustatory perception*. Available at http://ssrn.com/abstract=1030197 (accessed January 2014).

Genschow, O., Reutner, L. and Wanke, M. (2012) The color red reduces snack food and soft drink intake. *Appetite*, **58**, 699–702.

Goetz, K. O., Borisy, A. R., Lynn, R. and Eysenck, H. J. (1979) A new visual aesthetic sensitivity test: I. Construction and psychometric properties. *Perceptual and Motor Skills*, **49**, 795–802.

Gopnik, A. (2011) *Sweet revolution*. Available at http://www.newyorker.com/reporting/2011/01/03/110103fa_fact_gopnik (accessed January 2014).

Halligan, M. (1990) *Eat my Words*. Angus and Robertson, London.

Hardy, S. (2007) Creative generalist Q&A: Homaro Cantu. Available at http://creativegeneralist.com/2007/11/creative-generalist-qa-homaro-cantu/ (accessed January 2014).

Harrar, V. and Spence, C. (2013) The taste of cutlery. *Flavour*, **2**, 21.

Harrar, V., Piqueras-Fiszman, B. and Spence, C. (2011) There's more to taste in a coloured bowl. *Perception*, **40**, 880–882.

Hekkert, P. and Leder, H. (2008) Design aesthetics. In: *Product experience* (eds H. N. J. Schifferstein and P. Hekkert), pp. 259–285. Elsevier, London.

Horowitz, J. and Singley, P. (eds) (2004) *Eating Architecture*. MIT Press, Cambridge, MA.

Hudson, K. and Pettifer, J. (1979) *Diamonds in the Sky: A Social History of Air Travel*. Bodley Head, London.

Hultén, B., Broweus, N. and van Dijk, M. (2009) *Sensory Marketing*. Palgrave Macmillan, Basingstoke, UK.

Hutchings, J. B. (1994) *Food Colour and Appearance*. Blackie Academic and Professional, London.

Imram, N. (1999) The role of visual cues in consumer perception and acceptance of a food product. *Nutrition and Food Science*, **99**, 224–230.

Kildegaard, H., Olsen, A., Gabrielsen, G., Møller, P. and Thybo, A. K. (2011) A method to measure the effect of food appearance factors on children's visual preferences. *Food Quality and Preference*, **22**, 763–771.

King, S. (1980) Presentation and the choice of food. In: *Nutrition and Lifestyle* (ed M. Turner) pp. 67–78. Applied Science Laboratories, London.

Leibowitz, H., Myers, N. A. and Chinetti, P. (1955) The role of simultaneous contrast in brightness constancy. *Journal of Experimental Psychology*, **50**, 15–18.

Levitsky, D. and Youn, T. (2004) The more food young adults are served, the more they overeat. *Journal of Nutrition*, **134**, 2546–2549.

Liang, P., Roy, S., Chen, M.-L. and Zhang, G.-H. (2013). Visual influence of shapes and semantic familiarity on human sweet sensitivity. *Behavioural Brain Research*, **253**, 42–47.

Lyman, B. (1989) *A Psychology of Food, more than a Matter of Taste*. Avi, van Nostrand Reinhold, New York.

Maga, J. A. (1974) Influence of color on taste thresholds. *Chemical Senses and Flavor*, **1**, 115–119.

Marchiori, D., Corneille, O. and Klein, O. (2012) Container size influences snack food intake independently of portion size. *Appetite*, **58**, 814–817.

Marinetti, F. T. (1932) *La Cucina Futurista (The Futurist Kitchen)*. Sonzogno, Milan.

Masters, S. (2012) Top chef quit sold-out restaurant 'after clash over burger and chips'. *The Independent*, 24th December, 12.

Meiselman, H. L. (ed) (2000) *Dimensions of the Meal: The Science, Culture, Business, and Art of Eating*. Aspen Publishers, Gaithersburg, MA.

Michel, C., Velasco, C., Gatti, E. and Spence, C. (2013) *A taste of Kandinsky: Enhancing expectations and experience through the use of art-inspired food presentation*. Poster presentation given at the *10th Pangborn Sensory Science Symposium*. 11–15 August 2013, Rio de Janeiro, Brazil.

Mielby, L. H., Kildegaard, H., Gabrielsen, G., Edelenbos, M. and Kistrup Thybo, A. (2012) Adolescent and adult visual preferences for pictures of fruit and vegetable mixes – Effect of complexity. *Food Quality and Preference*, **26**, 188–195.

Mitchell, S. E. (1999) *Arranging Food Beautifully*. John Wiley, New York.

Monroe, D. (2007) Can food be art? The problem of consumption. In *Food and Philosophy* (eds F. Allhoff and D. Monroe), pp. 133–144. Blackwell Publishing, Oxford.

Neill, A. and Ridley, A. (2002) The art of food. In: *Arguing about Art: Contemporary Philosophical Debates* (2nd edition) (eds A. Neill and A. Ridley), pp. 5–8. Routledge, London.

Nielsen, S. J. and Popkin, B. M. (2003) Patterns and trends in portion size. *Journal of the American Medical Association*, **289**, 450–453.

Nolan, S. (2013) *The Heston effect? List of food trends moves away from the traditional and embraces the unusual. Daily Mail Online*, 2 April. Available at http://www.dailymail.co.uk/news/article-2302999/The-Heston-effect-List-food-trends-moves-away-traditional-embraces-unusual.html (accessed January 2014).

O'Mahony, M. (1983) Gustatory responses to nongustatory stimuli. *Perception*, **12**, 627–633.

Oppenheim, M. (1936) Object (Le Déjeuner en fourrure). *Journal of the American Psychoanalytic Association*, **44S**, 22.

Oshima, A. and Cwiertka, K. J. (2006) *Yamazato: Kaiseki Recipes: Secrets of the Japanese Cuisine*. Stichting Kunstboak, West-Vlaanderen.

Park, M. Y. (2013) *A history of how food is plated, from medieval bread bowls to Noma*. Available at http://www.bonappetit.com/trends/article/a-history-of-how-food-is-plated-from-medieval-bread-bowls-to-noma (accessed February 2014).

Piqueras-Fiszman, B. and Spence, C. (2011) Crossmodal correspondences in product packaging: Assessing color-flavor correspondences for potato chips (crisps). *Appetite*, **57**, 753–737.

Piqueras-Fiszman, B. and Spence, C. (2012a) The weight of the container influences expected satiety, perceived density, and subsequent expected fullness. *Appetite*, **58**, 559–562.

Piqueras-Fiszman B. and Spence C. (2012b) The influence of the feel of product packaging on the perception of the oral-somatosensory texture of food. *Food Quality and Preference*, **26**, 67–73.

Piqueras-Fiszman, B., Harrar, V., Alcaide, J. and Spence, C. (2011) Does the weight of the dish influence our perception of food? *Food Quality and Preference*, **22**, 753–756.

Piqueras-Fiszman, B., Alcaide, J., Roura, E. and Spence, C. (2012) Is it the plate or is it the food? The influence of the color and shape of the plate on the perception of the food placed on it. *Food Quality and Preference*, **24**, 205–208.

Piqueras-Fiszman, B., Giboreau, A. and Spence, C. (2013) Assessing the influence of the colour/finish of the plate on the perception of the food in a test in a restaurant setting. *Flavour*, **2**, 24.

Reisfelt, H. H., Gabrielsen, G., Aaslyng, M. D., Bjerre, M. S. and Møller, P. (2009) Consumer preferences for visually presented meals. *Journal of Sensory Studies*, **24**, 182–203.

Rolls, B. J., Roe, L. S., Halverson, K. H. and Meengs, J. S. (2007) Using a smaller plate did not reduce energy intake at meals. *Appetite*, **49**, 652–660.

Rozin, P., Kabnick, K., Pete, E., Fischler, C. and Shields, C. (2003) The ecology of eating: Smaller portion sizes in France than in the United States help explain the French Paradox. *Psychological Science*, **14**, 450–454.

Schneider, I. K., Rutjens, B. T., Jostmann, N. B. and Lakens, D. (2011) Weighty matters: Importance literally feels heavy. *Social Psychological and Personality Science*, **2**, 474–478.

Seremetakis, C. N. (ed) (1994) *The Senses Still*. The University of Chicago Press, Chicago.

Siple, M. and Sax, I. (1982) *Foodstyle: The Art of Presenting Food Beautifully*. Crown Publishers, New York.

Sobal, J. and Wansink, B. (2007) Kitchenscapes, tablescapes, platescapes, and food-scapes: Influences of microscale built environments on food intake. *Environment and Behavior*, **39**, 124–142.

Spence, C. (2012) Managing sensory expectations concerning products and brands: Capitalizing on the potential of sound and shape symbolism. *Journal of Consumer Psychology*, **22**, 37–54.

Spence, C. and Gallace, A. (2011) Multisensory design: Reaching out to touch the consumer. *Psychology and Marketing*, **28**, 267–308.

Spence, C. and Piqueras-Fiszman, B. (2012) The multisensory packaging of beverages. In: *Food Packaging: Procedures, Management and Trends* (ed. M. G. Kontominas), pp. 187–233. Nova Publishers, Hauppauge NY.

Spence, C., Levitan, C., Shankar, M. U. and Zampini, M. (2010) Does food color influence taste and flavor perception in humans? *Chemosensory Perception*, **3**, 68–84.

Spence, C., Ngo, M., Percival, B. and Smith, B. (2013a) Crossmodal correspondences: Assessing the shape symbolism of foods having a complex flavour profile. *Food Quality and Preference*, **28**, 206–212.

Spence, C., Hobkinson, C., Gallace, A. and Piqueras Fiszman, B. (2013b) A touch of gastronomy. *Flavour*, **2**, 14.

Stewart, P. C. and Goss, E. (2013) Plate shape and colour interact to influence taste and quality judgments. *Flavour*, **2**, 27.

Stuckey, B. (2012) *Taste What You're Missing: The Passionate Eater's Guide to Why Good Food Tastes Good*. Free Press, London.

Styler, C. and Lazarus, D. (2006) *Working the Plate: The Art of Food Presentation*. John Wiley, New York.

Tannahill, R. (1973). *Food in History*. Stein & Day, New York.

Titchener, E. B. (1908) *Lectures on the Elementary Psychology of Feeling and Attention*. Macmillan, New York.

Van Kleef, E., Shimizu, M. and Wansink, B. (2012) Serving bowl selection biases the amount of food served. *Journal of Nutrition Education and Behavior*, **44**, 66–70.

Wansink, B. (1996) Can package size accelerate usage volume? *Journal of Marketing*, **60**, 1–15.

Wansink, B. (2006) *Mindless Eating: Why We Eat More Than We Think*. Hay House, London.

Wansink, B. (2010) From mindless eating to mindlessly eating better. *Physiology and Behavior*, **100**, 454–463.

Wansink, B. and Cheney, M. M. (2005) Super bowls: Serving bowl size and food consumption. *Journal of the American Medical Association*, **293**, 1727–1728.

Wansink, B., van Ittersum, K. and Painter, J. E. (2006) Ice cream illusions: Bowl size, spoon size, and self-served portion sizes. *American Journal of Preventive Medicine*, **31**, 240–243.

Wansink, B., Just, D. R. and Payne, C. R. (2009) Mindless eating and healthy heuristics for the irrational. *American Economics Review: Papers and Proceedings*, **98**, 165–169.

Wei, S.-T., Ou, L.-C., Luo, M. R. and Hutchings, J. B. (2012) Optimization of food expectations using product colour and appearance. *Food Quality and Preference*, **23**, 49–62.

Weinstein, S. (1968) Intensive and extensive aspects of tactile sensitivity as a function of body part, sex, and laterality. In: *The Skin Senses* (ed D. R. Kenshalo), pp. 195–222. Thomas, Springfield, Ill.

Wheeler, E. (1938) *Tested Sentences that Sell*. Prentice and Co. Hall, Inc., New York.

Williams, J. M. (1976) Synesthetic adjectives: A possible law of semantic change. *Language*, **52**, 461–478.

Wilson, A. and Chatterjee, A. (2005) The assessment of preference for balance: Introducing a new test. *Empirical Studies of the Arts*, **23**, 165–180.

Wilson, C. A. (1991) *The Appetite and the Eye: Visual Aspects of Food and its Presentation within the Historical Context*. Edinburgh University Press, Edinburgh, Scotland.

Yang, J. (2011) The art of food presentation. Available at http://www.cravemag.com /features/the-art-of-food-presentation/ (accessed January 2014).

Young, L. R. and Nestle, M. (2002) The contribution of expanding portion sizes to the US obesity epidemic. *American Journal of Public Health*, **92**, 246–249.

Youssef, J. (2013) *Molecular Cooking at Home: Taking Culinary Physics out of the Lab and into your Kitchen*. Quintet, London.

Zampollo, F., Wansink, B., Kniffin, K. M., Shimuzu, M. and Omori, A. (2012) Looks good enough to eat: How food plating preferences differ across cultures and continents. *Cross-Cultural Research*, **46**, 31–49.

Zellner, D. A., Lankford, M., Ambrose, L. and Locher, P. (2010) Art on the plate: Effect of balance and color on attractiveness of, willingness to try and liking for food. *Food Quality and Preference*, **21**, 575–578.

Zellner, D. A., Siemers, E., Teran, V., Conroy, R., Lankford, M., Agrafiotis, A., Ambrose, L. and Locher, P. (2011) Neatness counts. How plating affects liking for the taste of food. *Appetite*, **57**, 642–648.

5
Getting Your Hands on the Food: Cutlery

"The eating utensils that we use daily are as familiar to us as our own hands. We manipulate knife, fork and spoon as automatically as we do our fingers, and we seem to become conscious of our silverware only when right- and left-handers cross elbows at a dinner party." (Petroski 1994, p. 3)

"One of the more spectacular triumphs of human 'culture' over 'nature' is our determination when eating to avoid touching food with anything but metal implements." (Visser 1991, p. 167)

5.1 Introduction

By this point of the day most of us will already have picked up several pieces of cutlery, likely giving them no thought. Using cutlery probably seems as natural as breathing, right? According to the Oxford English Dictionary, cutlery (also referred to as flatware) defines those utensils (knives, forks and spoons) that are used for eating and serving food. However, the word itself comes from the old French *coutellerie*, a term that refers to the art of making knives as well as to the place in which they were once made and sold. This is why the word is primarily used in many regions when talking about cutting utensils. In this chapter we will take a brief look at the history of flatware, highlighting how people used to eat (even just a couple of centuries ago) with cutlery that was made from a far wider range of materials than today. The discussion will then lead on to a review of the latest research looking at whether the material qualities of the utensils we use to eat with can affect the taste of the food.

The Perfect Meal: The Multisensory Science of Food and Dining, First Edition.
Charles Spence and Betina Piqueras-Fiszman.
© 2014 John Wiley & Sons, Ltd. Published 2014 by John Wiley & Sons, Ltd.

This in turn raises a number of questions, such as: what exactly is so special about a 'silver spoon'? Why do we all, at least in the west, eat using 'standard' cutlery made of either stainless steel or silver? And why did the knife, spoon and eventually the fork turn out in the way that they eventually did?

Of course, no chapter on cutlery would be complete without some discussion of chopsticks. We will therefore also take a look at their material properties, how they are used for particular dishes in different parts of Asia and how some researchers believe that the obesity crisis that is currently afflicting the west might be alleviated simply by encouraging more people to eat with chopsticks rather than with the more traditional knife, fork and spoon.

In this chapter, we will also review the latest research to have addressed the question of whether the weight of the cutlery influences the taste and flavour of the food, as well as looking at whether we really do eat more if we are given a larger spoon or fork. The reality is that cutlery is still very much in a state of development; what is particularly exciting at the present time is that a growing number of designers and chefs are starting to experiment with a bewildering variety of new materials, sizes and formats.[1] However, rather than trying to minimize the impact of the material properties of the cutlery on a diner's perception of the food (as was the case previously), the aim now is very much to try and enhance the sensory properties of the food (such as by changing its perceived texture, temperature, taste and/or flavour, etc.) or else perhaps to convey specific feelings to the diner. One can think of this as using the cutlery to season the food! By the end of this chapter, we hope to have convinced you that getting the cutlery right represents a much more important component than you might have thought when it comes to delivering the perfect meal.

As mentioned in Chapter 4, some chefs have not only been inspired by the visual arts but they have even started to apply artistic techniques to their culinary creations. Take the following quote from Andoni Luis Aduriz: "*Since the coating gave the potato a stony feel, dry, and clean to the touch, this quality suggested to the guest to dispense with knife and fork to eat it with one's hands*" (Aduriz *et al.* 2012, p. 76). Here, the culinary team from the Mugaritz restaurant in San Sebastian, Spain used the *trompe l'oeil* or 'fool the eye' technique in order to give their food a novel appearance which, in turn, changes the way in which the diner approaches the dish. In this case, chefs can use the visual presentation of the dish in order to influence the way in which their diners interact with it. Over the last few years, a growing number of chefs have started to seek alternative designs to the standard forms of cutlery in order to modify the eating behaviours and protocols of their diners. Forget about formal *etiquette* (and the advice one sometimes finds doled out in the newspapers, e.g. Anonymous 1982). Increasingly, you'll be better off following the instructions provided by the waiter instead. In fact, sometimes the

---

[1] In a way, bringing back the range of different materials that one might have found back in the sixteenth century.

diner is invited to forget about cutlery altogether – just use your hands and enjoy the *feeling* of the food between your fingers, and not just in your mouth. After all, who says that the texture of the food is something that can only be enjoyed orally? Certain foods (think hamburgers, sandwiches and wraps) have been intentionally designed to be enjoyed manually (as suggested by the term 'finger food'), while others have acquired that status as a result of convention (e.g. asparagus spears or artichokes; Crumpacker 2006). But just how much more enjoyable might our experience of a range of other foods be should we simply allow ourselves the pleasure of eating with our hands? And is it ever acceptable to do so, when dining in polite company? The good news here, at least for those who like to feel the food (or at least certain foods) between their fingers, is that the rules of etiquette are changing fast. Debrett's (see http://en.wikipedia.org/wiki/Debrett's) recently caused something of a stir when they acknowledged that it is now OK to eat certain foods with one's hands (see Section 5.8 for more).

5.2 The story of cutlery

When we humans first began to cook our food hundreds of thousands of years ago (Wrangham 2010), sharpened stones and sticks were used to help break down and consume recently hunted meals. Shells and hollowed-out animal horns were also common, leading to the early development of the spoon. In fact, both the Greek and Latin words for spoon are derived from *cochlea*, meaning a spiral-shaped snail shell (Wilson 2012, p. 246). Spoon technology seems to have hit something of a standstill in prehistoric times however, and the knife soon became the primary tool used by our ancestors to help them to eat. In fact, it's possible to trace human evolution through this humble instrument made first with bone or stone, then later with bronze and finally with iron in around 1000 BC (see Wilson 2012).

For the wealthy living in medieval Europe, knives were often elaborately carved and decorated with bone or ivory handles. At large parties, the host certainly couldn't be expected to furnish such a costly piece of equipment for each and every member of a large group of guests, and so they would have been expected to show up with their own knives in tow. They were still not deemed necessary for everyone, but diners would sometimes use them just to impress those with whom they were dining. Early table knives had sharp, pointy ends that were used to spear the food and to bring it direct to the diner's mouth. One can certainly imagine how this practice must have led to quite a few punctured palates! Sometimes diners would use two knives, one to hold the meat still while cutting it with the other. Finally, after the introduction of the fork at the table in 1637, it was no longer necessary to spear the food with the point of one's knife; cutlers began to make blunt-ended knives instead (Visser 1991; Wilson 2012).

France's Cardinal Richelieu is reputed to have demanded that the tips of knives be rounded after tiring of seeing his men stabbing their knives into the meat on the table and then using the tip of their knives to pick their teeth after the meal was over (Ward 2009); that is how the modern knife was born. What is more, this also led people to start holding their knives as they do nowadays: modern manners dictate that the knife should never be held in a way suggesting that one is about to start a fight (Fanshawe 2005, pp. 130–131).

The spoon is the oldest of all eating utensils used by humans. The first spoon-shaped utensils are thought to have consisted of shells found along the shoreline to which, at some point, wooden or bone handles were attached to make them more convenient to use. The earliest spoons (from Ancient Egypt) were made of wood, and the oldest examples include some made from slate and ivory as well. In other parts of the world, metals such as bronze, gold, pewter and brass increasingly started to be used. The sixteenth and seventeenth centuries saw the arrival of copper and silver spoons (though silver only for the high nobility). Soon, the quality of one's spoons began to signal one's rank in society. Around this time, many silver craftsmen and jewellers concentrated on making spoons solely for the benefit of the wealthy (Wilson 2012; see also Emery 1976). The family with the most ornate spoons would rapidly gain prestige.

Apostle spoons – that is, spoons with an ornamental ending in the handle (known as the 'finial') that represented the Apostles – were fashionable gifts at christenings during Tudor and Stuart times. A complete set of twelve silver Apostle spoons would undoubtedly have been considered a valuable gift. Tiny spoons, especially crafted out of silver, were commissioned by rich parents for their newborn babies. This is the likely origin of the expression "born with a silver spoon in the mouth".[2,3]

The story of the fork is much more interesting. It has even shaped our eating action and led to us having an overbite (or so it has been claimed). Before the use of this utensil became widespread, people would have used their incisors in a guillotine-like fashion with the two rows of teeth bearing down against one another. This would eventually have led to people grinding their teeth down (Brace 1977; Brace *et al.* 1987; Wilson 2012). Since people started to use a fork,

---

[2] Certainly, no other item of cutlery has made its way into our everyday language as 'the silver spoon' has. Not only is this the title of the classic Italian cookbook *Il Cucchiaio d'Argento* (1950, 1997) that even today is still given to newlyweds in Italy (in fact, this bible of authentic Italian cooking has been the best-selling cookbook in Italy for over half a century), but it is also the name of the famous brand of sugar originally launched in 1972 as part of the retail arm of British Sugar (http://www.silverspoonpeople.co.uk/page/our_brands).

[3] Why should the teaspoon be so much more popular than the coffee spoon, especially when sales of coffee look set to exceed those of tea in the UK? While the teaspoon has become the ubiquitous item of cutlery to add sugar to our hot beverages, no matter whether it is tea, coffee or something else, writers seem only to talk about the coffee spoon. Take the following: "*The sun is mirrored even in a coffee spoon*" (Giedon 1948/1975, p. 3); or T. S. Eliot's line "*I have measured out my life with coffee spoons*" (in *The Love Song of J. Alfred Prufrock* 1917, cited in Horowitz and Singley 2004, p. 6). There is even a book filled with phrases about spoons (see Jünger 1993)!

there was no longer any need to tear the food with one's front incisors and the overbite slowly emerged. Interestingly, in China (where the introduction of chopsticks occurred around 900 years earlier) the overbite has been shown to have emerged nearly 1000 years earlier than it did in the west (Wilson 2012). The fork is a surprisingly modern invention. In fact, it was the last utensil to be introduced to the dinner table (its use only becoming widespread at the start of the nineteenth century). Humans got along just fine without forks for thousands of years. And yet, that said, it remains a bizarre object. People today still have difficulties in deciding how best to use it (consider the debate regarding the correctness of switching the fork from one hand to the other that goes on between Americans and Europeans; Visser 1991).

The form of the fork has been around a lot longer than the eating utensil itself (think of ancient Greek Poseidon with his trident or the Devil with his pitchfork); however, the fork had no place at the Greek table.[4] At that time, people only used spoons, the point of their knife or their hands. It was only much later in the eleventh century that records first document the use of forks as we are familiar with them nowadays, namely in the Byzantine Empire. The source of this account comes from the story of St Peter Damian, a hermit and ascetic who criticized the Venetian princess Maria Argyra for her (according to his way of thinking) excessive delicacy in using a fork to eat: "*Such was the luxury of her habits ... [that] she deigned [not] to touch her food with her fingers, but would command her eunuchs to cut it up into small pieces, which she would impale on a certain golden instrument with two prongs and thus carry to her mouth.*" Damian found Maria's table manners so offensively pretentious that when she later died of the plague, he is said to have regarded it as nothing less than a fair punishment from God! The second piece of evidence regarding the origins of the fork comes from an eleventh century illustrated manuscript from Monte Cassino that apparently shows a couple of men using a two-pronged fork (see Forbes 1956).

After this, not much is heard of the humble fork over the course of the next five centuries. As mentioned in Chapter 4, in the Middle Ages most people would eat off rounds of stale bread called trenchers. These could hold (or absorb) anything from soup to cooked meat and vegetables. Trenchers could serve both as plates and cutlery since they could be brought directly to the mouth; at that point in time, knives and spoons were basically used for those foods that one could not manage to manipulate with one's hands. Forks were brought from Byzantium to Italy[5] and arrived in France along with Catherine de Medici, who travelled from Italy to France in 1533 to marry Henry II.

---

[4] As Bee Wilson (2012) notes, fork-like instruments had been in use in kitchens for hundreds of years. It is only their use at the dinner table that had yet to be firmly established.

[5] As Wilson (2012) notes, forks must surely have been very handy when it came to eating spaghetti. The residents of Naples, where the modern version of this form of pasta originated, were known to eat spaghetti (although there was no sauce) with their hands in the nineteenth century (see Del Conte 1976; Visser 1991, pp. 17–18).

Catherine used massive public festivals in order to emphasize the power of the monarchy, and food was an integral part of this strategy of spectacle. Some years later, thanks to her tour of the countryside, this food etiquette and culture started to percolate down to the middle strata of society. In the late sixteenth and early seventeenth centuries, most forks were two-pronged, hefty and deemed so refined that they were only used occasionally, mainly to eat sweets at the end of the meal.[6]

However, the use of forks was still not universally accepted. It was said that Louis XIV banned his children from eating with them, despite the instructions of their tutor (Braudel 1992). Meanwhile, the English ridiculed forks as being effeminate and, what is more, unnecessary.[7] It wasn't until the late seventeenth and early eighteenth centuries that people began to purchase multiple sets of silverware for their homes. It was around that time that homes were just beginning to be equipped with a room that would have been set aside specifically for the purpose of dining. It was also around that time that forks with three and then four tines started to become more commonplace. The fork had become sufficiently widespread by the middle of the century, such that those who didn't use one (properly) were regarded as the odd ones out.

> "We sat down for dinner, which was served up after the French manner, on a circular table, with chairs set around it, spoons, forks, etc, and nothing wanting but to know how to make use of all these things. They seems nevertheless to be desirous not to omit any of our customs which begin to be in the same vogue, among the Greeks, which those of the English are with us; and I have seen a woman, at dinner, take olives up with her fingers, and afterwards put them on a fork, to eat them after the French fashion." (Baron de Tott 1786, pp. 98–99)

Prior to the eighteenth century, most utensils were made of silver (one of the most inert of metals and hence one that reacts the least with food)[8] but, as can be imagined, rare and rather expensive. By the beginning of the nineteenth century the fork made its way firmly onto the French table and beyond (in America they were called 'split spoons'). It was around this time that the meal started to become a centre of social life; not just for the aristocracy, but for the newly established middle class (or *bourgeoisie*) as well (Ward 2009).

> "The fork has now become the favorite and fashionable utensil for conveying food to the mouth. First it crowded out the knife, and now in its pride it has invaded

---

[6] Crumpacker (2006, p. 186) mentions the case of an Englishman who wrote home describing the forks that he had seen in Italy. When this fellow finally made his way back to England people apparently made fun, calling him 'Furcifer' or silly fork carrier for his affected ways.

[7] Bee Wilson (2012, p. 257) points out that up until around 1900, British sailors would still eat without using a fork to demonstrate their manliness.

[8] For this same reason, tastevins (very shallow, small cups) were increasingly made from silver as well. Winemakers, traders and sommeliers in France would use a tastevin to assess the quality of a wine. Silver was ideal for this purpose in not giving any taint to the wine (Burk and Bywater 2008, pp. 200–202).

the domain of the once powerful spoon. The spoon is now pretty well subdued also, and the fork, insolent and triumphant, has become a sumptuary tyrant. The true devotee of fashion does not dare to use a spoon except to stir his tea or to eat his soup with, and meekly eats his ice-cream with a fork and pretends to like it." (Hall 1887, p. 87)

The invention of silver-plating in the middle of the nineteenth century, accompanied by a strong expansion of the consumer market, resulted in the widening usage of cutlery for those from all levels of society (Himsworth 1953; Visser 1991; Brown 2001; Elias 2005). Think of the Victorian approach to cutlery: the upper classes would have had a bevvie of servants to polish their extensive silver dinner services, keeping them free of tarnish. Polishing the silver was obviously no mean feat given that, at the time, each set might contain anywhere up to around 87 specialized pieces of cutlery for each place setting. Such large dinner services might well include items such as a marrow or cheese scoop, a chocolate muddler, a tomato server and pickle and oyster forks (see Petroski 1994, pp. 130–131). They would no doubt have had a few spoons made of horn or mother of pearl in order to eat their soft-boiled eggs in the morning (mother of pearl being used because egg yolk tarnished the silverware; Wilson 2012, p. 246).[9] Although there was probably never a time when more than a few of these forks and utensils were in play at any one dinner, anxiety can't have been too far away from the nervous diner's mind (Coffin *et al.* 2006).[10] Nowadays, if shown a number of these unusual items of cutlery, most people would likely have little clue as to what they would originally have been used for (see Figure 5.1).

"There also appeared the likes of salad forks, lemon forks, pickle forks, asparagus forks, sardine forks, and more, each with its tines widened, thickened, sharpened, splayed, barbed, spread, joined, or somehow modified to reduce the faults that other forks exhibited in handling some very specific food." (Petroski 1994, p. 149, on the evolution of the fork in the closing years of the nineteenth century)

Even the distinction between the fork and other items of cutlery became somewhat blurred. For example, one entry in the *Collectors' Handbook for Grape Nuts* (see Petroski 1994, p. 130), is named a 'melon knife or fork' while another is labelled as an 'olive fork or spoon'. The equivalent today is perhaps the 'spork', that most curious of utensils one sometimes finds inside the packs of fruit salads to-go in the supermarket (though the name of this unusual item

---

[9] It's the sulphur in eggs (and in caviar) that tarnishes silver and adversely affects the taste of food (http://kaufmann-mercantile.com/horn-spoon/; Wilson 2012). When such foods were eaten with the aid of a silver spoon, they were said to taste and smell awful! At Thomas Keller's Per Se restaurant in New York, caviar is still served with spoons made of mother of pearl (Damrosch 2008).

[10] Crumpacker (2006, p. 190) describes a table service for 18 diners that was sold at Sotheby's auction house in 1994. Each setting had 18 pieces including an ice-cream fork and spoon and an oyster spoon. Six of the pieces in each setting had been fashioned from solid gold.

Figure 5.1 Range of Victorian period sterling cutlery set. From left to right: a large Durgin division of Gorham twisted handle fish server, Dominick & Haff 'twist handle' slotted spoon and olive fork, a twisted handle sugar shell and a Gorham bright cut serving spoon with a curved handle. *Source*: Reproduced with permission of Jeffrey Burchard, Burchard Galleries Inc., FL, USA

apparently first made its way into the English dictionary back in 1909; Wilson 2012, p. 271).[11]

Aluminium cutlery briefly became fashionable around this time: spoons made of this metal were for a time so rare that they were once displayed in the Fifth Avenue windows of Tiffany's (McGee 1990, p. 244). Meanwhile, over in France, Napoleon Bonaparte purportedly commissioned a fancy set for himself (Aldersey-Williams 2011). Silver still won the beauty contest, however. In his *The theory of the Leisure Class*, Thorsten Veblen (1899/1992) took an aluminium spoon and a silver spoon to illustrate his idea that the utility of objects that are valued for their beauty 'depends closely on the expensiveness of the articles'. At the time he was writing, aluminium spoons would have cost somewhere around 10–20 cents apiece while, by contrast, silver spoons would cost as much in dollars (Aldersey-Williams 2011).

Harry Brearley invented stainless steel in Sheffield in 1913 (just before the outbreak of the First World War) while trying to improve on the design of gun barrels (Himsworth 1953, p. ix; Wilson 2012, p. 4). One perhaps unexpected consequence of the success of this new metal was the rapid

---

[11] Nowadays, one can even find a 3-in-1 knife, fork and spoon (e.g. see http://www.figmenta.co.uk/).

reduction in the range of other materials that were used to manufacture cutlery.[12] Today, stainless steel is by far the most commonly used material for cutlery. By the early twentieth century, the great designers of the era (and here we are referring to legendary figures such as Henry van der Velde, Charles Mackintosh and Josef Hoffman) included cutlery design, along with the design of the tables, the chairs and the lamps, for their buildings. In fact, each decade of the twentieth century has had its style of cutlery to match the corresponding design/art movement that happened to be in vogue at the time[13]. From the opening years of the twenty-first century, home cutlery sets would normally have ranged in number from 4 to 6 pieces.

In the rest of this chapter, we'll see how at some restaurants unique pieces of cutlery are becoming the essential accompaniment for each dish! Who knows, it may not be too long before delivering the perfect meal will necessitate one purchasing several new items of cutlery, and the dinner services of the world's fancy restaurants may once again reach well into the double digits. Restaurateurs will however need to be careful not to commit the same mistake as Alan Ducasse when he opened his restaurant ADNY in New York back in 2000. According to Steinberger (2010, p. 168), the diners were rather perturbed to be offered a dozen different ornate knives with which to cut their meat. As the New York magazine critic Gael Greene put it: "*I'm not really amused being forced to choose my knife or my pen just so the house can show off how many it has assembled … It's vulgar*" (Greene 2000).

5.3　The material qualities of the cutlery

Historically, wood, bone and ceramic spoons were especially common as were spoons made from iron, brass, bronze and pewter because they were the only metals that were affordable. However, these would often give an unpleasant taint to food (Miodownik 2008; Wilson 2012). The most common material used to make cutlery today is stainless steel, followed by silver. Of course, one can also find plenty made from plastic and sometimes wood. The decision to choose one material over another is normally conditioned by the intended context of use. For example, you certainly wouldn't expect to find

---

[12] One problem with the traditional steel knife was that acidic fish juices (often aggravated by the addition of lemon juice) would corrode the blades, not to mention giving the fish an unpalatable taste (see Petroski 1994, p. 149). By contrast, stainless steel is made of a mixture of steel and chromium. A thin layer of inert chromium oxide forms on the outer surface of the spoon, ensuring that this material is completely tasteless (Wilson 2012). Scratch the surface and a new layer of chromium oxide forms making this, according to Mark Miodownik (from The Institute of Making in London), a kind of self-cleaning cutlery; the most important fact, at least for those of us without servants, is that it doesn't tarnish.

[13] Actually, the creation of cutlery (and in particular the evolution of the fork) is used by many theorists as a paradigm case of the design process of industrial products (see Weber 1992; Fabian 2011).

plastic cutlery in the setting of a formal restaurant, or to find fine sterling silver cutlery in a fast-food joint.[14]

However, among the spectrum of possible options offered by different material properties, shapes and sizes, other factors should perhaps also be taken into account in order to facilitate the selection of the most appropriate material from which to fashion the cutlery. Just as we saw with the plateware in Chapter 4 (see also McGee *et al.* 1984), the wide variety of different cutlery designs that are available nowadays serve not only to meet the functional needs of the diner, but also to transmit a certain feeling. In Thailand, whether or not the diner is given chopsticks to use depends on the nature of the dish being served (Visser 1991). Going way back in time, one finds the ancient Fijians apparently eating pretty much everything with their hands except for human flesh, for which delicacy they reserved a special wooden fork (Visser 1991, p. 13).

"We know the lighter aluminium spoon is easier to use, yet we prefer the silver because it 'gratifies our taste'". (Aldersey-Williams 2011, p. 257)

5.3.1   The quality of the cutlery

One might reasonably ask if having cutlery made from different materials also affects the perceived quality or taste of the food that is consumed from it. This is just the sort of question we like to investigate. We therefore conducted an experiment in which people were given identical samples of yoghurt to try with two different spoons. When tasted with a heavier metallic spoon, yoghurt was rated as significantly more pleasant and perceived to be of higher quality (with a difference of 16% in both cases) than when tasted with a metallic-looking plastic spoon (Piqueras-Fiszman and Spence 2011).[15]

According to Piqueras-Fiszman and Spence (2011), the increased pleasantness ratings for the food tasted with the aid of the stainless steel spoon may have been attributable to the widespread belief that stainless steel spoons are of higher quality than those made of plastic. The significant effects observed in terms of perception of quality and liking could be taken to support the assertion that, in many product categories, heavier objects are generally perceived

---

[14] The airlines would prefer their passengers to use lightweight cutlery on board rather than heavy metallic utensils. A few years ago, a fine lightweight titanium cutlery set was custom designed for use on Concorde (when it was still flying). Unfortunately, this cutlery ended up being so light that passengers simply did not like eating with this flatware when it was trialled, despite the obvious high quality of the design. That said, even this lightweight cutlery was likely more pleasant to eat with than the bizarre combination of metal fork and plastic knife that one can find on Asiana Airlines. And have you ever wondered why it is that the blade of the knife on planes has become almost triangular in shape in recent years? According to Visser (1991), this development allows the passengers seated at the back of the plane to cut their meat even though their arms may be squished tightly by their sides.
[15] Given that both the weight and material properties of the spoons varied in this study, it wasn't possible to determine the relative contributions of each factor to the overall perception of the food eaten from them.

to be of higher quality (see also Ariely 2008; Lindstrom 2005; Spence and Piqueras-Fiszman 2011). However, these findings also suggest that the sense of higher quality that the stainless steel spoon might have engendered in a diner's mind could be transmitted to the food itself. One can even think of this in terms of the notion of sensation transference that we came across in Chapter 1. Any such effect might well be expected to result in people perceiving the food to be of higher quality and consequently positively affecting their overall hedonic judgments.

5.3.2 Tasting the cutlery

Until very recently, little research had been conducted in order to investigate whether the material properties of the cutlery would influence a diner's multisensory perception of a meal, and ultimately how acceptable they find a particular food to be. However, material scientists working with psychologists and sensory scientists have now started to conduct research in order to more systematically investigate the taste that different metals have (Laughlin *et al.* 2009, 2011; Miodownik 2008). For example, in one study Laughlin *et al.* (2011) investigated the effects on people's sensory-discriminative and hedonic ratings of metallic tastes arising from seven spoons plated with different metals: gold, silver, zinc, copper, tin, chrome and stainless steel. The blindfolded participants who took part in this study evaluated the taste of each of the spoons (note that there was actually nothing on the spoons themselves) and then rated the following attributes on a 7-point scale: cool, hard, salty, bitter, metallic, strong, sweet and unpleasant.

The results demonstrated that the gold, chrome, tin, silver and stainless steel spoons did not differ significantly for any of the attributes that were rated, but they were significantly different from the zinc and copper spoons. Meanwhile, gold and chrome were rated as the least metallic, least bitter and least strong tasting of all the spoons,[16] whereas the zinc and copper spoons had the strongest, most metallic, most bitter and least sweet taste. The fact that the gold and chrome-plated spoons were rated as being the least metallic and bitter doesn't necessarily mean that we should all trade in our silver/stainless steel spoons for these other metallic spoons, but simply that these metals may correlate with a 'preference' for different metallic tastes.[17]

Piqueras-Fiszman *et al.* (2012) extended this line of research in order to investigate the transfer of taste qualities from these metal spoons to the food that was consumed from them. Their participants evaluated sweet, sour, bitter,

---

[16] Put simply, those metals that are highly susceptible to oxidation (e.g. aluminium and copper) will give off a noticeable 'metallic' taste, whereas gold and silver are almost tasteless (hence the status attached to silver cutlery; see Laughlin *et al.* 2009).

[17] Note here that consumers apparently came to like the metallic taste of tinned tomatoes so much that many of them complained when innovations in technology removed this distinctive metallic taint from one of their favourite tinned products (see Rosenbaum 1979, p. 81).

salty or plain (i.e. unadulterated) cream samples using spoons that had been plated with one of four different metals: gold, copper, zinc and stainless steel. Once again, the spoons had exactly the same shape, size and weight. The participants were blindfolded in order to eliminate any effects that might have been associated with the spoons' visual attributes (think only of the sensation transference effects that were mentioned earlier). In addition to transferring a somewhat metallic and bitter taste to the food, the zinc and copper spoons were also found to enhance each cream's dominant taste (by as much as 25% in the case of bitterness). Surprisingly, the presence of a metallic taste didn't influence the participants' ratings of the pleasantness of the taste to any great extent. The gold and stainless steel spoons did not exert any significant effect on the taste of the different creams.

Taken together, the results reported so far in this chapter suggest that manufacturing spoons from a wider range of materials could, in the future, be used to enhance (or, at the very least, to alter) the taste and/or flavour of foods. That said, given that bitterness is not an attribute of food that is necessarily always appreciated by diners (indeed, more often than not, people tend to avoid those foods having a pronounced bitter taste), the ability of the cutlery to enhance bitterness might only be useful for a restricted number of foodstuffs (e.g. when eating foods that have been flavoured with coffee, Italian aperitifs such as Aperol or Campari or very dark chocolate). By contrast, the increased saltiness associated with eating salty foods with the aid of zinc and copper spoons could perhaps be expected to have a more widespread application (e.g. for those on restricted sodium diets).

> "The sight of 15 adults sucking their spoons like babies was an unusual start to a dinner party, but they had surprisingly different flavours. Copper and zinc were bold and assertive, with bitter, metallic tastes; the copper spoons even smelt metallic as they gently oxidised in the air. The silver spoon, despite its beauty, tasted dull in comparison, while the stainless steel had a faintly metallic flavour that is normally overlooked." (Dunlop 2012)

But what would one expect if these findings were to be applied to a real restaurant serving more complex foods (or at least, flavours)? On 16 April 2012, an eclectic mix of people was invited to dine at Quilon, a Michelin-starred Indian restaurant in central London (see http://www.quilon.co.uk/). The guests (who included Harold McGee and Heston Blumenthal, among other luminaries) were presented with a seven-course tasting menu. Each lucky diner was provided with an array of seven freshly polished spoons made from copper, gold, silver, tin, zinc, chrome and stainless steel (see Figure 5.2a). The diners were told to start by tasting each of the spoons. Using the various spoons to taste the different dishes that were presented led to some surprising observations. According to Fuchsia Dunlop, the guest journalist invited to the event, baked black cod with zinc was

(a)   (b)

Figure 5.2 (a) Range of seven spoons of different metals used at a dinner at Quilon restaurant in London. (b) Edible dough kitchen utensil set designed by Andere Monjo. *Source*: Reproduced with permission of Andere Monjo

extremely unpleasant, as was grapefruit eaten with the copper spoon. But both metals combined well with a mango relish, their loud, metallic tastes somehow harmonized with the food's sweet-sour flavour. Tin turned out to be a popular match for pistachio curry, as was the gold-plated spoon for sweet foods. That said, it is worth remembering that the taste of the spoon can be fairly subtle, and so can easily be overpowered by any strong-tasting food on the menu (cf. Dunlop 2012).

The idea that the meal is a multisensory experience is nothing new; what is revolutionary here is the science that is slowly starting to illuminate the complexity behind our everyday perception of food. Might chefs one day really consider the taste of their cutlery to be an integral part of the flavour of a dish? In the past, Heston Blumenthal has been known to serve edible cutlery made of chocolate dusted with silver. After the Quilon dinner, he was quoted as saying:

> "*I can imagine a spoon being part of a dish, … I've been surprised at the range of metal flavours we've tasted, and at the way some sit quite well with certain sour notes in food, like the zinc and copper with mango. I've always been sensitive to metallic tastes and had thought of the cutlery as interfering with the food; but here, the metallic note can, with some flavours, be more enjoyable than otherwise.*"

Indeed, as Mark Miodownik, one of the spoons' creators pointed out: "*we were not just tasting the spoons but actually eating them, because with each lick we were consuming 'perhaps a hundred billion atoms'*" (quoted in Dunlop 2012)

Meanwhile, over in Switzerland, Denis Martin serves a dish that goes by the name of '*Rien*', which means 'nothing' in English. The waiter brings you an empty bowl with a spoon resting inside: "Just a little something?" the waiter says mischievously. You look at your empty bowl sitting in front of you and then shift your gaze to the waiter, who then proceeds to pour nothing from a sauceboat into the bowl with something of a conspiratorial air. Once your plate is filled with 'nothing', you look puzzled. "You do not taste your food?" the waiter might ask. Not knowing what else to do, you decide to bring the empty spoon to your mouth. And then, surprise, surprise, an explosion of flavours floods your palate: tomatoes, cucumbers, *gazpacho* … It turns out that the seemingly tasteless spoon has actually been coated with a transparent film of a very tasty invisible flavour (see Martin 2007).

Talking of cutlery that has a taste, take a look at the dinnerware from the Spanish designer Andere Monjo created in collaboration with the master pastry chef Lionel Bethaz. These items are made entirely from flour, water and lots of seeds (Figure 5.2b; see Cottrell 2012). Although they were created specifically for an artistic installation, one might ask why this idea couldn't be introduced in the wacky world of modernist cuisine (if the culinary team working behind the scenes can make the pieces tasty enough, that is).

To the best of our knowledge, the effects of cutlery on people's perception of food have so far only been tested with spoons (but see Harrar and Spence 2013 for the latest data on knives and toothpicks). One might reasonably ask whether similar effects would also be observed for the foods that are normally eaten with the aid of forks, knifes and/or chopsticks? Since forks present a much smaller surface area to the mouth/tongue, and knives are rarely inserted into the mouth (at least in polite company), they might well *a priori* be expected to exert less of an effect on the taste/flavour of the food.[18]

Chopsticks tend to be manufactured from a fairly restricted range of materials including cheap wood and plastic, lacquered wood (popular in Japan and Korea) and metal (especially in Korea).[19] It would therefore be interesting in future research to determine what effect the weight of the chopsticks (and the material from which they happen to have been made) has on people's perception of the taste/flavour of foods eaten with them. It would also be intriguing to follow-up on any cross-cultural differences that there might be here too. Another question that arises in this area is why many Chinese feel the need to use disposable (wooden) chopsticks?[20] This may highlight just how strange

---

[18] You would be well advised to keep a close eye on your teaspoons, given that they are apparently the most frequently stolen item of cutlery according to the results of an observational study by Australian researchers who obviously had a little too much spare time on their hands (Lim *et al.* 2005)! Members of the House of Representatives in the US are apparently no more trustworthy when it comes to the mysterious disappearance of institutional cutlery (Fahrenthold and Sonmez 2011).

[19] Wilson (2012, p. 262) notes that silver chopsticks were used at the imperial Chinese table both for their obvious luxury but also because they could aid in the detection of poison – the idea was that arsenic would turn the silver black. However, silver is less useful given its (lack of) gripping power and because of the fact that the chopsticks would likely transfer too much heat to the diner's hand if one tried to eat a dish that was too hot.

[20] While disposable chopsticks may well be a popular option in many countries, a number of the world's forests may be in danger given how many pairs of them the world's population gets through.

our relationship to cutlery really is. If you were asked whether you would mind using someone else's toothbrush, the answer would undoubtedly be a resounding NO WAY. So why is it all right to stick cutlery that has been in who-knows-how-many other people's mouths into your own? While Westerners don't seem to mind doing this, many Chinese diners do. For them, the very idea of sharing chopsticks can trigger thoughts of contamination.[21]

Now that we know a bit more about the evolution of flatware and about the effect (and connotations) of the different materials from which the cutlery may have been made (or coated) on people's perception of food, let's move on to explore other parameters of the cutlery that the designer, chef or psychologist can play with.

5.4 Size matters

In addition to the material properties (i.e. taste, weight, etc.) of the cutlery affecting our perception of food, its size can also make a difference and, as one can imagine, mostly in terms of how much food is served and how much is consumed. Mishra *et al.* (2012) recently demonstrated that the size of the cutlery impacts on how much people eat in a restaurant setting. Counter-intuitively however, they found that those who ate with the aid of smaller forks tended to consume *more* than those diners eating with a larger fork. These findings were explained in terms of 'the goal of satiation'. That is, when people go to a restaurant, the cost and effort typically involved in the dining experience causes the diner to demand an appropriate benefit: in other words, they want a greater number of forkfuls of food in order to satisfy their predetermined satiation goal. Under laboratory conditions however, where the participants typically do not have to pay for their food (more likely, they will have been paid to come into the lab to eat), Mishra and colleagues found, as one might expect, that people ate less with a smaller fork than with a larger fork (that is, the *opposite* effect to that seen in the restaurant setting; see also Geier *et al.* 2006).

These results can perhaps be related to those observed by Bolhuis *et al.* (2013). These researchers pumped tomato soup into their participants' mouths via a tube. The tube could deliver different sip sizes: small (5 g, 60 g/min), large (15 g, 60 g/min) and freestyle (where the sip size was determined by the participants themselves). The *ad libitum* intake in the small-sip condition was around 30% lower than in the large-sip and free-sip conditions, regardless of whether the participants were focused on the taste and flavour

---

Bo Guangxin, head of a major forestry group, told Chinese parliamentarians that the Chinese alone get through as many as 80 billion pairs a year, equivalent to 20 million trees (Visser 1991; Moore 2013). Wilson (2012) notes that the demand for disposable chopsticks is so great that the Chinese have been forced to start importing them from North America.

[21] Part of the reason as to why cutlery but not toothbrushes can be shared may relate to the relative ease of cleaning.

while sipping or else distracted by watching a short animation film (*Pat and Mat*). It would therefore appear that with smaller portions of food being taken into the mouth each time, people eat less than with larger portions (regardless of whether the food is pumped directly to one's mouth or served by means of a spoon/fork); however, since the results from real-world settings differ from those observed under laboratory conditions, the conclusion that one should draw is still rather uncertain and likely depends on a number of factors.

Wansink *et al.* (2006) reported that when offered free ice-cream, participants who were given bigger spoons served themselves nearly 15% more than those who were left holding a small spoon.[22] The amount consumed by a diner would therefore appear to be based on at least two factors: (1) the size of the cutlery; and (2) some kind of cost/benefit analysis related to the cost of the food whereby people tend to eat less when the food is free and they are given a small utensil to eat with.

In his fascinating book *Mindless Eating*, Brian Wansink (2006, p. 86) also reports that diners tend to eat more slowly and to eat less per bite when they use chopsticks (we assume he is referring to Western consumers who may have been relatively less familiar with using this type of cutlery). He also notes that of the 33 people whom he informally observed eating with chopsticks in three Chinese buffets in the US, 26 were of normal weight and only 7 were obese, an observation that (if verified with a much larger sample) would appear to suggest that people who tend to eat a lot wouldn't choose chopsticks as an eating utensil, since they would just make the activity more frustrating. Perhaps the problem of overeating (WHO 1998) could be addressed in some small part by encouraging more people to eat with chopsticks instead of with knife and fork (but not disposable chopsticks, of course). Although this seems unlikely to happen on a daily basis if one has grown up in a non-chopstick-wielding culture, we might nevertheless still want to consider what impact such a move would have on our overall multisensory eating experiences.

5.5   On the texture/feel of the cutlery

Standard modern cutlery doesn't normally have any special texture or feel. It is likely to be entirely smooth, apart perhaps from a subtle embossed pattern on the handles if decorated. However, many practitioners are currently starting to use the insights of laboratory-based research on multisensory design in the context of real dining. For example, when the House of Wolf restaurant

---

[22] It is worth noting that the effect of the spoons observed in Mishra *et al.* (2012) and Wansink *et al.* (2006) should not necessarily be seen as contradictory. There were several methodological factors that might have been predicted to give rise to differing results reported in these two studies: the food was presented differently (already served on the plates versus self-service, respectively) and the contextual conditions were also very different (real restaurant versus invitation to attend a social event).

(a)   (b)   (c)

Figure 5.3   (a) The textured spoons presented to diners for one of the touch courses organized by Caroline Hobkinson at the House of Wolf restaurant. *Source*: Reproduced with permission of Caroline Hobkinson. (b) The Osetra "dish" served at Alinea, in which the bottom surface of the spoon is textured to give a nice contrast to the upper elements of the dish. (c) Spoon set designed by Jinhyun Jeon in order to enhance the different textures, temperatures and flavours of foods, as part of a project entitled 'Tableware as Sensorial Stimuli'. *Source*: Reproduced with permission of Jinhyun Jeon. *See colour plate section*

opened in London in 2012 (see http://houseofwolf.co.uk/), the inaugural chef (the chef rotates every month or so) and artist Caroline Hobkinson created a series of courses designed to sequentially stimulate each of the diner's senses. Of particular relevance here are the dishes that were specifically designed to engage the diner's sense of touch. These included the use of small whittled tree branches as cutlery, while another dish (again designed to more effectively stimulate the sense of touch) consisted of a 'Hendrick's gin infused cucumber Granita'. This dish was eaten with spoons that had been treated with rose water crystals and Maldon sea salt to give them a distinctive and unusual gritty texture (see Figure 5.3a). Meanwhile, the Osetra 'dish' that initiates a meal at Alinea is served on nothing but a single spoon whose bowl part is bathed in a delicious cream which serves to bind some crumbs to the lower surface. This then creates a playful crackle to balance the bright pop of the caviar and the smoothness of the jelly sphere (see Figure 5.3b).

Similarly, Jinhyun Jeon, an industrial designer based in Eindhoven, Holland, designed 'Tasty Formulas'. This project consisted of a set of cutlery (forks and spoons) with each piece designed to engage more than just the sense of taste through the combined use of temperature, colour, texture, volume and form (in both the handle and the bowl; see Figure 5.3c).

> "*Via exploring 'synesthesia' if we can stretch the borders of what tableware can do, the eating experience can be enriched in multi-cross-wiring ways. The tableware we use for eating should not just be a tool for placing food in our mouth, but it should become extensions of our body, challenging our senses even in the moment when the food is still on its way to being consumed. Each of the designs have been created to stimulate or train different senses – allowing more than just our taste buds to be engaged in the act and enjoyment of eating as sensorial stimuli.*"
> (Jinhyun Jeon, taken from http://jjhyun.com)

As we have seen in this chapter, there are so many more parameters to play with when a chef wants to convey specific sensations to the diner. Using the cutlery for this purpose provides a very convenient route for this, since it can easily be used to transfer specific sensations to our hands and mouths and even to season the food itself! One restaurant that has recently started to capitalize on playing with the temperature of the cutlery is Nerua in Bilbao (Spain, http://www.nerua.com). When the service starts, the diners find themselves seated at an empty table: no cutlery, no glasses and no plates. Next, the waiter brings each diner a warm napkin. The purpose here, according to the restaurant's founder and chef Josean Alija, is to transmit tenderness and care (cf. Williams and Bargh 2008). The cutlery is then brought to the table and, depending on its temperature, the attentive diner can infer whether the plate that will be served next is going to be warm or cold – congruency even at the level of the flatware![23] Meanwhile, at Per Se in New York, the bouillon spoons are presented to diners chilled (Damrosch 2008). Of course, given what we have already seen in this chapter, one might at this point be wondering whether changing the temperature of the cutlery would impact on the taste of the food (just as changing the texture or weight has been shown to do). Although we haven't seen anything published on this topic as yet, we have heard informally that Peter Barham and his colleagues in Bristol have tried just such an experiment on behalf of an unnamed restaurant. However, these researchers apparently failed to demonstrate any appreciable effect of the temperature of the cutlery on people's rating of food.

5.6   Colourful cutlery

Crumpacker (2006, p. 181) reports that, once upon a time, "*Some knifes were made in different colours, to match church holidays: There were black knives for Lent and White knives for Easter and, in England, checkered knives for Whitsun, made of black horn and mother-of-pearl.*" Surely, no one at the time would have given a thought as to whether the food eaten with these colourful items of cutlery would have tasted any different. However, the latest research suggests that it may very well have done. The Crossmodal Research Laboratory in Oxford recently published a study demonstrating the influence of the colour of the cutlery on the perception of two differently coloured yoghurts (Harrar and Spence 2013). The participants in their study tried two samples of yoghurt (once with its natural white colour, and once dyed to give the yoghurt a pinkish colour) with five different spoons (red, blue, green, black and white; see Figure 5.4). For each, they had to rate the yoghurt in terms of expense, sweetness, saltiness and overall liking on a 9-point scale. The samples from

---

[23] Similarly, the presence of the fish knife on the table again signals to the diner that there will be a fish course.

Figure 5.4 A selection of the cutlery and foods used in Harrar and Spence's (2013) study looking at the effects of cutlery on taste/flavour perception. *Source*: Harrar and Spence 2013. Reproduced with permission of Elizabeth Willing, photographer. *See colour plate section*

the black spoon were rated as tasting significantly less sweet and pleasant. Furthermore, the white yoghurt seemed to be slightly more pleasant on the white spoon than on the black spoon, while the opposite was true for the pinkish yoghurt sample.

5.7 Cutlery that is not

Up to this point, we've seen how modifying any of the parameters of standard cutlery can (to a greater or lesser extent) affect how much we decide to eat, how we perceive the taste and flavour of the food, how much we like it and even our appraisal of the food's overall quality. Most of the examples that have been discussed so far have concerned (more or less) standard items of cutlery. Recently however, researchers have started to turn their attention towards assessing the consequences of changes to the design of cutlery on the overall multisensory experience of eating (e.g. Piqueras-Fiszman and

Spence 2011; Spence and Piqueras-Fiszman 2011; Harrar and Spence 2012, 2013; Piqueras-Fiszman *et al.* 2012). As we have seen above, a number of insights from this relatively novel area of research have already started to appear in the context of modernist dining. So how exactly do consumers react when they are given small branches instead of 'proper' cutlery to eat with? How does changing the cutlery in such a radical departure from tradition impact on the diner's overall experience of a dish or meal?

Over the last couple of years, culinary artist Caroline Hobkinson (see http://www.stirringwithknives.com/) has been providing diners with a variety of unusual eating utensils to work with. In one of her dishes, the diners were invited to use hand-carved branches in order to 'spear' the foraged Chanterelles and the wild Venison from the table. Hobkinson specifically designed the utensils in order to enhance the feeling of the unaltered nature of the wild, gamey food. This example shows that trends are indeed cyclical (even if that cycle happens to run over the course of some millions of years).

Meanwhile, Homaro Cantu's Moto sample menu items include 'flat'-ware with Toro and Caviar that comes with curly-cue flatware handles that are filled with a sprig of fragrant thyme. The unsuspecting diner gets a burst of colour, perhaps even a hint of aromatherapy, from the fresh sprigs of herbs inserted in the cutlery's spiral handles (see Figure 5.5). Similarly, designer Luki Huber created a spoon for the elBulli restaurant that had an integrated clip in the handle to allow the fresh herbs to be clipped just next to the bowl and the spoon, thereby providing a complementary aroma with each and every spoonful that the diner takes. After all, why not directly use natural ingredients in order to deliver the aroma direct to the diner's nostrils? At the end of a meal at Alinea, the diners are brought a course consisting of a tempura of caramel with Meyer lemon and cinnamon sugar. The nugget of caramel was stuck to the end of a stick of natural cinnamon bark. The diner had to grab the bark in order to eat the caramel part, while the cinnamon bark gave the dish its distinctive aroma.

At this point, it is interesting to consider the future of utensils. At his restaurant Alinea, chef Grant Achatz has been working closely with Crucial Detail (see Chapter 4; www.crucialdetail.com) to devise a range of novel utensils for modernist dining. For instance, Figure 5.6a depicts the 'antenna' dish, the 'spike' of which support a single morsel of food. But is it a plate or is it cutlery? A similar question can be asked of certain of the pieces created by Andreas Fabian in his PhD work investigating 'spoons and spoonness' (see Fabian 2011). Perhaps that doesn't really matter. Rather, the question that you should be asking yourself is whether the utensil has genuinely been designed to make the food more interesting to eat. And before you imagine yourself incapable, or at least unwilling, of leaning forwards, hands-free, to

Figure 5.5 'Aromatic fork' served at Homaro Cantu's Moto. *Source*: Reproduced with permission of Homaro Cantu. *See colour plate section*

(a)   (b)

Figure 5.6 (a) One of the unusual utensils used at Alinea, ideal for encouraging the diner to grab 'a bite' even if they have no cutlery (or don't wish to use their hands). *Source*: Photograph by Lara Kastner. Reproduced with permission of Crucial Detail LLC. (b) Spoons designed for elBulli, especially to address their culinary needs when having to serve one of their molecular gastronomy creations. *Source*: Reproduced with permission of Julio Arias Pérez. *See Colour plate section*

snap the morsels of food from the thin stems of metal with your mouth, just remember how silly it must have felt in previous centuries when we humans were first learning to use the humble fork.

> "*Monkeys with knitting needles would not have looked more ludicrous than some of us did*' commented one of those present on the first recorded occasion of Americans eating Chinese food in China in 1819." (quoted in Wilson 2012, p. 260)

Another top chef who has been collaborating closely with industrial cutlery designers is Ferran Adrià. Together, these creative minds have come up with a collection of serviceware specifically for the chef's cuisine and for other companies such as Lavazza (e.g. the Èspresso spoon designed in 2003, with a hole in the centre of the bowl to emphasize the solid texture of their *espresso* thick coffee). Among a most curious collection of utensils, Ferran requested that the straining spoon (see Figure 5.6b) should be designed specifically to serve the restaurant's signature 'olives' made from olive juice (by means of the spherification technique). One cannot help but smile at the irony of these innovative olives served, as they were, in old-fashioned jars with oil and herbs (see also Barba 2013).

The writing spoon is one of our personal favourites from the elBulli FACES collection (http://faces-usa.com/writing.html). The aim here is to encourage the playfulness of stirring your coffee at the end of a meal and having 'espresso ink' to doodle away on the paper tablecloth or on one's immaculately clean plate. It suddenly turns the ingredient into a medium for the diner to take a turn at being creative at the table, an interactive twist on the classic end to what has hopefully been a perfect meal. We are reminded of what we discussed in Chapter 4 about being presented with plates of food that have been artistically plated. The tables have been turned however, and it is the creative and adventurous diners who can start to make their own art. What is more, they are encouraged by chefs and restaurateurs to do so (although we must surely all remember being admonished as children with the phrase: "Don't play with your food"; Dornenburg and Page 1996, p. xix; Visser 1991, p. 176).

Here at Oxford University's Crossmodal Research Laboratory, and in collaboration with the young Franco-Colombian chef Charles Michel, we have recently been experimenting with serving the Kandinsky on a plate dish, which itself is served on a white artist's canvas (as described in Chapter 4), with a paintbrush as cutlery. While this may strike the diner as a little peculiar – one never thinks of eating with a paintbrush – the wider thin brushes are actually just as effective at wiping up the colourful puree from the plate or canvas as they are at painting the living room. What is more, they deliver a much more textural experience to the diner's mouth. They also offer the opportunity for the chef to impregnate the bristles with the scent of a wonderfully aromatic virgin olive or truffle oil.[24]

Others who are experimenting with scenting their cutlery at the moment include Louise Bloor who runs the Fragrant Supper Club (see http://www.louisebloor.com/fragrant.htm). She recently provided diners with wooden forks and a fragrance that they could apply to the fork's handle. The idea is that the porous wood would absorb the fragrance, and the diner's

---

[24] Although the audience at culinary events and gastronomy shows are more than willing to trade in their forks for paintbrushes, at least when eating this dish, individual participants tested under laboratory conditions tended to be rather less adventurous.

Figure 2.3 The literally tasty 'tasting menu' served at Homaro Cantu's Moto restaurant printed with food-based ink. *Source*: Reproduced with permission of Homaro Cantu

Figure 3.1 'Snail Porridge', as served at Heston Blumenthal's The Fat Duck restaurant. The name of the dish deliberately juxtaposes two seemingly incongruous ingredients. *Source*: Photograph by Ashley Palmer-Watts. Reproduced with permission of Lotus PR and The Fat Duck

The Perfect Meal: The Multisensory Science of Food and Dining, First Edition.
Charles Spence and Betina Piqueras-Fiszman.
© 2014 John Wiley & Sons, Ltd. Published 2014 by John Wiley & Sons, Ltd.

Figure 3.2 The 'meat fruit' dish from Heston Blumenthal's Dinner. In this case, the name of the dish makes it unclear whether the meat or the fruit is the dominant element

Figure 3.3 'Spherical egg of white asparagus with false truffle', a dish served at elBulli before it closed. Just by reading the name of the dish on the menu, one knows that most probably it will be useless to search for the egg or the truffle (or even the asparagus). *Source*: Reproduced with permission of Blake Jones

Figure 4.2 Fried deer moss from the Noma restaurant in Copenhagen. *Source*: Reproduced with permission of Sonali Fenner

Figure 4.5 Heinz released a limited number of these Beanz Flavour Experience packs which included a matching bowl with the texture and shape of its main ingredient for each flavour. Left: the Beanz Garlic and Herbs Flavour Experience bowl; right: the Beanz Cheese Flavour Experience bowl. Pictures taken by Nathan Pask, *Source*: Photograph by Nathan Pask Photography. Reproduced with permission of Ann Charlott Ommedal, Bompas and Parr, London, and NathanPask

(a)　　　　　　　　　　　　　　　(b)

Figure 4.7 (a) Bacon hung in a steel bow designed by Crucial Detail for the Alinea restaurant in Chicago. *Source*: Photograph by Lara Kastner. Reproduced with permission of Martin Kastner, Crucial Detail LLC, Chicago. (b) Morsels of food tied to virtually invisible fishing wire and suspended from helium balloons that nestled under the ceiling. *Source*: Reproduced with permission of Caroline Hobkinson

(a)　　　　　　　　　　　　　　　(b)

Figure 4.8 Two examples of olfactorily enhanced plateware: (a) a Blanch & Shock dish served on burnt wood. *Source*: Reproduced with permission of Josh Pollen, Blanch and Shock, London and (b) clam chowder served on top of sizzling hot rocks designed to give the diner something of the impression of being at a beach, where this ancient tradition would have taken place, at Eleven Madison Park

(a)   (b)

Figure 4.9 (a) One of Noma's camouflaged starters: radish and carrots in edible soil. Basically, the empty pot is the only thing that the diner should leave on the table. *Source*: Reproduced with permission of Jannie Vestergaard. (b) Alinea's most impressive dessert, served using the table as a canvas

(a)   (b)

Figure 4.11 (a) One of the dishes served at Eleven Madison Park in New York: beef, roasted onion, cherries and ginger, aesthetically simple and asymmetric and (b) a typical multi-colourful Michel Bras dish. *Source*: Reproduced with permission of Andy Hayler

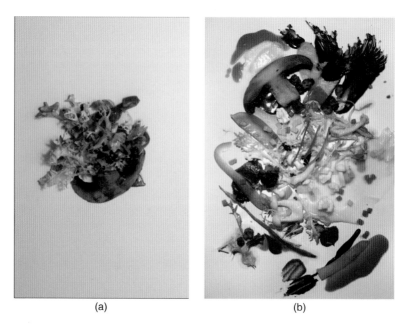

Figure 4.12 (a) 17 ingredients disposed as in a tossed salad and (b) 'Kandinsky on a plate' created by Charles Michel: an experimental dish with the same 17 ingredients arranged to look like one of Kandinsky's paintings

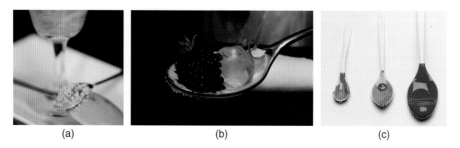

Figure 5.3 (a) The textured spoons presented to diners for one of the touch courses organized by Caroline Hobkinson at the House of Wolf restaurant. *Source*: Reproduced with permission of Caroline Hobkinson. (b) The Osetra "dish" served at Alinea, in which the bottom surface of the spoon is textured to give a nice contrast to the upper elements of the dish. (c) Spoon set designed by Jinhyun Jeon in order to enhance the different textures, temperatures and flavours of foods, as part of a project entitled 'Tableware as Sensorial Stimuli'. *Source*: Reproduced with permission of Jinhyun Jeon

Figure 5.4 A selection of the cutlery and foods used in Harrar and Spence's (2013) study looking at the effects of cutlery on taste/flavour perception. *Source*: Harrar and Spence 2013. Reproduced with permission of Elizabeth Willing, photographer

Figure 5.5 'Aromatic fork' served at Homaro Cantu's Moto. *Source*: Reproduced with permission of Homaro Cantu

(a)   (b)

Figure 5.6 (a) One of the unusual utensils used at Alinea, ideal for encouraging the diner to grab 'a bite' even if they have no cutlery (or don't wish to use their hands). *Source*: Photograph by Lara Kastner. Reproduced with permission of Crucial Detail LLC. (b) Spoons designed for elBulli, especially to address their culinary needs when having to serve one of their molecular gastronomy creations. *Source*: Reproduced with permission of Julio Arias Pérez.

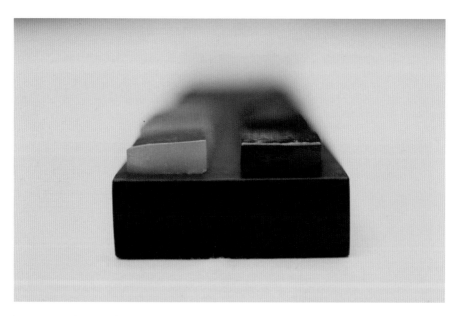

Figure 7.2 The Beetroot and Orange Jelly dish served at The Fat Duck restaurant in Bray, UK. *Source*: Reproduced with permission of Lotus PR and The Fat Duck

Figure 7.3 Olive oil jelly candy. If you'd expect this to taste like lime, you're in for a surprise!

Figure 7.4 *'Pan con tomate'* (developed by Denis Martin). *Source*: Reproduced with permission of Denis Martin

(a)   (b)   (c)

Figure 7.5 Appetisers served at elBulli: (a) Parmigiano marshmallows; (b) spring rolls made out of lightly sugared cotton; and (c) black olive 'Oreos'. *Source*: Reproduced with permission of Blake Jones

Figure 7.6 Tartufo al cerdo Ibérico y aceituna verde (Truffles with Iberian pork aroma, and green olive). Photography taken by Francesc Guillamet. *Source*: Reproduced with permission of Francesc Guillamet

Figure 7.7 An oyster leaf as served at Alinea. A bit incongruent? Yes, but incongruency of the most natural sort

Figure 7.8 These natural pine berries (hint: look at the right side of the image) are often served in restaurants as a means to amaze the diner, who very likely does not come across such fruits very often. For example, this fruit is found in the dessert Sotobosque served at 41°, the bar founded by Adrià in Barcelona. *Source*: Reproduced with permission of Miguel Angel Castillo & Rocio García Martin, http://la-cocina-creativa.blogspot.com.es

(a)   (b)

Figure 8.1 (a) 'Halibut black pepper, coffee, lemon' served by Grant Achatz in his restaurant Alinea. Note that all of the black food elements are presented in white form. *Source*: Reproduced with permission of Ron Kaplan/LTHForum.com. (b) Detail of one of Marije Vogelzang's White funeral meals. *Source*: www.marijevogelzang.nl. Reproduced with permission of Marije Vogelzang

Figure 8.3 (a) A rare steak for a normal-sighted person on the left, compared to how a red/green colour-blinded person would see the same steak (i.e. as well cooked) on the right. (b) A plate of cooked spinach for a normal-sighted person (on the left) compared to how a green colour-blind diner might see it (as a meat stew, perhaps?)

Figure 9.1 Urban picNYC table, design by Haiko Cornelissen Architecten. *Source*: Reproduced with permission of Haiko Cornelissen Architecten, WY, USA

Figure 9.2 Image of the interior of the Icebar located in Oslo

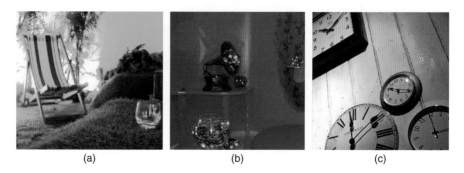

Figure 9.3 Stills taken from the three environments used in the study of Velasco *et al.* (2013) demonstrating the impact of the atmosphere on the whisky drinking experience. (a) the 'grassy' room; (b) the 'sweet' room; and (c) the 'woody' room. *Source*: Velasco et al. 2013

Figure 10.5 Eating direct from a tablet computer is one of the future plateware possibilities that may be well worth considering. *Source*: Reproduced with permission of Adam Scott

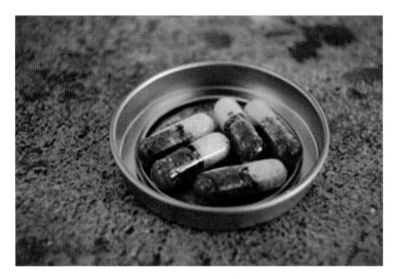

Figure 11.1 'Perfected pizza': this dish consists of dehydrated tomatoes, fresh basil, Parmigiano–Reggiano and sea salt in a vegetable capsule. It was served at The Tactile Dining Car, a participatory dining and performance installation held at the Flashpoint Gallery, Washington DC, 9–24 September 2011. *Source*: Reproduced with permission of banished? productions/Carmen C. Wong

Figure 11.2 'Aerofood': a mylar balloon holding food scent, one of the dishes served at The Tactile Dining Car, a participatory dining and performance installation held at the Flashpoint Gallery, Washington DC, 9–24 September 2011. *Source*: Reproduced with permission of banished? productions/Carmen C. Wong

Figure 11.3 Chimp stick dish. A carved down liquorice stick coated in juniper-wood-infused honey, with two species of ant (Lasius fuliginosus and Formica rufa) stuck to it with shiso, buckwheat, flax and freeze-dried raspberries. *Source*: Photograph by Chris Tonnesen. Reproduced with permission of Nordic Food Lab

Figure 11.5 The workshop of Paco Roncero: restaurant or experimental laboratory? The introduction of technology is increasingly starting to blur this boundary. *Source*: Photograph by Gerald Kiernan. Reproduced with permission of Estudio Baselga and Gerald Kieran

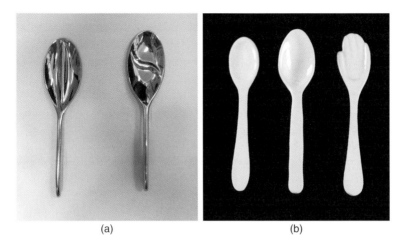

(a)   (b)

Figure 5.7 (a) A pair of spoons designed by Studio William for Heston Blumenthal. These divided spoons enable two different ingredients to be placed in the mouth separately in an original way. (b) Set of ceramic spoons with porous handles for adding essences designed by Elizabeth Willing. *Source*: Reproduced by permission of Elizabeth Willing

nostrils would be reminded of the scent every time they brought their fork to their mouth.

Continuing with the thread of specialized spoons, industrial designer William Welsh (http://www.studiowilliam.com) recently designed a unique set of 'Taster spoons'. These special spoons have a divider in the bowl part that enables two ingredients or morsels of food to be kept separate until they meet in the diner's mouth (see Figure 5.7a). Elizabeth Willing has created a slightly more upmarket version of the same concept. Her solution was to create a set of three ceramic spoons with porous handles for adding essences, the suggestion being that diner guests could then experience a bowl of vanilla ice-cream with each of the three spoons (see Figure 5.7b).

Here these spoons would seem to offer up the possibility of delivering sequential flavour sensations in much the same way that Denis Martin does in Switzerland. One of his dishes, served on a porcelain appetizer spoon, involves four different flavour experiences (white chocolate, wasabi, raw tuna and peanut oil) hitting your palate sequentially (see Martin 2007).

Finally, sometimes the design of the spoon has to be changed for basic functional reasons. How many times have you tried to get a spoonful of ice-cream from a scoop sitting on the plate, only to find it sliding across the plate as soon as you try to insert the spoon into the ice-cream? The solution for this problem can also be found in the modified design of the spoon! At the Tabla restaurant in New York, a serrated spoon was introduced in order to stop their frozen kulfi dessert from sliding across the plate (Meyer 2010, pp. 165–166; though see Fabricant 2010).

5.8   Finger food

In western countries, eating with one's hands is often seen as uncultured or uncivilized in today's day-and-age (the historians among us might be reminded of Henry VIII eating uncouthly with his paws; Wilson 2012), especially in those contexts in which cutlery is available/provided to the diner. It is important to note that our eating behaviours (and rituals; Visser 1991) have changed quite dramatically over the years, however (and continue to evolve). Now, high-end meals have changed in order to enable this format of eating (what's often referred to as 'finger food'; Crumpacker 2006). For instance, have you ever thought of eating yoghurt directly with your fingers? Probably not, but we bet many of you like to lick the last residues of the mixture out of the cake bowl with your finger. Nothing else could ever pick up the very last smear quite like your index finger!

> *"To people who eat with their fingers, hands seem cleaner, warmer, more agile than cutlery. Hands are silent, sensitive to texture and temperature, and graceful – providing, of course, that they have been properly trained."* (Visser 1991, pp. 167–168)

For many eastern cultures, the practice of eating with one's fingers offers a genuinely multisensory approach to dining and one that provides the opportunity for a more sensual, tactile/haptic experience (see the quote below). Of course, the foods that are eaten at such establishments are not piping hot and hence can be easily handled manually. As fine dining becomes ever more adventurous and playful however, a growing number of chefs are starting to consider course-specific alternatives to the knife and fork, that is, to traditional cutlery. The culinary/design teams working with certain of the world's top chefs (e.g. Grant Achatz) invest a great deal of thought and planning into the customization of serving vessels for each of the micro-courses on their menus. Indeed, one of Achatz's most famous dishes was a shrimp tempura skewered on a vanilla pod. Diners were instructed to tilt their heads back and simply take the shrimp into their mouths in one go (note that this dish is another example in which the eating element is used to deliver scent).

> *"The first time I went to an Indian restaurant in Canada I used my fingers. The waiter looked at me critically and said, 'Fresh off the boat, are you?' I blanched. My fingers, which a second before had been taste buds savouring the food a little ahead of my mouth, became dirty under his gaze. They froze like criminals caught in the act. I didn't dare lick them. I wiped them guiltily on my napkin. He had no idea how deeply those words wounded me. They were like nails being driven into my flesh. I picked up the knife and fork. I had hardly ever used such instruments. My hands trembled. My sambar lost its taste."* (Martel 2001, p. 7)[25]

---

[25] Or take the following email from Peter Zwart in California on reading some of our cutlery research: *"I was delighted by your research on the effects of cutlery on the taste of food. One thing that my wife*

Leaving such extravagant creations to one side for a moment, the fundamental truth is that many people simply find it more enjoyable to eat certain foods with their hands rather than using cutlery. Although it may not be everyone's favourite, the hamburger constitutes perhaps the perfect example of such a food. In fact, it is interesting to note that a number of restaurants have recently started to appear where the diner isn't provided with any cutlery at all (e.g. Il Giambellino in Milan, Italy).[26] That is, the diners are expected to eat using their hands and nothing else. As Jo Bryant, etiquette advisor at Debrett's explains: "*The influence of other cultures and new foods, such as calzone, means eating with our hands is a growing trend*" (quoted in Furness 2012). In fact, the latest version of this classic guide to etiquette has, for the first time, stated that eating with one's hands[27] is now acceptable practice (at least for certain foods such as pizza, calzone and ice-cream cones), just as long as the diners remember not to lick their fingers afterwards.

Apart from the sensuous enjoyment that interacting with the food with our hands may provide in terms of the eating experience, it is important to note that people can also evaluate the texture of food by means of haptic information. To give but one example here, Barnett-Cowan (2010) demonstrated that the tactile feel of a pretzel held in the hand would influence people's oral perception of the pretzel when bitten from the other extreme. The softer the pretzel felt in the hand, the staler it was perceived in the mouth, thus suggesting something of a contrast effect (cf. Zampini *et al.* 2006; see also Chapter 6). While in this case it was the food itself that the participants were feeling, there would seem to be nothing preventing the texture of the cutlery in a diner's hands from conveying some meaning or setting up some expectation in the diner's mind regarding the food that is lodged, albeit temporarily, at the end of their fork.

5.9 Eating without hands

Another trend that is currently very popular relates not to the enhancement of specific sensory cues (through the cutlery) but to their very removal. How about removing the hands' touch entirely from the diner's experience of eating? One example of this comes from a most memorable experience that your first author had when dining at Heston Blumenthal's The Fat Duck

---

from South India always tells me is that the food she cooks (rasam, sambar, etc.) tastes better when she eats with her hands instead of a spoon. I have always (quietly) dismissed this as total nonsense, but your research could warrant an apology from my side!"

[26] For all those restaurateurs who may be thinking of removing the cutlery from their own dining establishments, this particular restaurant didn't last very long before it doors were closed to the public for good.

[27] As suggested by the title of Zachary Pelaccio's most recent cookbook: *Eat With Your Hands* (Pelaccio 2012).

restaurant over a decade ago. For one of the courses on the tasting menu, the waitress arrived at the table and instructed your author to open wide. She then proceeded to insert a spoonful of Heston's latest gastronomic creation (on this occasion, lime gelee) into his mouth. Meanwhile, restaurants such as Madeleines Madteater in Denmark sometimes dispense with the need for a plate and instead place the food directly into the diner's mouth with a spoon or perhaps a pipette. These items of cutlery are often used instead of knives and forks. Note that by so doing, many of the tactile and haptic elements that are normally associated with eating, such as the holding and wielding of cutlery, have been removed from the diner's grasp. Certainly the pipetting of aromatic liquids over the food provides an intriguing means of ensuring the targeted delivery of aromatic flavours to a dish. Erica Duffy does this very effectively with the smoked salmon starter that comes with a side pipette of Highland malt whisky that she serves at various gastronomy events (see also Caroline Hobkinson's menu in Chapter 8 for another example).[28]

When the waitress at The Fat Duck asked your first author to open wide and swallow more than a decade ago, it was an incredibly powerful/emotional moment. The flavour of the food itself became secondary to the theatre of the whole experience. (As Elizabeth Carter, editor of the Good Food Guide, described the experience at the restaurant: "*It is food as theatre*"). Take away the hands, and the diner is immediately taken back to early childhood, to experiences of being helpless and spoon-fed while sitting on a caregiver's lap.

Another recent example of eating without the aid of cutlery comes from Caroline Hobkinson. In one dish/experience, morsels of food were tied to fishing wire that was virtually invisible and suspended from helium balloons that nestled under the ceiling; even the bread was suspended from the ceiling with string. The diners were then encouraged to eat the food using only their mouths to, as it were, 'catch' a bite. Similarly, the current menu at Alinea includes a candy balloon (as the first sweet course) which is full of helium and left on the table so that the inflated part is at the level of the diner's head. The diner is then instructed to approach the balloon only with the mouth (no hands!) and suck the air in while trying to eat the balloon at the same time (and avoiding getting the candy balloon stuck all over their face).[29]

We wonder whether the modernist chefs will soon start to plan dishes consisting of nothing more than a gas or vapour (one utensil perfectly suited for this purpose is Le Whaf, http://labstoreparis.com/#&p=products&products=le-whaf); there would be no need for cutlery whatsoever, perhaps only a straw. Only the future will tell (but see Chapters 10 and 11 on the future of technologically enhanced cutlery).

---

[28] Presumably we should think of the pipette as an item of unusual plateware, rather than as a new piece of cutlery.

[29] The important point about this dish is to lose the fear of looking ridiculous. When your second author tried it somewhat shyly, the last thing she expected was to end up sounding like a Smurf!

5.10 Conclusions

As the evidence reviewed in this chapter has made clear, the cutlery that we use to eat has shaped the way (and even the food) that we all eat nowadays. It can even exert a profound impact on our overall experience of a dish, not to mention leading to us all now having an overbite. We have seen that the connotations that we all have concerning materials can be transferred to the food (mostly in terms of quality), silver being the material that, for centuries, has been associated with high-quality flatware. The material properties of the cutlery can also be used to transfer a subtle taste and/or flavour to the food; even in those cases where it doesn't, the innovative chef can undoubtedly figure out a way in which to cover the spoon with a tasty film of invisible flavour (e.g. Martin 2007). In this chapter we have seen how the size of the cutlery can sometimes influence how much we choose to eat. Counter-intuitive as it may seem, smaller cutlery appears to make us eat more by forking or spooning more frequently in some contexts; in other cases (when we're invited to the laboratory), it seems that grabbing hold of a bigger item of cutlery will make the shovelling easier and hence we are also likely to gobble down more food.

So, what does the future of flatware look like? As we will see in Chapter 10, designers are currently exploring how a range of digital technologies can be merged with cutlery in order to help us to eat more slowly. The idea is that, in doing so, we should all be better able to control our calorie intake. In the meantime however, we see a growing trend toward diners eating with their hands. We also see the creation of customized flatware that can convey some sensory or hedonic attributes to the dinner/diner by means of its smell, temperature and form and a multitude of novel materials that are now at the disposal of the designer with an interest in cuisine.

Ironically, while previously everyone seemed to want to manufacture tasteless cutlery (in the gustatory, rather than aesthetic, meaning of that term), we are now starting to see some of the most innovative chefs thinking about integrating the taste of cutlery into the diner's multisensory experience of a dish (rather than something to be avoided). If things keep going as they are at the moment, might the top chefs one day be known as much for their signature cutlery as for their signature dishes? What does seem clear is that the cutlery (or better said, the eating utensils) will likely come to play an increasingly important role in the perfect meal in the years to come.

References

Aduriz, A. L., Vergara, J., Lasa, D., Oliva, O. and Perisé, R. (2012) Culinary trompe-l'oeil: A new concept in coating. *International Journal of Gastronomy and Food Science*, **1**, 70–77.

Aldersey-Williams, H. (2011) *Periodic Tales: The Curious Lives of the Elements*. Viking, London.

Anonymous (1982) The spoon question: Or how to eat pasta like an expert. *The New York Times*, 19 May. [Online], Available: http://www.nytimes.com/1982/05/19/garden/the-spoon-question-or-how-to-eat-pasta-like-an-expert.html?pagewanted =2 (accessed January 2014).

Ariely, D. (2008) *Predictably Irrational: The Hidden Forces that Shape our Decisions.* Harper Collins Publishers, London.

Barba, E. D. (2013) My cuisine is tradition in evolution. Available at http://www.swide.com/food-travel/chef-interview/michelin-starred-chef-an-interview-with -massimo-bottura/2013/4/23 (accessed January 2014).

Barnett-Cowan, M. (2010) An illusion you can sink your teeth into: Haptic cues modulate the perceived freshness and crispness of pretzels. *Perception*, **39**, 1684–1686.

Baron de Tott, F. (1786) Memoirs of Baron de Tott: Containing the state of the Turkish empire and the Crimea, during the late war with Russia. With numerous anecdotes, facts *and observations, on the manners and customs of the Turks and Tartars (Vol. 1)*. Available at http://books.google.co.uk/books?id=WIc2AAAAMAAJ (accessed January 2014).

Bolhuis, D. P., Lakemond, C. M. M., de Wijk, R. A., Luning, P. A. and de Graaf, C. (2013) Consumption with large sip sizes increases food intake and leads to underestimation of the amount consumed. *PLoS ONE*, **8**(1), e53288.

Brace, C. L. (1977) Occlusion to the anthropological eye. In: *The Biology of Occlusal Development* (ed. J. McNamara), pp. 179–209. Centre for Human Growth and Development, Ann Arbor, Michigan.

Brace, C. L., Rosenberg, K. R. and Hunt, K. D. (1987) Gradual change in human tooth size in the Late Pleistocene and Post-Pleistocene. *Evolution*, **41**, 705–720.

Braudel, F. (1992) *The Structure of Everyday Life*. University of California Press, Berkeley, CA.

Brown, P. (2001) *British Cutlery: An Illustrated History of its Design, Evolution and Use*. Philip Wilson Publishers, London.

Burk, K. and Bywater, M. (2008) *Is This Bottle Corked? The Secret Life of Wine*. Faber & Faber, London.

Coffin, S., Lupton, E., Goldstein, E. and Bloemink, B. (eds) (2006) *Finding Desire: Design and the Tools of the Table, 1500–2005*. Assouline in collaboration with Smithsonian Cooper-Hewitt, New York.

Cottrell, C. (2012) Are you going to eat that chair? Yes, and my pencil is cheese, and this candle is chocolate. *The Atlantic Magazine*, 27 November. Available at http://www.theatlantic.com/health/archive/2012/11/are-you-going-to-eat-that-chair -yes-and-my-pencil-is-cheese-and-this-candle-is-chocolate/265629/ (accessed January 2014).

Crumpacker, B. (2006) *The Sex Life of Food: When Body and Soul Meet to Eat*. Thomas Dunne Books, New York.

Damrosch, P. (2008) *Service Included: Four-Star Secrets of an Eavesdropping Waiter*. William Morrow, New York.

Del Conte, A. (1976). *Portrait of Pasta*. Paddington Press, London.

Dornenburg, A. and Page, K. (1996) *Culinary Artistry*. John Wiley & Sons, New York.

Dunlop, F. (2012) *Spoon fed: How cutlery affects your food*. Available at http://www.ft.com/cms/s/2/776ba1d4-93ee-11e1-baf0-00144feab49a.html#axzz2Ufqd6mmL (January 2014).

Elias, N. (2005) On medieval manners. In *The Book of Touch* (ed. C. Classen), pp. 266–272. Berg, Oxford.

Emery, J. (1976) *European Spoons Before 1700*. John Donald Publishers, Edinburgh.

Fabian, A. (2011) *Spoons and spoonness: A philosophical inquiry through creative practice*. Thesis submitted for the degree of Doctor of Philosophy, Brunel University.

Fabricant, F. (2010) Tabla is closing. *The New York Times*, 30 September. Available at http://dinersjournal.blogs.nytimes.com/2010/09/30/tabla-is-closing/ (accessed January 2014).

Fahrenthold, D. A. and Sonmez, F. (2011) *Stick a fork in Hill's 'green' cutlery*. The *Washington Post*, 5 March, A1, A4.

Fanshawe, S. (2005) *The Done Thing: Negotiating the Minefield of Modern Manners*. Arrow Books, London.

Forbes, R. J. (1956) Food and drink. In: *A History of Technology vol. II: The Mediterranean Civilization and the Middle Ages* (eds C. Singer, E. J. Holmyard, A. R. Hall and T. I. Williams), pp. 103–146. Oxford University Press, New York.

Furness, H. (2012) *How to eat with one's fingers: The Debrett's guide to very modern etiquette*. Available at http://www.telegraph.co.uk/foodanddrink/foodanddrinknews/9696223/How-to-eat-with-ones-fingers-the-Debretts-guide-to-very-modern-etiquette.html (accessed January 2014).

Geier, A. B., Rozin, P. and Doros, G. (2006) Unit bias: A new heuristic that helps explain the effect of portion size on food intake. *Psychological Science*, **17**, 521–525.

Giedion, S. (1948/1975) *Mechanization Takes Command: A Contribution to Anonymous History*. W. W. Norton, New York.

Greene, G. (2000) *Gold-plate special. New York Magazine*. Available at http://nymag.com/nymetro/food/industry/features/3647/index2.html (accessed January 2014).

Hall, F. H. (1887) *Social Customs*. Estes and Lauriat, Boston.

Harrar, V. and Spence, C. (2012) A weighty matter: The effect of spoon size and weight on food perception. *Seeing and Perceiving*, **25**(Suppl), 199.

Harrar, V. and Spence, C. (2013) The taste of cutlery. *Flavour*, **2**, 21.

Himsworth, J. B. (1953) *The Story of Cutlery: From Flint to Stainless Steel*. Ernest Benn, London.

Horowitz, J. and Singley, P. (eds) (2004) *Eating Architecture*. MIT Press, Cambridge, MA.

Jünger, H. (1993) *Herbei, herbei, was Löffel sei …* (Come on, come, what is a spoon …). Anabas-Verlag, Giessen.

Laughlin, Z., Conreen, M., Witchel, H. and Miodownik, M. A. (2009) The taste of materials: Spoons. In *MINET Conference: Measurement, Sensation and Cognition*, pp. 127–128. National Physical Laboratories, Teddington.

Laughlin, Z., Conreen, M., Witchel, H. J. and Miodownik, M. (2011) The use of standard electrode potentials to predict the taste of solid metals. *Food Quality and Preference*, **22**, 628–637.

Lim, M. S. C., Hellard, M. E. and Aitken, C. K. (2005) The case of the disappearing teaspoons: Longitudinal cohort study of the displacement of teaspoons in an Australian research institute. *British Medical Journal*, **331**, 1498–1500.

Lindstrom, M. (2005) *Brand Sense: How to Build Brands through Touch, Taste, Smell, Sight and Sound*. Kogan Page, London.

Martel, Y. (2001) *Life of Pi*. Harcourt Trade Publishers, New York.

Martin, D. (2007) *Evolution*. Editions Favre, Lausanne.

McGee, H. (1990) *The Curious Cook: More Kitchen Science and Lore*. Collier Books, New York.

McGee, H. J., Long, S. R. and Briggs, W. R. (1984) Why whip egg whites in copper bowls? *Nature*, **308**, 667–668.

Meyer, D. (2010) *Setting the Table: Lessons and Inspirations from one of the World's Leading Entrepreneurs*. Marshall Cavendish International, London.

Miodownik, M. (2008) The taste of a spoon. *Materials Today*, **11**, 6.

Mishra, A., Mishra, H. and Masters, T. (2012) The influence of the bite size on quantity of food consumed: A field study. *Journal of Consumer Research*, **38**, 791–795.

Moore, M. (2013) *Chinese 'must swap chopsticks for knife and fork'*. Available at http://www.telegraph.co.uk/news/worldnews/asia/china/9926599/Chinese-must-swap-chopsticks-for-knife-and-fork.html (accessed January 2014).

Pelaccio, Z. (2012) *Eat With Your Hands*. Ecco, New York.

Petroski, H. (1994) *The Evolution of Useful Things*. Vintage Books, New York.

Piqueras-Fiszman, B. and Spence, C. (2011) Do the material properties of cutlery affect the perception of the food you eat? An exploratory study. *Journal of Sensory Studies*, **26**, 358–362.

Piqueras-Fiszman, B., Laughlin, Z., Miodownik, M. and Spence, C. (2012) Tasting spoons: Assessing how the material of a spoon affects the taste of the food. *Food Quality and Preference*, **24**, 24–29.

Spence, C. and Piqueras-Fiszman, B. (2011) Multisensory design: Weight and multisensory product perception. In: *Proceedings of RightWeight2* (ed. G. Hollington), pp. 8–18. Materials KTN, London.

Steinberger, M. (2010) *Au Revoir to All That: The Rise and Fall of French Cuisine*. Bloomsbury, London.

Veblen, T. (1899/1992) *The Theory of the Leisure Class*. Transaction Publishers, New Brunswick, NJ.

Visser, M. (1991) *The Rituals of Dinner: The Origins, Evolution, Eccentricities, and Meaning of Table Manners*. Penguin Books, London.

Wansink, B. (2006) *Mindless Eating: Why We Eat More Than We Think*. Hay House, London.

Wansink, B., van Ittersum, K. and Painter, J. E. (2006) Ice cream illusions: Bowl size, spoon size, and self-served portion sizes. *American Journal of Preventive Medicine*, **31**, 240–243.

Ward, C. (2009) *The uncommon origins of the common fork*. Available at http://leitesculinaria.com/1157/writings-the-uncommon-origins-of-the-common-fork.html#comment-87562 (accessed January 2014).

Weber, R. J. (1992) *Forks, Phonographs and Hot Air Balloons*. Oxford University Press, New York.

Williams, L. E. and Bargh, J. A. (2008) Experiencing physical warmth promotes interpersonal warmth. *Science*, **322**, 606–607.

Wilson, B. (2012) *Consider the Fork*. Particular Books, London.

World Health Organization (1998) *Obesity: Preventing and Managing the Global Epidemic*. Geneva: World Health Organization.

Wrangham, R. (2010) *Catching Fire: How Cooking Made us Human*. Profile Books, London.

Zampini, M., Mawhinney, S. and Spence, C. (2006) Tactile perception of the roughness of the end of a tool: What role does tool handle roughness play? *Neuroscience Letters*, **400**, 235–239.

6
The Multisensory Perception of Flavour

6.1 Introduction

No book on the perfect meal would be complete without some consideration of the flavour of the food itself. Flavour turns out to be one of the most multisensory of our everyday experiences. Have you ever wondered what happens in our brain while we're eating? Understanding how diners experience flavour is an area that has seen a rapid growth of interest from scientists over the last few years (see Small 2004; Small and Prescott 2005; Verhagen and Engelen 2006; Auvray and Spence 2008; Stevenson 2009; Spence 2013 for reviews). In fact, cognitive neuroscience approaches are increasingly coming to complement the traditional techniques of food science in the design of great-tasting dishes. The scientists have certainly made much progress in terms of understanding the mechanisms underlying multisensory flavour perception.

In this chapter, we review the evidence illustrating the role of multisensory integration in flavour perception. We demonstrate how important each of the diner's senses (e.g. vision, taste, smell, sound and touch) is to the overall experience of flavour, which sense dominates and when, and we try to explain how the chef can better meet a diner's sensory expectations. We tackle the question of whether or not we all live in the same taste world, addressing questions such as 'What does it mean to be a supertaster?' and 'Would you even want to be one anyway?'

6.2 Perceiving flavours

According to the International Standards Organization (ISO 5492 2008), flavour can be defined as a "*Complex combination of the olfactory, gustatory*

The Perfect Meal: The Multisensory Science of Food and Dining, First Edition.
Charles Spence and Betina Piqueras-Fiszman.
© 2014 John Wiley & Sons, Ltd. Published 2014 by John Wiley & Sons, Ltd.

and trigeminal sensations perceived during tasting. The flavour may be influenced by tactile, thermal, painful and/or kinaesthetic effects" (see Delwiche 2004, p. 137). That is, the experts would have us believe that taste, smell, trigeminal and oral-somatosensory cues, are the *only* senses that contribute *directly* to the perception of flavour. This is not to say that what a diner sees and/or hears cannot modify a food's flavour; they most certainly do. It is just that they are not, at least according to the ISO definition, integral to it. This narrow definition of flavour is however currently being challenged by researchers (e.g. Auvray and Spence 2008; Stevenson 2009; Spence 2012a; Spence *et al.* in press; see also McBurney 1986). Part of the reason for this is the emergence of a growing body of evidence, some of which is reviewed later in this chapter, demonstrating just how profoundly the sound of food (and the sounds we make while eating) – not to mention its visual attributes (as we saw in Chapter 4) – affects our perception of both the sensory-discriminative attributes of food and our hedonic responses to a dish.

When talking about the role of olfactory cues in multisensory flavour perception, one critical point to note at the start is that there are actually two relatively distinct sensory systems: the older *orthonasal* system, associated with the inhalation of external odours (as when we sniff; for those who are old enough, remember the Bisto kid); and the somewhat newer *retronasal* system (involving the posterior nares), associated with the detection of the smells/aromas emanating from the food we have in our mouths while we are chewing and swallowing it. It turns out that odours are periodically forced out from the back of the nasal cavity during this process. In fact, it has been suggested that we may be the only creatures born with two distinct senses of smell (Gilbert 2008). A growing body of empirical research highlights a number of important differences between orthonasal and retronasal smell at the subjective/perceptual level (e.g. Rozin 1982; Diaz 2004). Interestingly, differences have now also started to show up in terms of the parts of the brain that preferentially process these two kinds of sensory input (e.g. Small *et al.* 2005, 2008; see also Heilman and Hummel 2004). Both orthonasal and retronasal smell are clearly important in terms of multisensory flavour perception.[1]

When thinking about the senses and their role in multisensory flavour perception, it can be helpful to distinguish between two categories: the *exteroceptive* senses of vision, audition and orthonasal olfaction are typically stimulated prior to (and sometimes during) the consumption of food; and the *interoceptive* senses are those that are stimulated while a diner

---

[1] It is striking how little thought researchers traditionally gave to whether the odours in their experiments were presented via the orthonasal or retronasal route; in fact, more often than not, olfactants were nearly always delivered orthonasally (presumably because it is simpler to do so). However, this approach is turning out to be increasingly problematic given the mounting empirical evidence that substantially different patterns of multisensory integration can sometimes be observed as a function of whether an odour is presented via the orthonasal or retronasal route (e.g. Koza *et al.* 2005; Pfeiffer *et al.* 2005; Bult *et al.* 2007).

is eating. In the latter case, the relevant senses are taste, retronasal smell, oral-somatosensation and the sounds associated with the consumption of food (see Figure 6.1). Different brain mechanisms may be involved in these two cases (see also Small *et al.* 2008). It would seem likely that the multisensory integration of interoceptive flavour cues is more automatic than the combination of cues that is involved in interpreting exteroceptive food-related signals (but see van der Klaauw and Frank 1996; Auvray and Spence 2008).

One of the most important means by which exteroceptive cues influence a diner's perception of a plate of food relates to expectancy effects (Deliza and MacFie 1997; Hutchings 2003; Cardello 2007; Shankar *et al.* 2010a, b). That is, visual appearance cues (especially the colour and visual form of the food on the plate; remember the popular expression mentioned in Chapter 4, 'eye appeal is half the meal'), orthonasal olfactory cues and distal food sounds (for example, the sizzle of the steak or the sound of the veggies being chopped) can all set up powerful expectations regarding the food that a diner is about to eat. When the food or drink is then evaluated (in terms of its flavour, aroma or taste), assimilation may occur if there is only a small discrepancy between what was expected and what was provided. However, if the discrepancy between a diner's expectations and the actual interoceptive information is too great, then contrast may occur instead (as we will see more in detail in Chapter 7).

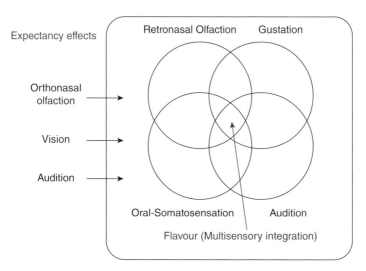

Figure 6.1 Multisensory flavour perception, emphasizing the important role played by food-eating sounds. Note that this view distinguishes between the unique roles played by orthonasal and retronasal olfaction. The four interoceptive sensory inputs (shown in the box) are likely combined through a process of multisensory integration to deliver flavour percepts. Meanwhile, exteroceptive cues (shown outside the box) can also influence flavour perception, primarily by means of setting up expectations concerning the flavour, aroma and/or taste of a food or drink

The assimilation/contrast model has been used by food science researchers in order to try and account for the results of multisensory integration on people's hedonic responses to (and perception of) food (Yeomans *et al.* 2008). There would seem to be no good reason why it should not be extended to the case of people's sensory-discriminative food judgments as well (see Shankar *et al.* 2010a, b). In the following, we review the key evidence concerning the role of each of the human senses in the sensory-discriminative aspects of multisensory flavour perception.

6.3   Taste

We are sensitive to a number of basic tastes including sweet, sour, salty, bitter, umami (Schiffman 2002) and metallic (Lindemann 1996; Lawless 2001; Lawless *et al.* 2005). However, anyone who has tried eating a Szechuan button (see www.koppertcress.com), included in the David Bellamy cocktail served at The House of Wolf (houseofwolf.co.uk) in Islington, London will know that we may need to add an electric taste as well (but see Keeley 2002). The key chemical in this case is hydroxyl-α-sanshool (Sugai *et al.* 2005a, b; Bautista *et al.* 2008). Take a bite of a Szechuan flower, or else crunch on one of the peppercorns, and your tongue will start to tingle at a frequency of about 50 Hz (Hagura *et al.* 2013). The sensation feels similar to the one you probably had as a child when you put a 9 volt battery on your tongue.

Traditionally, researchers thought that the receptors for each taste were distributed asymmetrically over the surface of the human tongue (see Guyton and Hall 1996, pp. 676–678; Mackenna and Callander 1997, p. 266), with the sweet receptors lying exclusively at the front, bitter receptors at the rear, sour receptors to the side of the tongue and so on. However, it turns out that the various receptors are actually more evenly spaced over the tongue's surface than had previously been thought.[2] That said, there is some evidence to suggest that the taste buds may be somewhat more densely packed towards the tip of the tongue than they are elsewhere (Todrank and Bartoshuk 1991; Duffy 2007). It is no longer clear that each basic taste is necessarily transduced (or coded) by an individual specialized receptor (Lindeman 2001), thus leading some researchers to query the very notion of basic tastes (see Delwiche 1996; Erikson 2008). Interestingly, the questions of how many 'basic' tastes there are, and whether they really deserve to be called 'basic' (or even 'taste' instead of 'sensation' as others would prefer), are still open questions among many of those working in the field of sensory science. In her book *Taste What You're Missing*, Stuckey (2012) interviewed a number of sensory

---

[2] The erroneous idea that the receptors were so nicely separated has been traced back to Boring's (1942) incorrect description of early research conducted by Hanig (1901).

scientists who believe that there may, in fact, be more than 20 distinct basic tastes! These include a fatty acid taste and a kokumi taste (a Japanese term that can perhaps best be translated as 'heartiness' or 'mouthfulness').

6.3.1 Are you a supertaster?

Taste (or, more technically, gustation) is certainly the sense in which one can see the largest individual differences in terms of the number of sensory receptors that people have. Each of the taste buds on the human tongue contains a number of taste cells. The taste buds themselves are located within structures known as fungiform papillae. Hard though it is to believe, there is a 16-fold variation in the density of taste buds between people. (Miller and Reedy 1990).[3] Perhaps unsurprisingly, people differ in terms of their sensitivity to certain gustatory stimuli. Those who are especially sensitive to bitter-tasting compounds, such as phenylthiocarbamide (PTC) or the chemically-related 6-*n*-propylthiouracil (PROP), are referred to as supertasters. Approximately 25% of the population fall into this category; 50% are 'medium tasters' while the remainder are 'non-tasters' (Bartoshuk 2000). While the population clearly differs quite dramatically in terms of their sensitivity to these bitter-tasting chemicals, the tri-partite division of the population isn't so obviously correct.

While supertasters typically have more taste buds than non-tasters (e.g. Bartoshuk *et al.* 1994; Hayes and Duffy 2007, 2008; Bajec and Pickering 2008), researchers are still uncertain as to the exact relationship between a diner's taster status and the density of taste papillae on their tongue (e.g. Prescott *et al.* 2004; Hayes *et al.* 2008). Supertasters also seem to be more sensitive to (or at least rate more intensely the sensation elicited by) other (non-bitter) taste stimuli (Bajec and Pickering 2008; Bajec *et al.* 2012). These individual differences in taster status are especially interesting here in that they have been shown to influence various aspects of our multisensory flavour perception (e.g. Ishiko *et al.* 1978; Essick *et al.* 2003; Bajec and Pickering 2008; Zampini *et al.* 2008; Eldeghaidy *et al.* 2011; Bajec *et al.* 2012; Hayes and Pickering 2012). While we hear anecdotally that chefs are more likely to be supertasters than the rest of the population, the evidence suggests that any such tendency, if indeed it exists, doesn't extend to foodies (Hayes and Pickering 2012). Certainly, one imagines that many chefs must have come up against this in the restaurant setting when a specific dish tastes great to one diner but is simply not liked by another.[4]

---

[3] While the incidence of supertasters appears to vary from region to region, there is only one group living in the Amazon who are all supertasters (Allen 2012).

[4] For anyone who finds themselves wondering whether they are a supertaster or not, the easiest way to check is to go online and order yourself some tasting strips (e.g. Fisher Scientific; http://www.fisher.co.uk/).

6.4   Olfactory–gustatory interactions

"*Smell and taste are in fact but a single sense, whose laboratory is in the mouth and whose chimney is in the nose.*" (Jean Anthelme Brillat-Savarin 1835)

While it is certainly true that taste plays an important role in the multisensory perception of flavour, smell is also critical. In fact, olfactory cues may actually be far more important than taste cues, with some researchers estimating that the sense of smell actually contributes as much as 80–90% to what people normally report as flavour (e.g. Martin 2004; Ge 2012; Stuckey 2012; Roach 2013). Just think of how dull even your favourite dish tastes when you have a head cold! Perhaps you also remember how, when you were still a child, your parents would tell you to pinch your nose whenever you had to swallow that awful syrupy medicine. That said, while the journalists (not to mention a few scientists, who should know better) love to spout this 80–90% factoid in their articles, it is far from clear whether there is actually any good evidence to support the claim. The closest one comes to an answer can be found in an article published by Murphy *et al.* in 1977, but their evidence and discussion fall some way short of substantiating such a claim. Indeed, one would think that the answer would depend on the food that one likes to eat. Smell would seem to play a far more important role for someone who enjoys eating stinky French cheeses, say, than for someone who loves to eat sushi and sashimi. While there may be no easy answer to the question of the relative importance of taste and smell to flavour perception, what is certainly true is that people are often unsure of the relative contributions of smell and taste to flavour perception (e.g. Davidson *et al.* 1999; Rozin 1982; Stevenson *et al.* 1999; but see Spence *et al.* in press). Such confusions have even led some researchers to argue that we may all be synaesthetic,[5] at least when it comes to flavours (Stevenson and Boakes 2004; Verhagen and Engelen 2006; Stevenson and Tomiczek 2007; Auvray and Spence 2008).

Some of the most convincing evidence regarding the multisensory integration of orthonasal olfactory and gustatory cues in human flavour perception has been reported by Dalton *et al.* (2000). The participants in this study were given four bottles to sniff, each containing a liquid. One bottle contained benzaldehyde, which has an almond-cherry-like odour, while the other three bottles only contained an odourless diluent. On each trial, the participants had to determine which bottle the benzaldehyde had been added to. The concentration of the olfactant was varied on a trial-by-trial basis in order to determine each participant's detection *threshold* for the smell. In one experiment, people performed this olfactory discrimination task while holding a

---

[5] Synaesthesia is a neurological condition in which the stimulation of one sensory or cognitive pathway leads to an automatic, involuntary experience in a second sensory or cognitive pathway (Cytowic 2002). In short, it often presents as a confusion of the senses.

10 mL sub-threshold concentration solution of saccharin in their mouths at the same time. The saccharin solution had no detectable taste or, importantly, odour. Surprisingly, under such conditions, the smell of almond was perceived as significantly more intense when assessed relative to a baseline condition in which no tastant was present (see Figure 6.2).

The results of a follow-up study demonstrated that holding a little water or monosodium glutamate (MSG) on the tongue did not give rise to any change in people's ability to smell the almond aroma. Taken together, these results demonstrate that orthonasal olfactory and sub-threshold gustatory stimuli are easily integrated. However, it turns out that this multisensory effect is specific to the particular combination of smell and taste stimuli that are used.

Dalton *et al.*'s (2000) study provides convincing evidence that gustatory cues that people are not aware of tasting (because they were presented at a level that happened to fall below awareness) can nevertheless still enhance their sense of smell when sniffing a food or drink. Similar results have now been reported in several subsequent studies (e.g. Delwiche and Heffelfinger 2005; Pfeiffer *et al.* 2005). For instance, Pfeiffer *et al.* demonstrated a 50% lowering in the olfactory threshold (that is, they observed complete additivity) in the majority of the participants that they tested when the taste and smell stimuli were presented simultaneously. Similar effects were observed when the odour was delivered retronasally (i.e. from the back of the mouth/nose). Equivalent multisensory integration effects have now been demonstrated when people actually taste solutions as well (i.e. in a situation that is closer to real life; Delwiche and Heffelfinger 2005).

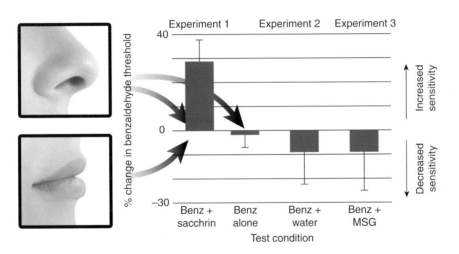

Figure 6.2   Results of a series of experiments conducted by Pam Dalton and her colleagues, showing the integration of orthonasal olfactory and gustatory cues. *Source*: Based on data from Dalton *et al.* (2000)

6.4.1   Cross-cultural differences in multisensory flavour perception

Researchers have subsequently demonstrated that this form of multisensory integration effect depends both on the particular combination of stimuli used and on where in the world the participants (or diners) happen to have been brought up. For instance, the Japanese typically show these kinds of multisensory integration effects when experiencing MSG and benzaldehyde (see Breslin *et al.* 2001; Spence 2008), but not for the combination of saccharine and benzaldehyde. Note that this is the *opposite* pattern of results to those shown by the North American participants tested in Dalton *et al.*'s (2000) original study. Why should this be so? Well, while the typical Western consumer is frequently exposed to almond-cherry odour and sweet taste together in desserts such as Bakewell tart, the combination of an almond smell and a salty taste occurs far less frequently. By contrast, the reverse is true in Japan. There, the combination of an almond odour with a salty taste is common in pickled condiments, whereas sweet almond desserts are rarely eaten. Given that olfactory stimuli that happen to be presented subliminally (that is, at a level that falls below awareness) can enhance sweetness perception (Labbe *et al.* 2007), any account of these effects solely in terms of cognitive/expectancy-based effects can be ruled out. Instead, the results would appear to be more consistent with an explanation in terms of multisensory integration (cf. Labbe *et al.* 2007).

These results with sub-threshold gustatory stimuli, highlighting the importance of stimulus congruency, are also linked to earlier results demonstrating similar multisensory interactions between congruent *supra-threshold* combinations of odour and taste stimuli. For example, Frank and Byram (1988) reported some years ago that the presence of a strawberry odour enhanced the sweetness of sucrose delivered in a whipped cream base (the participants in this particular study swallowed the cream in order to enhance the retronasal olfactory hit). By contrast, these researchers reported that adding the odour of peanut butter had no such effect. That is, stimulus congruency (defined by Schifferstein and Verlegh 1996 as the extent to which two stimuli are likely to be combined in a food product) turns out to play an important role at both the sub- and supra-threshold levels. Meanwhile, Labbe *et al.* (2006) have reported that odour congruency also plays a role in people's perception of bitterness in real foods such as cocoa beverages (see also Cliff and Noble 1990; Schifferstein and Verlegh 1996; Stevenson and Boakes 2004).

Taken together, the results reported in this section support the view that no matter where in the world a diner happens to have been born (or where they grew up), their brain will use the same rules of multisensory integration in order to combine the smell, taste and presumably also oral-somatosensory textural cues that give rise to the perception of flavour. However, the tastes and smells (and, as we will see in Section 6.8, colours) which are integrated will depend on the particular ingredients that tend to be combined in the particular cuisine that the diner is familiar with. Interestingly, we start to acquire our preferences for specific flavours while still in the womb. Newborns seem to express a preference (as measured by their head-turning responses) toward certain odours as a function of the foods that were consumed by their mothers during pregnancy (Schaal *et al.* 2000; Ganchrow and Mennella 2003; Schaal and Durand 2012; Spence 2012b).

6.5 Oral-somatosensory contributions to multisensory flavour perception

The oral-somatosensory system plays a much more important role in multi-sensory flavour perception than many of us realize. Such cues enable a diner to determine the temperature of a food as well as its texture, not to mention informing them about the burning sensation associated with eating chilli (see Lawless *et al.* 1985). The grippiness and astringency of the tannins in a young oaked red wine or in an overstewed cup of black tea, say, also result from the stimulation of the sense of touch. The latest research suggests that tactile cues play an important role in helping a diner to localize their flavour experiences within the oral cavity (Todrank and Bartoshuk 1991; Lim and Green 2008; Lim and Johnson 2011, 2012; Stevenson *et al.* 2011). The oral texture or *mouthfeel*[6] of food and drink can also influence the way in which multisensory flavours are experienced (e.g. Szczesniak 2002; Weel *et al.* 2002; Bult *et al.* 2007; Frost and Janhoj 2007). While the results of early research (e.g. Christensen 1980) led to the suggestion that increased viscosity in a foodstuff reduced taste perception, it has for many years been difficult to determine whether such effects had a physicochemical or psychological origin (since, as any food chemist will tell you, increased viscosity is likely to reduce volatility at the food–air interface; see Delwiche 2004). However, in recent years technological advances have meant that it is now much easier to demonstrate the genuinely psychological nature (of at least a part) of this cross-modal effect.

Gaining a better knowledge of how the senses are combined to deliver the rich range of flavour experiences that we all know and love has also been very important for a number of the world's largest flavour houses and food com-panies. In one particularly nice study, a creamy odour was presented either orthonasally or retronasally using a computer-controlled smell delivery device or what is called an olfactometer in the trade (Bult *et al.* 2007). At the same time, milk-like substances with different viscosities were delivered to the par-ticipant's mouth. The participants had to rate the intensity of the flavour, as well as the thickness and creaminess of the resulting flavour. The key finding to emerge from this study was that participants' ratings of the intensity of the

---

[6] Note that texture is a property of the food itself, whereas the mouthfeel refers to the sensations that arise in the oral cavity.

flavour decreased as the viscosity of the liquid increased, regardless of whether the odour was presented orthonasally or retronasally. Given the independent control of texture and odour delivery, these results therefore highlight the important role that texture (mouthfeel) can sometimes play in the multisensory perception of flavour. They also suggest that the presence of a retronasal odour can alter how thick a foodstuff is felt to be in the mouth (see also Sundqvist *et al.* 2006; Tournier *et al.* 2009). Roudnitzky *et al.* (2011) reported that when milk (either thickened or not) was placed in the mouth, people rated a butter aroma as being more intense, although the odour intensity was not affected by the concentration of the thickener.[7] Results such as these certainly open up a number of possibilities for chefs in terms of matching aromas with different flavours and textures in order to enhance one or the other.

> *"Thickness has a strong impact on taste, as every chef intuitively knows. Thicker sauces remain in the mouth longer and release their flavors more slowly. When a sauce lingers in the mouth, taste receptors for sour, sweet, bitter, salty, and umami are under the influence of the taste molecules for a longer period."* (Vilgis 2012, p. 143)

Everyone has heard of the ventriloquist, the illusionist who projects his voice to the articulated lips of his dummy. This illusion provides a beautiful example of how where we think that a sound is coming from can be captured by a concurrently presented visual stimulus (Alais and Burr 2004). The evidence now suggests that a very similar effect may also be taking place in our mouths whenever we eat. It turns out that the perceived localization of the taste of a food in the mouth follows the location of a tactile stimulus drawn across the tongue (Todrank and Bartoshuk 1991; Green 2002; see also Lim and Green 2008). The same may also be true in the case of olfactants (Murphy and Cain 1980). What is more, given the pronounced differences in transduction latencies (that is, in the conversion of a sensory stimulus from one form of energy to another) between the senses, the tactile sensations associated with eating and drinking will normally arrive centrally in the diner's brain before either the associated gustatory or retronasal olfactory stimuli. This 'prior entry' of the tactile signals may also play a key role in situating the combined multisensory flavour experience in the mouth rather than in the nose say, where the majority of the information apparently originates (see also von Békésy 1964; Pfeiffer *et al.* 2005; Small *et al.* 2005; Spence and Parise 2010; Gotow and Kobayakawa 2014). Thinking about the timing and sequencing of flavour sensations is something that is becoming more and more important to many chefs.

---

[7] While these results would appear to demonstrate that the intensity of the flavour of a cream, say, decreases as its thickness increases, it's also true that thicker substances remain in the mouth for longer; the perception of the flavour might be less intense, but it should at least last longer! Indeed, the bitter astringent notes in a red wine may help to prolong the flavour, or length, of a wine in just this way.

6.5.1 Are you a thermal taster?

Another kind of cross-modal interaction involving oral-somatosensation takes place between *temperature* and *taste*. Everyone has likely had the experience of tasting a warm soft drink (like Coca-Cola) that is unbearably sweet. There are, however, marked individual differences here too: roughly 33–50% of the population experience what is known as the 'thermal-taste' illusion (see Cruz and Green 2000; Green and George 2004). Simply by raising or lowering the temperature at various points on the tongue, it is possible to elicit sensations of sweetness, sourness, saltiness and bitterness in these individuals. Furthermore, those diners who experience the thermal-taste illusion will also tend to experience other tastes as more intense (Bajec and Pickering 2008). If you want to know whether you are one of the thermal tasters or not, all you need to do is take an ice cube from the freezer and place it against the side of your tongue. Assuming that the water from which the ice cube was made was tasteless, if you can taste something, anything at all (say, sweetness), then you know you are a thermal taster. Should you be wondering, supertasters are also likely to be thermal tasters.

On occasion, we judge a food's texture (and quality and freshness) by touching it. Just think about pushing gently against the flesh of a fruit, say, in order to check how ripe it is. It turns out that the haptic cues we feel when eating food with our hands can also influence the perceived texture of that food in the mouth. In one study, a group of blindfolded participants had to rate the freshness/staleness and the crispness/softness of a series of pretzels while biting into either the fresh or stale end of the pretzel. Barnett-Cowan (2010) manipulated whether the tactile/haptic information provided to the participants' hand matched that provided to their mouth. In half of the trials, the participants were given a half-fresh/half-stale pretzel (the incongruent conditions); in the remainder of the trials, they were given either a whole fresh or stale pretzel (the congruent conditions). In the incongruent conditions, the stale half of the pretzel was rated as being significantly fresher and crispier in-mouth because the hands held what felt like a fresh pretzel, and vice versa when the hand felt a stale pretzel. Such results therefore suggest that the perceived texture of food in-mouth can be altered simply by changing the haptic information provided to the consumer's hands. (As we will see in Section 6.7, the texture of the food in our mouths can also be modified by changing the sounds that foods make in the mouth.)

6.6 Interim summary

We have reviewed the evidence concerning the multisensory integration of gustatory, olfactory (both orthonasal and retronasal) and oral-somatosensory cues that *directly* contribute to (or influence) the multisensory perception of flavour. The results reported so far in this chapter support the view that the diner's brain uses certain rules of multisensory integration in order to combine the relevant olfactory, gustatory and oral-somatosensory cues. By now, the reader is hopefully starting to understand the complexity that is involved in preparing a perfect mousse, say, to delight diners (think of all the parameters involved: sweetness, foaminess, thickness and aroma, all interacting but doing so differently as a function of the taster status of the diner).

In the remainder of this chapter, we look at the cross-modal influence of auditory and visual cues on the perception of flavour. Why? Because we all know that part of the enjoyment of a perfect mousse is appreciating the sound it makes, both when our spoon gently breaks into the surface and when breaking its bubbles in our mouth, and its appearance (colour, airy texture, etc.).

6.7   The sound of food

Heston Blumenthal famously attempted to introduce popping candy onto the menu at the Little Chef restaurant chain, before they finally went under (Fleming 2013). That said, the majority of papers on the topic of flavour perception tend to remain silent when it comes to hearing. Any researcher who does refer to this 'forgotten' flavour sense will typically provide only the briefest of mentions (e.g. Amerine and Roessler 1976; Verhagen and Engelen 2006; Shepherd 2012). It is therefore all too easy to come away from such reviews of the literature with the distinct impression that what we hear plays no role in our perception of flavour or, for that matter, in modulating what a plate of food will taste like. Indeed, when Delwiche (2003) questioned 140 researchers working in the field a few years ago, their responses suggested that such a view was in fact widespread. These experts rated 'sound' as the *least* important aspect of flavour, coming in well behind taste, smell, temperature, texture appearance and colour (see Schifferstein 2006 for similar results). That said, the situation is slowly starting to change. A number of researchers have recently included sound as one of the senses that can impact on the consumer's experience of food and drink (Spence and Shankar 2010; Spence *et al.* 2011; Knight 2012, p. 80). Furthermore, a growing number of chefs are now starting to make their dishes more audible by using everything from popping candy to the latest in digital technology (see Chapter 10). Indeed, we would argue that the last few years have seen something of a renaissance of interest in this previously neglected 'flavour' sense (see Spence 2012c for a review).

> "*I would consider sound as an ingredient available to the chef.*" (Heston Blumenthal, interviewed on BBC Radio 4)

Interest in the role of audition in multisensory flavour (and texture) perception actually goes back a surprisingly long way (see Spence 2012b for a review). Back in the 1960s, Drake was already analysing the information conveyed by food-crushing sounds. A few years later, Zata Vickers and her colleagues published an extensive body of research in which they investigated the factors that contribute to the perception of, and consumer distinction between, *crispiness* and *crunchiness* in various different dry food products (see Duizier 2001 for a review). Basically, this research revealed that foods that gave rise to higher-pitched biting sounds were more likely to be described by participants/consumers as 'crispy' than as 'crunchy' (see also Dacremont 1995; Dijksterhuis *et al.* 2007; Varela and Fiszman 2012 for more recent work in this area).

Try eating a crisp (or potato chip) without making a noise and you'll see that it is simply impossible! The question therefore arises as to whether such food-related eating sounds exert any influence on a diner's perception of food. Zampini and Spence (2004) demonstrated that the sounds made by people when they bite into a dry food product, such as a crisp, can contribute as much as 15% to their perception of crispness and/or freshness.[8] Zampini and Spence's participants had to bite into nearly 200 Pringles potato chips (all more or less identical in shape, size and weight, and hence ideal for psychophysical investigation) and rate each one in terms of its crispness and freshness. Real-time auditory feedback of the noise made by the participants while they were biting into each crisp was played back over closed-ear headphones.[9] The auditory feedback was sometimes altered in terms of its frequency composition and/or overall loudness. That is, when biting into each potato chip, the sound that participants heard could either be attenuated by 0, 20 or 40 dB across the entire frequency range, or else only those frequencies above 2 kHz could either be boosted, cut (by 12 dB) or left unaltered.

The potato chips were rated as tasting both significantly crisper and significantly fresher when the overall sound level was increased and/or when just the high frequency sounds were amplified. By contrast, people rated the crisps as being both significantly staler and significantly softer when the overall loudness of the sounds made by their biting into the potato chips was reduced and/or when the high-frequency sounds were suppressed. These results therefore highlight the important role that food-eating sounds play in terms of people's perception and evaluation of foods (at least for those foods that make a 'noise' when we bite into them).[10]

It should however be noted that it is not only the overall loudness or frequency composition of food-eating sounds that modulate the perception of a food's texture; the temporal qualities of food sounds, such as how uneven or

---

[8] Max Zampini and Charles Spence were awarded the 2008 IG Nobel prize for nutrition for this ground-breaking research (http://news.bbc.co.uk/1/hi/sci/tech/7650103.stm).

[9] The participants actually perceived the sounds to have come from the potato chips themselves (i.e. rather than from the headphones) due, presumably, to the audiotactile ventriloquism effect (Caclin *et al.* 2002).

[10] Zampini and Spence (2005) subsequently demonstrated that people's ratings of the carbonation of a fizzy drink can also be modified by changing the loudness of the popping sounds that they hear when evaluating a beverage presented in a cup.

discontinuous they are, can also influence how 'crispy' a food appears to be (e.g. Vickers and Wasserman 1979; de Liz Pocztaruk *et al.* 2011; Varela and Fiszman 2012).[11]

What does all this mean for the chef? A number of chefs have recently started to pay a lot more attention to how their dishes sound, providing everything from embedding a layer of popping candy in an otherwise silent chocolate mousse through to adding some sonic explosions to the crispy potato topping on a shepherd's pie (both examples from Heston Blumenthal's kitchens).[12] Ironically, even deaf diners apparently like the oral-somatosensory feeling associated with popping candy (Stuckey 2012).

6.8   Visual flavour

As we have seen at several points throughout this book, visual cues play an important role in multisensory flavour perception. In fact, well over 200 studies have been published on this topic since the original report by Moir (1936) first documented the fact that changing the colour of a food can alter its taste and flavour (see Clydesdale 1993; Spence *et al.* 2010 for reviews). By-and-large, the research that has been published to date demonstrates that judgments of the identity of a food's taste, aroma and flavour can all be influenced by changing the colour (such that it is appropriate, inappropriate or absent) of the foodstuff that people happen to be evaluating. By contrast, those studies that have investigated the effect of varying the intensity of the colour added to a food (typically coloured beverages) have revealed rather mixed results (see Spence *et al.* 2010 for a review). As yet, far less research has attempted to investigate the influence of other visual appearance cues such as the opacity etc. of a food or beverage item on multisensory flavour perception (but see Hutchings 1977; Okajima and Spence 2011).

6.8.1   How does colour influence flavour perception?

One of the classic studies demonstrating the influence of colour on *taste* sensitivity was conducted by Maga (1974). He investigated the consequences of

---

[11] Given the important role of auditory cues on our perception of food, one open question for future research concerns what happens in the case of those individuals who are deaf. It turns out to be surprisingly hard to find any published evidence on the topic (but see Srinivasan 1955 for an early discussion of the issue and Stuckey 2012 for some anecdotal reports). This contrasts with the large body of research assessing the effect of either short- or long-term visual deprivation on people's perception of flavour (see Chapter 8). It is presently unclear whether the multisensory flavour experiences of the deaf are similar to those of normal-hearing adults who eat food while listening to loud white noise, say (which masks food sounds; Masuda *et al.* 2008), or whether the relative contribution of each of the residual senses is somewhat different for deaf individuals due to the consequences of cortical plasticity (see also Spence *et al.* in press).

[12] To see what restaurateurs are doing in terms of background sounds to enhance the perception of food, see Chapter 9.

colouring aqueous solutions red, yellow or green on perceptual thresholds for the four basic tastes (sweet, sour, bitter and salty). In several cases, Maga found that the concentration of the tastant had to be increased in order for participants to correctly detect its presence in a coloured, as compared to in an uncoloured, drink. For example, the addition of yellow colouring to a sweet solution significantly decreased people's sensitivity to sweetness while adding green colour increased it. By contrast, the addition of red food colouring had no effect on the sweetness threshold (see also Frank *et al.* 1989), but did result in a significant lowering of participants' sensitivity to the bitter-tasting solution. The addition of yellow and green colouring had no such effect on the threshold for bitterness. With respect to people's sensitivity to the presence of a sour taste, colouring a solution either yellow or green significantly decreased participants' sensitivity, with red colouring again having no significant effect. Interestingly, taste detection thresholds for the salt solutions were unaffected by colouring, perhaps because salty foods can come in so many colours.

What do findings such as these mean for the modernist chef? The suggestion here is that they might be able to reduce the concentration of less healthy ingredients in a dish by simply changing the colour of the food (not forgetting, of course, the role of the colour and shape of the plate; see Chapter 4).

Several other studies have also demonstrated a similarly impressive influence of visual cues on people's olfactory discrimination performance (Stevenson and Oaten 2008) and on their identification of both fruit- and non-fruit-flavoured beverages (e.g. Davis 1981; Zellner *et al.* 1991; Blackwell 1995; Zellner and Durlach 2003; Shankar *et al.* 2010c; Spence 2010)[13] as well as on their ratings of the intensity of food odours (Zellner and Kautz 1990; Zellner and Whitten 1999). Who is more fooled by what they see: the expert (e.g. the chef) or the average diner? What little evidence there is would appear to suggest that experts are, if anything, more influenced by colour cues than non-experts, at least if judging in their domain of expertise (Parr *et al.* 2003; Lelièvre *et al.* 2009).

The story regarding the influence of what we see on what we smell is, however, made more complicated by the results of a study reported by Koza *et al.* (2005). These researchers demonstrated that colour had a qualitatively different effect on the perception of orthonasally versus retronasally presented odours associated with a commercial fruit-flavoured (tangerine-pineapple-guava) water drink (see also Christensen 1983; Zellner and Durlach 2003). In particular, they found that colouring the solutions red led to odour enhancement in those participants who sniffed the odour orthonasally, while leading to a reduction in perceived odour intensity when it was presented retronasally! Koza *et al.* (2005) accounted for this

---

[13] If you are wondering why there is so much focus on coloured liquids, it is by far the most commonly studied food stimulus that has been used by researchers interested in cross-modal interactions between colour and taste. Colours and tastes/flavours diluted in water all have the same texture, so it's a way of simplifying the procedure as the influence of all other visual cues can be ruled out.

surprising pattern of results by suggesting that it may be more important for us to correctly evaluate foods once they have entered our mouths, since that is when they pose a greater risk of poisoning. By contrast, the threat of poisoning from foodstuffs located outside the mouth is less severe. It may well be that people simply attend more to the stimuli within their bodies as compared to those situated externally (cf. Spence *et al.* 2001), and that this influence biased the pattern of sensory dominance that was reported. Should the striking results of Koza *et al.* be replicated (preferably using a within-participants experimental design), it would certainly add further weight to the argument that qualitatively different patterns of multisensory integration (and hence of multisensory perception) are observed following the presentation of orthonasal as compared to retronasal odours (see also Small *et al.* 2008).

Many times during conferences we are asked whether there is any sense that dominates over human perception. Many psychologists talk about visual dominance, the idea that what we see normally dominates the other senses. Given the significant effect of colour on both taste sensitivity (Maga 1974) and on various aspects of odour perception (e.g. Davis 1981; Blackwell 1995; Koza *et al.* 2005; Shankar *et al.* 2010b), it should come as little surprise that it also exerts a robust effect on people's identification of flavours. For example, the participants in a classic study reported by DuBose *et al.* (1980) had to identify the flavours of a variety of differently coloured fruit-flavoured soft drinks. Certain of the colour–flavour pairings used in this study were deemed 'appropriate' (e.g. a cherry-flavoured drink coloured red), while others were deemed 'inappropriate' (e.g. as when a lime-flavoured drink was coloured red).[14] The participants in this study misidentified the flavours of a number of the drinks when they were coloured inappropriately. In fact, their incorrect answers often seemed to be driven by the colours of the drinks themselves. That is, participants often made what might be classed as *visual-flavour* responses. For example, 26% of the participants reported that a cherry-flavoured drink tasted of lemon/lime when coloured green, as compared to no lime-flavour responses when the drink was coloured red instead. This phenomenon is referred to by psychologists as visual dominance.

In fact, similar results have been reported in many other studies over the years (see Spence *et al.* 2010 for a review). However, an interesting (and until recently, unanswered) question in this area asks why it is that only a proportion of participants' responses are seemingly influenced by the changing of the colour of a drink (as in DuBose *et al.* 1980). One recent suggestion has been that people may differ in the degree to which particular colours induce specific flavour expectations (see Zampini *et al.* 2007; Shankar *et al.* 2010a, b). It is possible that such individual differences in the flavour expectations elicited by particular colours may explain why certain colours

---

[14] But see Shankar *et al.* (2010a) on the problematic notion of colour 'appropriateness'.

have a more pronounced effect than others on people's flavour identification responses. What is more, and complicating matters further, different groups of consumers have on occasion been shown to generate different taste and flavour expectations in response to the sight of a food or beverage of a particular colour (Garber *et al.* 2001). This means that what may be judged as *incongruent* by one group of observers might actually be perceived as *congruent* by another (see Shankar *et al.* 2010c; Wan *et al.* 2014). The existence of such cross-cultural differences in the meaning of colour can cause a headache for those chefs interested in playing with sensory incongruity at the dining table (see Chapter 7).

When people were asked to report what flavour they would expect a red and a blue drink (presented in transparent plastic cups) to have, young Taiwanese adults typically suggested cranberry and mint, respectively (Shankar *et al.* 2010a). By contrast, young British consumers reported that they would expect the drinks to taste of strawberry or raspberry instead (yes, raspberry for the blue drink due, presumably, to the presence of various blue raspberry-flavoured drinks in the UK). In other words, exactly the same colour in a drink can give rise to very different flavour expectations in the mind of a diner (or drinker).[15] The modernist chef interested in surprising the diners in their restaurant will therefore need to take such insights on board in order to make sure that their dinner guests don't end up experiencing a negative disconfirmation of expectation response.

Given that we all live in very different taste worlds, as we saw in Sections 6.2 and 6.3, one might wonder here whether our taster status has anything to do with the extent to which vision dominates over what we perceive as the flavour of a food in our mouth? One particularly interesting result to have emerged in this regard comes from Zampini *et al.* (2008). These researchers found that supertasters showed less visual dominance over their perception of flavour than medium tasters who, in turn, showed less visual dominance than non-tasters. The participants in this particular study had to identify the flavour of a large number of fruit-flavoured drinks presented among other flavourless drinks. The drinks could be coloured red, orange, yellow, grey or else presented as clear and colourless solutions. Overall, the non-tasters correctly identified 19% of the solutions, the medium tasters 31% and the supertasters 67% of the drinks that they were given to taste. People certainly found it much easier to identify the orange-flavoured solutions when they were correctly coloured. Similarly, colouring the blackcurrant-flavoured solutions a greyish purple also led to a significant improvement in their correct flavour identification (note here that many blackcurrant-flavoured products such as yoghurt are typically coloured a greyish-purple). What is more, the participants in the study of Zampini *et al.* also rated the congruently coloured drinks as having a

---

[15] Such individual differences in people's visual flavour expectations may also help to explain the inconsistent effect of colour manipulations reported in some of the previous studies that have looked at the effect of colour on taste and flavour perception (see Spence *et al.* 2010).

more intense flavour than when presented either with no colour or else with an incongruent colour instead. By contrast, no such effect of congruent colouring was reported on participants' sourness or sweetness intensity judgments. Interestingly, the addition of colouring had the largest effect (in terms of determining whether or not the participants' correctly identified the flavour) on the performance of the non-tasters, less of an effect on the medium-tasters and very little effect on the colour identification responses of the supertasters.

6.8.2   Summary of research on visual flavour

In summary, the effect of colour (be it appropriate, inappropriate or absent) has a profound consequence on people's flavour identification responses. That said, manipulating the *intensity* of the colour added to a food does not appear to have such a clear-cut effect on a diner's expectations, and hence judgments, about a food's flavour (or, for that matter, taste) intensity (Petit *et al.* 2007; see Spence *et al.* 2010 for a review). One general point to note here is that if the effect of colour on flavour perception is more cognitive in nature than the multisensory integration effects that appear to govern oral–somatosensory–gustatory–orthonasal–olfactory interactions (that is, if the effect of colour depends on an expectancy effect set-up by the presence of a particular colour), then the exact testing protocol may well influence the results of studies in ways that, as yet, have not been fully thought out (see Frank *et al.* 1989; Spence *et al.* 2010).

One problem with many of the studies reported in this chapter is that only a single kind of food (e.g. jelly, cake or chocolate) or drink was presented at a given time. Of course (as we saw in Chapter 4), when it comes to the visual aspects of plating the perception of a meal becomes much more complex (e.g. the mysterious Lamb dish served at Alinea in Chicago was composed of 86 ingredients laid side by side). Most dishes on a restaurant's menus likely include multiple colours and elements. In the future, researchers will therefore need to move from studying the response of consumers to coloured drinks to studying their response to more realistic presentations of a variety of food elements.

6.9   The cognitive neuroscience of multisensory flavour perception

Here comes what is perhaps the most theoretical part of the book: the faint of heart may wish to skip this section. We hope however that the reader is eager to follow since, in order to better understand how our senses interact to create what we think of as the flavour of a food, a short section on how the brain integrates all this information is needed. The last few years have seen a rapid growth in our understanding of the neural networks that underlie multisensory flavour perception (Shepherd 2012). We now know, for example, that gustatory stimuli project from the tongue to the primary taste cortex (more specifically, the anterior insula and the frontal or parietal operculum; see Simon *et al.* 2006), while olfactory stimuli project directly to the primary olfactory (i.e. piriform) cortex (but see Johnson *et al.* 2000; Veldhuizen *et al.* 2009). From there, the inputs from both senses project directly to the orbitofrontal cortex (OFC), a small brain structure located behind the eyes. Gustatory stimuli appear to the project to caudolateral OFC whereas olfactory stimuli project to the caudomedial OFC. Currently, the available evidence suggests that the OFC plays a central role in mediating multisensory interactions in flavour perception (e.g. Small 2012). In fact, the consensus view is that the pleasantness (and reward value) of a food or drink is represented there (e.g. Rolls and Baylis 1994; Small *et al.* 1997, 2001; Small 2004; Small and Prescott 2005).

The participants in one influential study had to lie in a functional magnetic resonance imaging (fMRI) brain scanner while rating the pleasantness and congruency of various different pairings of orthonasal olfactory and gustatory stimuli (de Araujo *et al.* 2003). The olfactory stimuli consisted of methianol (which smells like chicken broth) and strawberry odour, while the tastants used were sucrose and monosodium glutamate (MSG). The participants were presented with both congruent (e.g. sucrose and strawberry odour) and incongruent (e.g. sucrose and chicken broth odour) combinations of orthonasal olfactory and gustatory stimuli. Crucially, increased OFC activity was correlated with increased ratings of the pleasantness and congruency of the olfactory–gustatory stimulus pairing that they happened to be evaluating. So now you know which area of the brain likely 'lights up' when tasting your favourite food!

In another study, Dana Small and her colleagues presented familiar/unfamiliar combinations of retronasal olfactory and gustatory stimuli to their participants (Small *et al.* 2004). Superadditive neural interactions (cf. Stein and Meredith 1993; Stein and Stanford 2008) were observed in the OFC for familiar (or congruent, sweet–vanilla), but not for unfamiliar (or incongruent) combinations of stimuli (such as for the salty–vanilla combo). Additionally, several other brain areas – including the dorsal insula, the frontal operculum and the anterior cingulate cortex – also lit up in what could be thought of as a 'flavour network' (Shepherd 2012; Small 2012). Thus, it would appear as though the presentation of familiar combinations of olfactory (both orthonasal and retronasal) and gustatory flavour stimuli can lead to enhanced neural responses in those parts of the brain that code for the hedonic (i.e. pleasantness) and reward value of food. This ultimately means that the chef really ought to be thinking about how exactly to light up his or her dinner guests' OFC. Although knowing this isn't yet of much practical use, knowing how the senses interact really ought to be a key part of a chef's training.

Behaviourally, the presentation of familiar (and/or congruent) stimulus combinations results in the perception of enhanced flavour intensity and increased pleasantness. By contrast, unfamiliar (and/or incongruent) combinations of taste and smell (delivered either orthonasally or retronasally) will likely lead to reduced liking and the suppression of neural response in those parts of the brain coding for the intensity and pleasantness of food (De Araujo *et al.* 2003; Small *et al.* 2005; see also Skrandies and Reuther 2008). For example, Small *et al.* (1997) observed a significant decrease in the regional cerebral blood flow in the primary and secondary gustatory cortices as well as in the secondary olfactory cortex during the simultaneous presentation of incongruent orthonasal olfactory and gustatory stimuli, as compared to when the stimuli were presented individually. That said (and as we will see in Chapter 7), sensory incongruity can often be a very desirable thing, at least when served in the context of a modernist restaurant rather than, say, the cramped and noisy confines of the brain scanner.

Oral-somatosensory information regarding whatever food we happen to have in our mouths is transferred to the brain via the trigeminal nerve. This projection carries information concerning touch, texture (mouthfeel; Christensen 2004), temperature, proprioception, nociception and chemical irritation from the receptors in the mouth directly to the primary somatic sensory cortex of the diner's brain (Simon *et al.* 2006). Given what we have seen so far in this chapter, it should come as little surprise to find that oral texture (including the perception of fattiness in food) is also represented in the OFC (Eldeghaidy *et al.* 2011; see also Cerf-Ducastel *et al.* 2001). Congruent combinations of colour and orthonasally presented odours also lead to enhanced activation in the OFC (Österbauer *et al.* 2005). To the best of our knowledge, neuroimaging techniques have not yet been applied to the case of somatosensory–olfactory interactions (but see Stevenson 2012).

6.10   Conclusions

By now, the reader can hopefully better understand why it is that flavour perception in one of the most multisensory of our everyday experiences, involving as it does the direct contribution of taste, smell and oral-somatosensory (e.g. mouthfeel and trigeminal) cues. That said, auditory and visual cues also influence our multisensory flavour perception although they have typically been excluded from traditional definitions of flavour (e.g. Delwiche 2004; ISO 2008). However, we believe that such definitions are overly restrictive and should actually be expanded to include sounds elicited by food mastication (see Figure 6.1). For instance, the reader enjoying his/her next creamy soup with croutons will hopefully remember just how important the crunchy sound of the croutons is to their overall perception of a dish, and that the multisensory flavour of the meal (and his/her enjoyment) will be based on everything from the thickness of the cream to its temperature, smell and taste.

The question of whether one should also include exteroceptive cues, such as those provided by the distal sounds associated with foods (e.g. the sizzling of the steak, the chopping of vegetables), visual appearance cues and orthonasal olfaction, in one's definition of flavour (see also Small *et al.* 2004) remains something of an open question. The answer here may depend upon what exactly is meant by the term 'flavour' (see also Auvray and Spence 2008; Stevenson 2009; Spence *et al.* in press). Nevertheless, however one chooses to define the term, the key point remains that our brains appear to effortlessly bind all of the available sensory cues (regardless of their modality of occurrence) into a single unitary multisensory flavour percept (or *Gestalt*; Spence in press).

The evidence reviewed in this chapter demonstrates just how important multisensory integration is to the multisensory perception of flavour. The available research shows that multisensory interactions appear to play just as important a role in people's hedonic responses to food and drink as they do in their sensory-discriminative responses (Yeomans *et al.* 2008). That said, the results of many a cognitive neuroscience study now converge on the conclusion that the OFC plays a central role in the multisensory integration of gustatory, olfactory, tactile and visual (and presumably also auditory) cues. The OFC appears to code for the reward value of food; it is however important to remember that it is just one node in a complex 'flavour network' that spans multiple sites in the human brain (Small *et al.* 2004; Small and Prescott 2005; Small 2012). Our better understanding of the neural substrates underlying flavour perception in humans does not however mean that we should necessarily jump on the bandwagon and say that flavours are located in the brain (see Small 2012; see also Shepherd 2012). Curiously, there are a few rare patients who, after suffering from a lesion in a particular part of the right anterior (front) part of their brains (specifically involving cortical areas, basal ganglia and/or limbic structures), suddenly develop a fascination with fine food. The condition is known as Gourmand's syndrome (see Regard and Landis 1997; Steingarten 1998, 2002; Uher and Treasure 2005).

While it has been known for many years that taster status influences the experience of certain bitter tastes, there is now evidence to suggest that a diner's taster status can also affect their responses to other tastants, as well as affecting how they perceive the oral-somatosensory attributes of food and drink. Surprisingly, a person's taster status might also influence how much emphasis they put on what they see when trying to identify a food or drink's flavour (Zampini *et al.* 2008). In addition, we have also documented the growing interest that surrounds the individual and cultural differences that pervade the field of multisensory flavour perception (e.g. Shankar *et al.* 2010a, b). All of these results highlight the challenges that chefs (and even those of us who cook at home) have to face when trying to design the perfect meal. Indeed, it might seem an impossible task to make a meal that would be judged as perfect by everyone, given the profound individual differences that have been identified. So, having seen how the diner's brain deals with the *congruent* combinations of sensory cues (with most of the examples coming from the laboratory), in the following chapter we look at what happens when the chefs and restaurateurs start to deliver *incongruent* combinations of sensory stimuli.

References

Alais, D. and Burr, D. (2004) The ventriloquist effect results from near-optimal bimodal integration. *Current Biology*, **14**, 257–262.

Amerine, M. A. and Roessler, E. B. (1976) *Wines: Their Sensory Evaluation.* W. H. Freeman & Company, San Francisco.

Auvray, M. and Spence, C. (2008) The multisensory perception of flavor. *Consciousness and Cognition*, **17**, 1016–1031.

Bajec, M. R. and Pickering, G. J. (2008) Thermal taste, PROP responsiveness, and perception of oral sensations. *Physiology and Behavior*, **95**, 581–590.

Bajec, M. R., Pickering, G. J. and DeCourville, N. (2012) Influence of stimulus temperature on orosensory perception and variation with taste phenotype. *Chemosensory Perception*, **5**, 243–265.

Barnett-Cowan, M. (2010) An illusion you can sink your teeth into: Haptic cues modulate the perceived freshness and crispness of pretzels. *Perception*, **39**, 1684–1686.

Bartoshuk, L. M. (2000) Comparing sensory experiences across individuals: Recent psychophysical advances illuminate genetic variation in taste perception. *Chemical Senses*, **25**, 447–460.

Bartoshuk, L. M., Duffy, V. B. and Miller, I. J. (1994) PTC/PROP tasting: Anatomy, psychophysics, and sex effects. *Physiology and Behavior*, **56**, 1165–1171.

Bautista, D. M., Sigal, Y. M., Milstein, A. D., Garrison, J. L., Zorn, J. A., Tsuruda, P. R., Nicoll, R. A. and Julius, D. (2008) Pungent agents from Szechuan peppers excite sensory neurons by inhibiting two-pore potassium channels. *Nature Neuroscience*, **11**, 772–779.

Blackwell, L. (1995) Visual clues and their effects on odour assessment. *Nutrition and Food Science*, **5**, 24–28.

Boring, E. G. (1942) *Sensation and Perception in the History of Experimental Psychology*. Appleton, New York.

Breslin, P. A., Doolittle, N. and Dalton, P. (2001) Subthreshold integration of taste and smell: The role of experience in flavour integration. *Chemical Senses*, **26**, 1035.

Brillat-Savarin, J. A. (1835) *Physiologie du Goût* (The Philosopher in the Kitchen/The Physiology of Taste). J. P. Meline, Bruxelles. *A Handbook of Gastronomy* (translated by A. Lalauze, 1884), Nimmo & Bain, London.

Bult, J. H. F., de Wijk, R. A. and Hummel, T. (2007) Investigations on multimodal sensory integration: Texture, taste, and ortho- and retronasal olfactory stimuli in concert. *Neuroscience Letters*, **411**, 6–10.

Caclin, A., Soto-Faraco, S., Kingstone, A. and Spence, C. (2002) Tactile 'capture' of audition. *Perception and Psychophysics*, **64**, 616–630.

Cardello, A. V. (2007) Measuring consumer expectations to improve food product development. In: *Consumer-led Food Product Development* (ed. H. J. H. MacFie), pp. 223–261. Woodhead Publishing, Cambridge, UK.

Cerf-Ducastel, B., Van dc Moortele, P.-F., Macleod, P., Le Bihan, D. and Faurion, A. (2001) Interaction of gustatory and lingual somatosensory perceptions at the cortical level in the human: A functional magnetic resonance imaging study. *Chemical Senses*, **26**, 371–383.

Christensen, C. M. (1980) Effects of solution viscosity on perceived saltiness and sweetness. *Perception and Psychophysics*, **28**, 347–353.

Christensen, C. (1983) Effects of color on aroma, flavor and texture judgments of foods. *Journal of Food Science*, **48**, 787–790.

Christensen, C. M. (1984) Food texture perception. In: *Advances in Food Research* (ed. E. Mark), pp. 159–199. Academic Press, New York.

Cliff, M. and Noble, A. C. (1990) Time-intensity evaluation of sweetness and fruitiness and their interaction in a model solution. *Journal of Food Science*, **55**, 450–454.

Clydesdale, F. M. (1993) Color as a factor in food choice. *Critical Reviews in Food Science and Nutrition*, **33**, 83–101.

Cruz, A. and Green, B. G. (2000) Thermal stimulation of taste. *Nature*, **403**, 889–892.

Cytowic, R. E. (2002) *Synesthesia: A Union of the Senses (2nd ed)*. MIT Press, Cambridge, MA.

Dacremont, C. (1995) Spectral composition of eating sounds generated by crispy, crunchy and crackly foods. *Journal of Texture Studies*, **26**, 27–43.

Dalton, P., Doolittle, N., Nagata, H. and Breslin, P. A. S. (2000) The merging of the senses: Integration of subthreshold taste and smell. *Nature Neuroscience*, **3**, 431–432.

Davidson, J. M., Linforth, R. S. T., Hollowood, T. A. and Taylor, A. J. (1999) Effect of sucrose on the perceived flavor intensity of chewing gum. *Journal of Agricultural and Food Chemistry*, **47**, 4336–4340.

Davis, R. G. (1981) The role of nonolfactory context cues in odor identification. *Perception and Psychophysics*, **30**, 83–89.

De Araujo, I. E. T., Rolls, E. T., Kringelbach, M. L., McGlone, F. and Phillips, N. (2003) Taste-olfactory convergence, and the representation of the pleasantness of flavour, in the human brain. *European Journal of Neuroscience*, **18**, 2059–2068.

de Liz Pocztaruk, R., Abbink J. H., de Wijk, R., da Fontoura Frasca, L. C., Duarte Gavião, M. B. and van der Bilt, A. (2011) The influence of auditory and visual information on the perception of crispy food. *Food Quality and Preference*, **22**, 404–411.

Deliza, R. and MacFie, H. J. H. (1997) The generation of sensory expectation by external cues and its effect on sensory perception and hedonic ratings: A review. *Journal of Sensory Studies*, **2**, 103–128.

Delwiche, J. (1996) Are there 'basic' tastes? *Trends in Food Science and Technology*, **7**, 411–415.

Delwiche, J. (2003) Attributes believed to impact flavor: An opinion survey. *Journal of Sensory Studies*, **18**, 437–444.

Delwiche, J. (2004) The impact of perceptual interactions on perceived flavor. *Food Quality and Preference*, **15**, 137–146.

Delwiche, J. and Heffelfinger, A. L. (2005) Cross-modal additivity of taste and smell. *Journal of Sensory Studies*, **20**, 137–146.

Diaz, M. E. (2004) Comparison between orthonasal and retronasal flavour perception at different concentrations. *Flavour and Fragrance Journal*, **19**, 499–504.

Dijksterhuis, G., Luyten, H., de Wijk, R. and Mojet, J. (2007) A new sensory vocabulary for crisp and crunchy dry model foods. *Food Quality and Preference*, **18**, 37–50.

DuBose, C. N., Cardello, A. V. and Maller, O. (1980) Effects of colorants and flavorants on identification, perceived flavor intensity, and hedonic quality of fruit-flavored beverages and cake. *Journal of Food Science*, **45**, 1393–1399, 1415.

Duffy, V. B. (2007) Variation in oral sensation: Implications for diet and health. *Current Opinion in Gastroenterology*, **23**, 171–177.

Duizier, L. (2001) A review of acoustic research for studying the sensory perception of crisp, crunchy and crackly textures. *Trends in Food Science and Technology*, **12**, 17–24.

Eldeghaidy, S., Marciani, L., McGlone, F., Hollowood, T., Hort, J., Head, K. *et al.* (2011) The cortical response to the oral perception of fat emulsions and the effect of taster status. *Journal of Neurophysiology*, **105**, 2572–2581.

Erikson, R. P. (2008) A study of the science of taste: On the origins and influence of the core ideas. *Behavioral and Brain Sciences*, **31**, 59–105.

Essick, G. K., Chopra, A., Guest, S. and McGlone, F. (2003) Lingual tactile acuity, taste perception, and the density and diameter of fungiform papillae in female subjects. *Physiology and Behavior*, **80**, 289–302.

Fleming, A. (2013) What makes eating so satisfying? *The Guardian*, 23 April. Available at http://www.guardian.co.uk/lifeandstyle/wordofmouth/2013/apr/23/what-makes-eating-so-satisfying (accessed January 2014).

Frank, R. A. and Byram, J. (1988) Taste-smell interactions are tastant and odorant dependent. *Chemical Senses*, **13**, 445–455.

Frank, R. A., Ducheny, K. and Mize, S. J. S. (1989) Strawberry odor, but not red color, enhances the sweetness of sucrose solutions. *Chemical Senses*, **14**, 371–377.

Frost, M. B. and Janhoj, T. (2007) Understanding creaminess. *International Dairy Journal*, **17**, 1298–1311.

Ganchrow, J. R. and Mennella, J. A. (2003) The ontogeny of human flavour perception. In: *Handbook of Olfaction and Gustation* (ed. R. L. Doty), pp. 823–846. Marcel Dekker, New York.

Garber, L. L., Jr., Hyatt, E. M. and Starr, R. G., Jr. (2001) Placing food color experimentation into a valid consumer context. *Journal of Food Products Marketing*, **7(3)**, 3–24.

Ge, L. (2012) *Why coffee can be bittersweet. FT Weekend Magazine*, 13/14 October, 50.

Gilbert, A. (2008) *What the Nose Knows: The Science of Scent in Everyday Life*. Crown Publishers, New York.

Gotow, N. and Kobayakawa, T. (2014) Construction of a measurement system for simultaneity judgment using odor and taste stimuli. *Journal of Neuroscience Methods*, **221**, 132–138.

Green, B. G. (2002) Studying taste as a cutaneous sense. *Food Quality and Preference*, **14**, 99–109.

Green, B. G. and George, P. (2004) 'Thermal taste' predicts higher responsiveness to chemical taste and flavor. *Chemical Senses*, **29**, 617–628.

Guyton, A. C. and Hall, J. E. (eds) (1996) *Textbook of Medical Physiology* (9th edition). W. B. Saunders Company, Philadelphia, PA.

Hagura, N., Barber, H. and Haggard, P. (2013) Food vibrations: Asian spice sets lips trembling. *Proceedings of the Royal Society B*, **280(1770)**, 20131680.

Hanig, D. P. (1901) Zur Psychophysik des Geschmackssinnes (On the psychophysics of taste). *Philosophische Studien*, **17**, 576–623.

Hayes, J. E. and Duffy, V. B. (2007) Revisiting sugar–fat mixtures: Sweetness and creaminess vary with phenotypic markers of oral sensation. *Chemical Senses*, **32**, 225–236.

Hayes, J. E. and Duffy, V. B. (2008) Oral sensory phenotype identifies level of sugar and fat required for maximal liking. *Physiology and Behaviour*, **95(1)**, 77–87.

Hayes, J. E. and Pickering, G. J. (2012) Wine expertise predicts taste phenotype. *American Journal of Enology and Viticulture*, **63**, 80–84.

Hayes, J. E., Bartoshuk, L. M., Kidd, J. R. and Duffy, V. B. (2008) Supertasting and PROP bitterness depends on more than the TAS2R38 gene. *Chemical Senses*, **33**, 255–265.

Heilman, S. and Hummel, T. (2004) A new method for comparing orthonasal and retronasal olfaction. *Behavioral Neuroscience*, **118**, 412–419.

Hutchings, J. B. (1977) The importance of visual appearance of foods to the food processor and the consumer. In: *Sensory Properties of Foods* (eds. G. G. Birch, J. G. Brennan and K. J. Parker), pp. 45–57. Applied Science Publishers, London.

Hutchings, J. B. (2003) *Expectations and the Food Industry: The Impact of Color and Appearance*. Plenum Publishers, New York.

Ishiko, N., Murayama, N., Hanamori, T. and Ito, H. (1978) Depression of lingual tactile sensation during chemical stimulation of the human tongue. *Neuroscience Letters*, **7**, 79–81.

ISO (2008) *Standard 5492: Terms relating to sensory analysis*. International Organization for Standardization. Vienna: Austrian Standards Institute.

Johnson, D. M., Illig, K. R., Behan, M. and Haberly, L. B. (2000) New features of connectivity in piriform cortex visualized by intracellular injection of pyramidal cells suggest that "primary" olfactory cortex functions like "association" cortex in other sensory systems. *Journal of Neuroscience*, **20**, 6974–6982.

Keeley, B. L. (2002) Making sense of the senses: Individuating modalities in humans and other animals. *Journal of Philosophy*, **XCIX**, 5–28.

Knight, T. (2012) Bacon: The slice of life. In: *The Kitchen as Laboratory: Reflections on the Science of Food and Cooking* (eds C. Vega, J. Ubbink and E. van der Linden), pp. 73–82. Columbia University Press, New York.

Koza, B. J., Cilmi, A., Dolese, M. and Zellner, D. A. (2005) Color enhances orthonasal olfactory intensity and reduces retronasal olfactory intensity. *Chemical Senses*, **30**, 643–649.

Labbe, D., Damevin, L., Vaccher, C., Morgenegg, C. and Martin, N. (2006) Modulation of perceived taste by olfaction in familiar and unfamiliar beverages. *Food Quality and Preference*, **17**, 582–589.

Labbe, D., Rytz, A., Morgenegg, C., Ali, C. and Martin, N. (2007) Subthreshold olfactory stimulation can enhance sweetness. *Chemical Senses*, **32**, 205–214.

Lawless, H. T. (2001) Taste. In: *Blackwell Handbook of Perception* (ed. B. E. Goldstein), pp. 601–635. Blackwell Publishers, Malden, MA.

Lawless, H., Rozin, P. and Shenker, J. (1985) Effects of oral capsaicin on gustatory, olfactory and irritant sensations and flavor identification in humans who regularly or rarely consume chili pepper. *Chemical Senses*, **10**, 579–589.

Lawless, H. T., Stevens, D. A., Chapman, K. W. and Kurtz, A. (2005) Metallic taste from electrical and chemical stimulation. *Chemical Senses*, **30**, 185–194.

Lelièvre, M., Chollet, S., Abdi, H. and Valentin, D. (2009) Beer-trained and untrained assessors rely more on vision than on taste when they categorize beers. *Chemosensory Perception*, **2**, 143–153.

Lim, J. and Green, B. G. (2008) Tactile interaction with taste localization: Influence of gustatory quality and intensity. *Chemical Senses*, **33**, 137–143.

Lim, J. and Johnson, M. B. (2011) Potential mechanisms of retronasal odor referral to the mouth. *Chemical Senses*, **36**, 283–289.

Lim, J. and Johnson, M. (2012) The role of congruency in retronasal odor referral to the mouth. *Chemical Senses*, **37**, 515–521.

Lindemann, B. (1996) Taste reception. *Physiology Review*, **76**, 719–766.

Lindemann, B. (2001) Receptors and transduction in taste. *Nature*, **413**, 219–225.

Mackenna, B. R. and Callander, R. (1997) *Illustrated Physiology* (6th edition). Churchill Livingstone, New York.

Maga, J. A. (1974) Influence of color on taste thresholds. *Chemical Senses and Flavor*, **1**, 115–119.

Martin, G. N. (2004) A neuroanatomy of flavour. *Petits Propos Culinaires*, **76**, 58–82.

Masuda, M., Yamaguchi, Y., Arai, K. and Okajima, K. (2008) Effect of auditory information on food recognition. *IEICE Technical Report*, **108(356)**, 123–126.

McBurney, D. H. (1986) Taste, smell, and flavor terminology: Taking the confusion out of fusion. In: *Clinical Measurement of Taste and Smell* (eds. H. L. Meiselman and R. S. Rivkin), pp. 117–125. Macmillan, New York.

Miller, I. J. and Reedy, D. P. (1990) Variations in human taste bud density and taste intensity perception. *Physiology and Behavior*, **47**, 1213–1219.

Moir, H. C. (1936) Some observations on the appreciation of flavour in foodstuffs. *Journal of the Society of Chemical Industry: Chemistry and Industry Review*, **14**, 145–148.

Murphy, C. and Cain, W. S. (1980) Taste and olfaction: Independence vs. interaction. *Physiology and Behavior*, **24**, 601–605.

Murphy, C., Cain, W. S. and Bartoshuk, L. M. (1977) Mutual action of taste and olfaction. *Sensory Processes*, **1**, 204–211.

Okajima, K. and Spence, C. (2011) Effects of visual food texture on taste perception. *i-Perception*, **2(8)**, http://i-perception.perceptionweb.com/journal/I/article/ic966 (accessed January 2014).

Österbauer, R. A., Matthews, P. M., Jenkinson, M., Beckmann, C. F., Hansen, P. C. and Calvert, G. A. (2005) Color of scents: Chromatic stimuli modulate odor responses in the human brain. *Journal of Neurophysiology*, **93**, 3434–3441.

Parr, W. V., White, K. G. and Heatherbell, D. (2003) The nose knows: Influence of colour on perception of wine aroma. *Journal of Wine Research*, **14**, 79–101.

Petit, C. E. F., Hollowood, T. A., Wulfert, F. and Hort, J. (2007) Colour-coolant-aroma interactions and the impact of congruency and exposure on flavour perception. *Food Quality and Preference*, **18**, 880–889.

Pfeiffer, J. C., Hollowood, T. A., Hort, J. and Taylor, A. J. (2005) Temporal synchrony and integration of sub-threshold taste and smell signals. *Chemical Senses*, **30**, 539–545.

Prescott, J., Bartoshuk, L. M. and Prutkin, J. (2004) 6-*n*-Propylthiouracil tasting and the perception of nontaste oral sensations. In: *Genetic Variations in Taste Sensitivity* (eds J. Prescott and B. Tepper), pp. 89–104. Marcel Dekker, New York.

Regard, M. and Landis, T. (1997) "Gourmand syndrome": Eating passion associated with right anterior lesions. *Neurology*, **48**, 1185–1190.

Roach, M. (2013) *Gulp: Adventures on the Alimentary Canal*. One World, London.

Rolls, E. T. and Baylis, L. L. (1994) Gustatory, olfactory, and visual convergence within the primate orbitofrontal cortex. *Journal of Neuroscience*, **14**, 5437–5452.

Roudnitzky, N., Bult, J. H., de Wijk, R. A., Reden, J., Schuster, B. and Hummel, T. (2011) Investigation of interactions between texture and ortho-and retronasal olfactory stimuli using psychophysical and electrophysiological approaches. *Behavioural Brain Research*, **216**, 109–115.

Rozin, P. (1982) "Taste-smell confusions" and the duality of the olfactory sense. *Perception and Psychophysics*, **31**, 397–401.

Schaal, B. and Durand, K. (2012) The role of olfaction in human multisensory development. In: *Multisensory Development* (eds A. J. Bremner, D. Lewkowicz and C. Spence), pp. 29–62. Oxford University Press, Oxford.

Schaal, B., Marlier, L. and Soussignan, R. (2000) Human foetuses learn odours from their pregnant mother's diet. *Chemical Senses*, **25**, 72–737.

Schifferstein, H. N. J. (2006) The perceived importance of sensory modalities in product usage: A study of self-reports. *Acta Psychologica*, **121**, 41–64.

Schifferstein, H. N. J. and Verlegh, P. W. J. (1996) The role of congruency and pleasantness in odor-induced taste enhancement. *Acta Psychologica*, **94**, 87–105.

Schiffman, S. S. (2002) Taste quality and neural coding: Implications from psychophysics and neurophysiology. *Physiology and Behavior*, **69**, 147–159.

Shankar, M. U., Levitan, C. and Spence, C. (2010a) Grape expectations: The role of cognitive influences in color-flavor interactions. *Consciousness and Cognition*, **19**, 380–390.

Shankar, M., Simons, C., Levitan, C., Shiv, B., McClure, S. and Spence, C. (2010b) An expectations-based approach to explaining the crossmodal influence of color on orthonasal olfactory identification: Assessing the influence of temporal and spatial factors. *Journal of Sensory Studies*, **25**, 791–803.

Shankar, M., Simons, C., Shiv, B., Levitan, C., McClure, S. and Spence, C. (2010c) An expectations-based approach to explaining the cross-modal influence of color on orthonasal olfactory identification: the influence of the degree of discrepancy. *Attention, Perception and Psychophysics*, **72**, 1981–1983.

Shepherd, G. (2012) *Neurogastronomy: How the Brain Creates Flavor and Why it Matters*. Columbia University Press, New York.

Simon, S. A., de Araujo, I. E., Gutierrez, R. and Nicolelis, M. A. L. (2006) The neural mechanisms of gustation: A distributed processing code. *Nature Reviews Neuroscience*, **7**, 890–901.

Skrandies, W. and Reuther, N. (2008) Match and mismatch of taste, odor, and color is reflected by electrical activity in the human brain. *Journal of Psychophysiology*, **22**, 175–184.

Small, D. M. (2004) Crossmodal integration - Insights from the chemical senses. *Trends in Neurosciences*, **27**, 120–123.

Small, D. M. (2012) Flavor is in the brain. *Physiology and Behavior*, **107**, 540–552.

Small, D. M., Jones-Gotman, M., Zatorre, R. J., Petrides, M. and Evans, A. C. (1997) Flavor processing: More than the sum of its parts. *Neuroreport*, **8**, 3913–3917.

Small, D. M. and Prescott, J. (2005) Odor/taste integration and the perception of flavour. *Experimental Brain Research*, **166**, 345–357.

Small, D. M., Zatorre, R. J., Dagher, A., Evans, A. C. and Jones-Gotman, M. (2001) Changes in brain activity related to eating chocolate: From pleasure to aversion. *Brain*, **124**, 1720–1733.

Small, D. M., Voss, J., Mak, Y. E., Simmons, K. B., Parrish, T. and Gitelman, D. (2004) Experience-dependent neural integration of taste and smell in the human brain. *Journal of Neurophysiology*, **92**, 1892–1903.

Small, D. M., Gerber, J. C., Mak, Y. E. and Hummel, T. (2005) Differential neural responses evoked by orthonasal versus retronasal odorant perception in humans. *Neuron*, **47**, 593–605.

Small, D. M., Veldhuizen, M. G., Felsted, J., Mak, Y. E. and McGlone, F. (2008) Separable substrates for anticipatory and consummatory food chemosensation. *Neuron*, **57**, 786–797.

Spence, C. (2008) Multisensory perception. In: *The Big Fat Duck Cook Book* (ed H. Blumenthal), pp. 484–485. Bloomsbury, London.

Spence, C. (2010) The color of wine – Part 1. *The World of Fine Wine*, **28**, 122–129.

Spence, C. (2012a) Multi-sensory integration and the psychophysics of flavour perception. In: *Food Oral Processing: Fundamentals of Eating and Sensory Perception* (eds. J. Chen and L. Engelen), pp. 203–219. Blackwell Publishing, Oxford.

Spence, C. (2012b) The development and decline of multisensory flavour perception. In: *Multisensory Development* (eds A. J. Bremner, D. Lewkowicz and C. Spence), pp. 63–87. Oxford University Press, Oxford.

Spence, C. (2012c) Auditory contributions to flavour perception and feeding behaviour. *Physiology and Behaviour*, **107**, 505–515.

Spence, C. (2013) Multisensory flavour perception. *Current Biology*, **23**, R365–R369.

Spence, C. (in press) Cross-modal perceptual organization. In: *The Oxford Handbook of Perceptual Organization* (ed J. Wagemans), Oxford University Press, Oxford.

Spence, C. and Parise, C. (2010) Prior entry. *Consciousness and Cognition*, **19**, 364–379.

Spence, C. and Shankar, M. U. (2010) The influence of auditory cues on the perception of, and responses to, food and drink. *Journal of Sensory Studies*, **25**, 406–430.

Spence, C., Kettenmann, B., Kobal, G. and McGlone, F. P. (2001) Shared attentional resources for processing vision and chemosensation. *Quarterly Journal of Experimental Psychology A*, **54A**, 775–783.

Spence, C., Levitan, C., Shankar, M. U., and Zampini, M. (2010) Does food colour influence flavour identification in humans? *Chemosensory Perception*, **3**, 68–84.

Spence, C., Shankar, M. U. and Blumenthal, H. (2011) 'Sound bites': Auditory contributions to the perception and consumption of food and drink. In: *Art and the Senses* (eds F. Bacci and D. Melcher), pp. 207–238. Oxford University Press, Oxford.

Spence, C., Smith, B. and Auvray, M. (in press) Confusing tastes and flavours. In: *The Senses* (eds M. Matthen and D. Stokes). Oxford University Press, Oxford.

Srinivasan, M. (1955) Has the ear a role in registering flavour? *Bulletin of the Central Food Technology Research Institute Mysore (India)*, **4**, 136.

Stein, B. E. and Meredith, M. A. (1993) *The Merging of the Senses*. MIT Press, Cambridge, MA.

Stein, B. E. and Stanford, T. R. (2008) Multisensory integration: Current issues from the perspective of the single neuron. *Nature Reviews Neuroscience*, **9**, 255–267.

Steingarten, J. (1998) *The Man Who Ate Everything: And Other Gastronomic Feats, Disputes, and Pleasurable Pursuits*. Headline Publishing, London.

Steingarten, J. (2002) *It Must've Been Something I Ate*. Adolf. A. Knopf, New York.

Stevenson, R. J. (2009) *The Psychology of Flavour*. Oxford University Press, Oxford.

Stevenson, R. J. (2012) Multisensory interactions in flavor perception. In: *The New Handbook of Multisensory Processes* (ed. B. E. Stein), pp. 283–299. MIT Press, Cambridge, MA.

Stevenson, R. J. and Boakes, R. A. (2004) Sweet and sour smells: Learned synaesthesia between the senses of taste and smell. In: *The Handbook of Multisensory Processing* (eds G. A. Calvert, C. Spence and B. E. Stein), pp. 69–83. MIT Press, Cambridge, MA.

Stevenson, R. J. and Tomiczek, C. (2007) Olfactory-induced synesthesias: A review and model. *Psychological Bulletin*, **133**, 294–309.

Stevenson, R. J. and Oaten, M. (2008) The effect of appropriate and inappropriate stimulus color on odor discrimination. *Perception and Psychophysics*, **70**, 640–646.

Stevenson, R. J., Prescott, J. and Boakes, R. A. (1999) Confusing tastes and smells: How odours can influence the perception of sweet and sour tastes. *Chemical Senses*, **24**, 627–635.

Stevenson, R. J., Mahmut, M. K. and Oaten, M. J. (2011) The role of attention in the localization of odors to the mouth. *Attention, Perception and Psychophysics*, **73**, 247–258.

Stuckey, B. (2012) *Taste What You're Missing: The Passionate Eater's Guide to Why Good Food Tastes Good*. Free Press, London.

Sugai, E., Morimitsu, Y., Iwasaki, Y., Morita, A., Watanabe, T. and Kubota, K. (2005a) Pungent qualities of sanshool-related compounds evaluated by a sensory test and activation of rat TRPV1. *Bioscience, Biotechnology and Biochemistry*, **69**, 1951–1957.

Sugai, E., Morimitsu, Y. and Kubota, K. (2005b) Quantitative analysis of sanshool compounds in Japanese pepper (Xanthoxylum piperitum DC.) and their pungent characteristics. *Bioscience, Biotechnology and Biochemistry*, **69**, 1958–1962.

Sundqvist, N. C., Stevenson, R. J. and Bishop, I. R. J. (2006) Can odours acquire fat-like properties? *Appetite*, **47**, 91–99.

Szczesniak, A. S. (2002) Texture is a sensory property. *Food Quality and Preference*, **13**, 215–225.

Todrank, J. and Bartoshuk, L. M. (1991) A taste illusion: Taste sensation localized by touch. *Physiology and Behavior*, **50**, 1027–1031.

Tournier, C., Sulmont-Rossé, C., Sémon, E., Vignon, A., Issanchou, S. and Guichard, E. (2009) A study on texture-taste-aroma interactions: Physico-chemical and cognitive mechanisms. *International Dairy Journal*, **19**, 450–458.

Uher, R. and Treasure, J. (2005) Brain lesions and eating disorders. *Journal of Neurology, Neurosurgery and Psychiatry*, **76**, 852–857.

van der Klaauw, N. J. and Frank, R. A. (1996) Scaling component intensities of complex stimuli: The influence of response alternatives. *Environmental International*, **22**, 21–31.

Varela, P. and Fiszman, S. (2012) Playing with sound. In: *The Kitchen as Laboratory: Reflections on the Science of Food and Cooking* (eds C. Vega, J. Ubbink and E. van der Linden), pp. 155–165. Columbia University Press, New York.

Veldhuizen, M. G., Nachtigal, D. J. and Small, D. M. (2009) Taste cortex contributes to odor quality coding. *Chemical Senses*, **34**, A8.

Verhagen, J. V. and Engelen, L. (2006) The neurocognitive bases of human multimodal food perception: Sensory integration. *Neuroscience and Biobehavioral Reviews*, **30**, 613–650.

Vickers, Z. M. and Wasserman, S. S. (1979) Sensory qualities of food sounds based on individual perceptions. *Journal of Texture Studies*, **10**, 319–332.

Vilgis, T. (2012) Ketchup as tasty soft matter: The case of xantham gum. In: *The Kitchen as Laboratory: Reflections on the Science of Food and Cooking* (eds C. Vega, J. Ubbink and E. van der Linden), pp. 143–147. Columbia University Press: New York.

von Békésy, G. (1964) Olfactory analogue to directional hearing. *Journal of Applied Physiology*, **19**, 369–373.

Wan, X., Velasco, C., Michel, C., Mu, B., Woods, A. T. and Spence, C. (2014) Does the shape of the glass influence the crossmodal association between colour and flavour? A cross-cultural comparison. *Flavour*, **3**, 3.

Weel, K. G. C., Boelrijk, A. C., Alting, P. J. J. M., van Mil, J. J., Burger, H., Gruppen, H., Voragen, A. G. J. and Smit, G. (2002) Flavor release and perception of flavored whey protein gels: Perception is determined by texture rather than by release. *Journal of Agricultural and Food Chemistry*, **50**, 5149–5155.

Yeomans, M., Chambers, L., Blumenthal, H. and Blake, A. (2008) The role of expectancy in sensory and hedonic evaluation: The case of smoked salmon ice-cream. *Food Quality and Preference*, **19**, 565–573.

Zampini, M. and Spence, C. (2004) Multisensory contribution to food perception: The role of auditory cues in modulating crispness and staleness in crisps. *Journal of Sensory Science*, **19**, 347–363.

Zampini, M. and Spence, C. (2005) Modifying the multisensory perception of a carbonated beverage using auditory cues. *Food Quality and Preference*, **16**, 632–641.

Zampini, M., Sanabria, D., Phillips, N. and Spence, C. (2007) The multisensory perception of flavor: Assessing the influence of color cues on flavor discrimination responses. *Food Quality and Preference*, **18**, 975–984.

Zampini, M., Wantling, E., Phillips, N. and Spence, C. (2008) Multisensory flavor perception: Assessing the influence of fruit acids and color cues on the perception of fruit-flavored beverages. *Food Quality and Preference*, **19**, 335–343.

Zellner, D. A. and Kautz, M. A. (1990) Color affects perceived odor intensity. *Journal of Experimental Psychology: Human Perception and Performance*, **16**, 391–397.

Zellner, D. A. and Whitten, L. A. (1999) The effect of color intensity and appropriateness on color-induced odor enhancement. *American Journal of Psychology*, **112**, 585–604.

Zellner, D. A. and Durlach, P. (2003) Effect of color on expected and experienced refreshment, intensity, and liking of beverages. *American Journal of Psychology*, **116**, 633–647.

Zellner, D. A., Bartoli, A. M. and Eckard, R. (1991) Influence of color on odor identification and liking ratings. *American Journal of Psychology*, **104**, 547–561.

7
Using Surprise and Sensory Incongruity in a Meal

7.1   Introduction

Whenever we are presented with a plate of food (the food itself or even just an image of it), we rapidly create an opinion (or a prediction) about how what we see in front of us might taste and smell as well as its oral-somatosensory and auditory properties (Cardello 1996). These food-centred expectations, or beliefs (Olson and Dover 1979), are based primarily on any previous experiences that we may have had with that dish or with similar dishes (Oliver and Winer 1987). Such visually derived expectations are sometimes referred to as 'visual flavour' (e.g. Masurovsky 1939; Spence *et al.* 2010). Indeed, as Apicius noted in the first century, "*We eat first with our eyes*" (Masurovsky 1939; Delwiche 2012). Particularly when hungry, the sight of food can lead to some quite dramatic changes in blood flow to various parts of the brain (e.g. Pelchat *et al.* 2004; Wang *et al.* 2004; van der Laan *et al.* 2011).

So why would a chef want to create a sensorially incongruent dish? Well, the immediate reaction to experiencing sensory incongruity in a meal setting is usually one of surprise (Desmet and Schifferstein 2008). Such surprise can be associated with either positive (i.e. joyful, pleasant, etc.) or negative (i.e. frustrating, unpleasant, shocking, etc.) feelings and emotions. The likely reaction of a diner to such an experience is going to depend on a number of factors, including: the nature of the sensory incongruity involved; whether it is visible (e.g. hinted at by a food label or description) or hidden (cf. Schifferstein and Spence 2008); the situation or context in which the incongruity is experienced (i.e. in a Michelin-starred restaurant versus in a student cafeteria or perhaps an airplane; e.g. Green and Butts 1945; see also Chapter 9); and on the person who happens to be experiencing it (i.e. whether they are

The Perfect Meal: The Multisensory Science of Food and Dining, First Edition.
Charles Spence and Betina Piqueras-Fiszman.
© 2014 John Wiley & Sons, Ltd. Published 2014 by John Wiley & Sons, Ltd.

neophobic or neophilic, young or old).[1] In this chapter, we step into the world of disconfirmed expectations. We look at what happens when what we perceive through our senses doesn't match up, and how chefs and restaurateurs can manage the experience so that it is more likely to be pleasant, and possibly memorable, for the diner.

7.2   How did sensory incongruity become so popular and why is it so exciting?

The intentional use of sensory incongruity has become increasingly common among experimental chefs and food developers in recent years. Why might this be so? We believe that there are several reasons for this emerging trend, discussed in the following sections.

7.2.1   The search for novelty

Ever since the seventeenth century, chefs have been searching for novelty (Beaugé 2012). The use of sensory incongruity can be seen as but one response to the search for the next new thing in contemporary cuisine, one which can help to demonstrate the skills of the culinary team who created the dish. Note here that when one asks what is novel in terms of gastronomy, the question can be addressed either from the viewpoint of the chef (or culinary team) who developed the dish or from the point of view of the diner consuming it.

7.2.2   The rise of molecular gastronomy/ modernist cuisine

As we have seen already, molecular gastronomy constitutes a novel and increasingly popular approach to cooking and culinary arts that has taken hold in recent decades. We would argue that molecular gastronomy is somehow implicitly linked to the use of surprise (Mielby and Bom Frøst 2010). Sensory incongruity can then be positioned as one of the principal means of achieving surprise in the restaurant context. The rise of internationally recognized chefs such as Ferran Adrià in Spain and Heston Blumenthal in the UK, who have occupied several of the top spots in many international restaurant rankings for some years, has certainly played its part in helping

---

[1] In his latest book 'Historic Heston Blumenthal', the UK chef has also unearthed some dramatic old recipes that involved surprise (Blumenthal 2013), for example a dish called 'How to roast a chicken and bring it back to life again'. The idea was to mix a live, but sleeping, plucked and basted chicken with a couple of roasted birds served at the Lord of the manor's table. When the carving of the animals begins, said chicken would wake up, make some surprised chicken noises and then run off, undoubtedly surprising the host in the process (Lathigra 2013). Of course, such a dish might not be everyone's taste.

to popularize sensory incongruity in the setting of the restaurant. (Note however that both Adrià and Blumenthal have been known to bristle at what they do being described as molecular gastronomy, as we saw in Chapter 1.) Increasingly, where the modernist chefs lead the cutting-edge molecular mixologists are starting to follow; many cocktail makers are now starting to take on board and utilize the tricks and techniques of the modernist chef (e.g. Uyehara 2011).

7.2.3 The rise of sensory marketing and multisensory design

One increasingly popular approach to product/experience design, partly linked to the growth of sensory marketing (e.g. Lindstrom 2005; Schifferstein and Hekkert 2008; Hultén *et al.* 2009; Krishna 2010, 2013; Hultén 2011; Spence *et al.* in press), has actively stressed the multisensory nature of the consumer's experience of the majority of products and services. According to the multisensory approach, designers should always try to stimulate as many of a consumer's senses as possible. What is more, they should try to ensure that all of the sensory cues in a given product or service go together (that is, that they are *congruent*). The intuitive claim here is that multisensory congruity ought to lead to experiences that the consumer finds preferable (e.g. Spence 2002; Lindstrom 2005; Hekkert 2006). In the wake of this mantra regarding the benefits of congruity, there is however always the temptation to be different, that is, to buck the trend. Thus, while acknowledging the importance of multisensory stimulation (and, by implication, multisensory congruity), chefs may also decide to play with a diner's expectations by introducing sensory (or should that be multisensory) incongruity instead.

7.2.4 Globalization

The growing awareness of sensory incongruity can also be thought of as a natural consequence of the increasing globalization that has taken place in society in recent years. For example, particular combinations of colours and flavours that may be entirely familiar to those living in one part of the world may not always match the combinations of sensory cues that are familiar to those in other parts of the world (cf. Rozin 1983; see also Chapter 6). In our own research here at Oxford University's Crossmodal Research Laboratory, we have found that when people smell cinnamon and are asked to pick the appropriate colour for a food having that smell, Europeans will typically choose brown (after the colour of the spice) whereas North Americans very often pick a dark red (after the hard-boiled sweets that are so popular in that part of the world; e.g. Demattè *et al.* 2006). Consider also the smell of lemons; while this citrus scent makes people in many parts of the world think of a yellow food, those in Colombia (including the first author's wife) immediately associate the smell with the colour green instead (Piqueras-Fiszman *et al.* 2012; Maric and Jacquot 2013). This can be seen as yet another kind of (in this case, unwanted or unintended) cross-cultural sensory incongruity which needs to be managed rather carefully, especially by the chef who is preparing dishes for an increasingly eclectic international clientele.

In this chapter, we discuss various examples of sensory incongruity that have been introduced in some of the world's most *avant-garde* restaurants in recent years. We look at sensory incongruity as it relates to both food and drink, although the focus will be squarely on the incongruity present on the plate.[2]

7.3   Defining sensory incongruity

Sensory incongruity can be defined as occurring when the sensory features present in one modality (e.g. vision) do not match up (that is, agree) with those present (or expected) in another sensory modality, as perceived by the majority of those individuals who are likely to come across a given dish. Here, it is crucial to note that the sensory features present in the food should have been *intentionally* designed to create incongruity in the mind of the diner; that is, it makes little sense to think of a hot meal that, for whatever reason, has accidentally been served cold as providing a meaningful example of incongruity.

Sensory incongruity can potentially occur between *any* combination of sensory modalities, or even between different attributes within the same modality (as in the 'Hot and Iced Tea' from The Fat Duck restaurant). This tea certainly provides an excellent example of the deliberate use of temperature incongruity. Diners are often perplexed as to why the tea, which is experienced at different temperatures on either side of the tongue/mouth, do not mix in the glass giving rise to a liquid that is uniformly tepid. The reason for this seeming physical impossibility is that the tea itself is actually made of a very finely chopped gel; hence the hot and cold parts resist mixing.[3] A similar technique was created by Ferran Adrià back in the early 1990s when he started serving 'Hot and Cold Pea Soup', one of his early signature dishes. Served in a shot glass, the soup was hot on top and cold on the bottom. As to which half was more incongruent, well that is really up to the diner's personal taste!

By far, the most common example of sensory incongruity occurs when the visual attributes of a food or beverage (most commonly its colour) do not match the actual taste/flavour, that is, when the expectations set up by what we see (what is referred to as the 'visual flavour') are not met in the actual taste

---

[2] Note that we have already come across numerous examples of verbal incongruity (that is sometimes found in the names/labels for food and beverage products) in Chapter 3.

[3] To make the dish, a barrier is initially placed in an empty glass, and then the hot gel is poured in one side, the cold gel in the other. Next, the barrier is removed, and the dish promptly presented to the expectant diner.

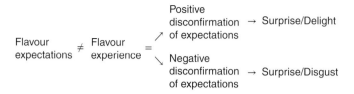

Figure 7.1   When a flavour experience doesn't match our expectations, we can either be pleasantly surprised or else find the surprise disgusting or shocking

of the dish. It is perhaps worth emphasizing the distinction between *flavour expectations* and *flavour experiences*. In particular, surprise and incongruity often occur when there is a noticeable mismatch between the initial flavour expectation and the subsequent experience (see Figure 7.1; Stevenson 2009; Spence *et al.* 2010). Note however that not all of our senses generate flavour expectations, and not all of our senses are capable of contributing directly to our flavour experiences (but this is currently something of a controversial issue). As we saw in Chapter 6, those sensory cues that are capable of eliciting robust flavour expectations prior to a diner tasting a dish consist of vision, orthonasal olfaction (as when we sniff; especially important if we happen to be trying one of the currently popular dine-in-the-dark restaurants described in the following chapter) and, to a lesser extent, sound (as captured by Elmer Wheeler's classic marketing phrase "*Sell the sizzle not the steak*"; Wheeler 1938) and temperature cues (e.g. as when we assess the warmth of a dish whose heat we feel at a distance). Note that the name and sensory description of the dish on the menu can also help to set up flavour expectations in the mind of the diner (see Chapter 3).[4]

7.4   Noticing sensory incongruity

It is important to point out here that the simple act of introducing sensory incongruity into a dish does not itself guarantee that a diner will notice it. The reason for this is that our brains normally try to integrate the inputs from the different senses that appear to belong to the same object/event (see Spence 2007; Auvray and Spence 2008). Hence, if the sensory incongruity between what a person sees and what he (or she) tastes is not too large, the diner's brain may well simply taste what it sees. That is, visual flavour (or rather visual flavour expectations) can all too easily end up dominating over the gustatorily and olfactorily determined flavour experiences that follow (see Chapter 6). The first scientific study to report that changing a food's colour could result in *sensory dominance* (and hence result in perceived congruity),

---
[4] By contrast, flavour *experience* consists of gustatory, retronasal olfactory and trigeminal cues, possibly as well as the self-generated sounds that are associated with food mastication (Stevenson 2009).

rather than the surprise of sensory incongruity, was published in 1936 by Moir (a chemist by training). When he gave his colleagues a range of coloured jellies to try, he noted that they would often taste the colours that they saw (that is, they tasted the flavours that they would normally associate with those colours) and not the flavour that was actually present in the food. (Note that they might have correctly identified the flavours had their eyes been closed; see the following chapter).

A few years later, the famous psychologist Karl Duncker (1939) conducted an intriguing study using white (rather than brown) chocolate, a novel product that had just been introduced into the marketplace in the US at the time. When his participants tasted the white chocolate (that they were allowed to see, and which was probably the first time that they had come across it) they reported that it was much milkier or else flavourless as compared to previous samples of common milk (brown) chocolate or a sample of white chocolate that they had tasted while blindfolded. Since these pioneering early studies, several hundred further experiments have been conducted on the topic and published in the academic journals (see Spence *et al.* 2010 for a review). Many of these studies have demonstrated the impact that food colouring has on people's flavour experiences (cf. Chapter 6). The key point to note here is simply that the existence of sensory dominance (that is, the brain's tendency to integrate sensory cues that it 'believes' belong together, and to use the most accurate sense in order to infer what is going on in the other senses; Spence 2013b) means that it is not as easy to introduce sensory incongruity at the dinner table as one might at first have thought.

What the inquisitive chef might be wondering at this point is: can the use of sensory incongruity as a means of surprising one's diners go wrong? The answer is that it most certainly can. For example, in order to count as incongruent, a chef needs to be confident that their diners will actually perceive the mismatching sensory impressions in the way that was intended. Ascertaining whether a diner will notice the discrepancy or not (i.e. whether they will register a certain disconfirmation of expectation) is not always easy to predict *a priori*. Hence, in order to introduce sensory incongruity and make it work as a surprising and attention-capturing device in the setting of the restaurant, one needs: (1) the technical means to deliver mismatching, yet nevertheless still harmonious, sensations to different senses; (2) an understanding of when exactly sensory incongruity is likely to be obscured by the neural processes that give rise to sensory dominance; and (3) plenty of time to devote to experimenting in the kitchen! It is only when the sensory incongruity between flavour expectation and flavour experience is sufficiently large that noticeable incongruity will be elicited and, hopefully, the diner pleasantly surprised. In this regard, collaboration between the chef and the psychologist can be useful, since the latter may well know some tricks that can help the former to ensure that the desired surprise on the part of the diner is, in fact, the end result.

7.4.1 Disconfirmed expectations

As just mentioned, sensory incongruity has to be perceived first. It should come as little surprise that as the magnitude of the sensory incongruity increases, so too does the likelihood that the diner will notice it. However, once the incongruity becomes noticeable, a *disconfirmation of expectation* response is likely to follow (e.g. Schifferstein 2001; Yeomans *et al.* 2008; Ludden *et al.* 2009). Back in 1963, Carlsmith and Aronson argued that such a disconfirmation of expectation usually leads to a negative hedonic appraisal of whatever it is that a person has been confronted with (see also Loken and Ward 1990). Cardello *et al.* (1985) supported these claims some years later, stating that giving consumers correct information was likely to increase their product acceptance. The latter study refers mainly to novel food products, however. Here, we would like to argue that there has been a dramatic shift in the mindset of many diners since these original studies on disconfirmed expectation were published. To illustrate the point, consider the following advice handed out by the Committee on Food Habits in the USA back in the 1940s, concerning what would make for an acceptable food: "*In its most basic form, an acceptable food must taste good; must be available; must be familiar; and must look, taste, and feel as expected*" (Wansink 2002, p. 92). Reading this advice now certainly hints at how peoples' attitudes (especially those of adventurous diners) to disconfirmed expectation have changed over the years.

The danger with cases such as those reviewed in this chapter (and where the sensory incongruity is *intentional* rather than accidental) is that consumers simply consider that the design/composition of the dish is bad. That is, they might come away thinking that the dish has been poorly thought out and/or executed and that the various elements do not go together, having no idea that the sensory incongruity was actually intentional on the part of the chef or the culinary team who developed it. However, as the degree of incongruity is increased still further, at some point there is a realization that it can *only* have been designed intentionally (in order to capture the attention of the diner, to surprise them and, hopefully, to delight them). At this point, the feeling of 'wrongness' decreases or, at least, is reappraised in a much more positive light. Of course, the context of dining in a modernist restaurant presumably helps here.

On the other hand, for more familiar foods a confirmation of expectation may not affect the consumer's hedonic assessment; instead, it may simply result in boredom (Schifferstein 2001). Introducing a level of sensory incongruity that is perceptible usually gives rise to culinary experiences that are more exciting overall. Nevertheless, it is fair to say that sensory incongruity has to be carefully managed in the restaurant setting (see Schifferstein 2001; Yeomans *et al.* 2008). In the case of familiar foods (such as a strawberry ice-cream), for example, increasing the degree of sensory incongruity (e.g. by colouring it green or blue) might be expected to have a negative effect in terms of people's sensory-discriminative and hedonic evaluation of it (see Yeomans *et al.* 2008). It therefore seems that you can always make familiar foods more exciting by adding a touch of incongruity, but don't mess around too much … better leave that for more novel foods!

7.4.2   Hidden and visible incongruity

With regard to the extent to which sensory incongruity is hinted at (and hence expected) or not, it can be argued that this can fall into either one of two categories: *visible* and *hidden* incongruity (Schifferstein and Spence 2008; see also Ludden and Schifferstein 2007). One might consider the raspberry flavour popular in certain blue fruit drinks currently sold in stores (think Cool Blue Gatorade or Slush Puppie) as providing an example of visible incongruity, since the name of the flavour is normally clearly visible on the packaging. Furthermore, the colour blue in a beverage is so unusual (unnatural even) that when most consumers first come across such a product they do not know what to expect.

Another example of what was meant to be visible incongruity was Clear Tab cola, since the label clearly stated that the drink was a 'cola'. However, it could easily become an example of hidden novelty. Imagine yourself picking up a glass of a clear fizzy liquid at a party; what would you expect it to taste like? Lemonade or sparkling water, presumably. It would seem likely that it was the negative disconfirmation of expectation reaction that consumers had under such conditions that led to the rapid demise of this particular product (no such problem for a blue drink though). Clear Tab cola was unceremoniously withdrawn from the stores within a year of its launch (Graves 2010).[5]

By contrast, a classic successful example of hidden sensory incongruity is the 'Beetroot and Orange Jelly' once served at The Fat Duck restaurant in Bray (see Figure 7.2). It can be argued that part of the enduring success of this dish related to the fact that the sensory incongruity was hidden from the diner (what is more, the colours had been achieved naturally, a point to which we will return in Section 7.12). Meanwhile, in the case of the 'Hot and Iced Tea' dish (described earlier in this chapter), despite the fact that the sensory incongruity is hinted at by the dish's name, it could also be thought of as an example of hidden incongruity. It is difficult to imagine the incongruent feeling of drinking tea that is hot on one side of your mouth but cool on the other, simply by looking at what appears to be a homogeneous liquid in the glass.

---

[5] One could perhaps also argue that another case of visible incongruity comes from the traditionally popular tendency to make copies of expensive dishes from cheaper ingredients as, for example, in the case of mock turtle soup, a dish that should be familiar to anyone who has read Alice in Wonderland (see Wilson 2009).

Figure 7.2 The Beetroot and Orange Jelly dish served at The Fat Duck restaurant in Bray, UK. *Source*: Reproduced with permission of Lotus PR and The Fat Duck. *See colour plate section*

In this case, the diner is confronted with an oral experience that seems to go against whatever they know or have experienced previously concerning the physics of liquids.

It is now worth pausing to consider the fact that if a diner were anywhere except a Michelin-starred temple to scientific cuisine (Sifton 2011), they might never realize the sensory incongruity that is present in the 'Beetroot and Orange Jelly' dish. However, since the cognizant diners know precisely where they are sitting, it is to be presumed that they arrive at the modernist chef's restaurant *expecting* to be amazed/surprised (see also Chapter 3). Consequently, it is the very normality of the appearance of the jelly that is itself what at first seems so surprising. This apparently straightforward dish somehow feels out of place; it is almost too simple. It is only when the inquisitive diner keeps on probing as to just what could be going on (looking for the scientific cuisine that surely must be hidden somewhere in the dish), or when the waiter suggests to the diners that they should try closing their eyes and then taste from the two sides of the plate again, that the hidden incongruity makes itself apparent. Time and again in this chapter we will come across various examples that illustrate the concept of sensory incongruity (of both the hidden and visible kind) being served at the dining tables of the modernist chefs. But how did we get to the stage where sensory incongruity has become, in some sense, the norm? Surely, it was not always thus.

7.5   A brief history of sensory incongruity at the dinner table

Historically, at least, sensory incongruity in the domain of food always seems to have been portrayed as undesirable (Wilson 2009). For instance, Alfred Hitchcock was on occasion known to deliberately try and wrong-foot his dinner guests (including Sir Gerald du Maurier) in a private dining room at the Trocadero restaurant in London by serving them food that had been artificially coloured blue.

> *"And all the food I had made up was blue! Even when you broke your roll. It looked like a brown roll but when you broke it open it was blue. Blue soup, thick blue soup. Blue trout. Blue chicken. Blue ice cream."* (Hitchcock and Gottlieb 2003, p. 76)

Another infamous dinner was served by a marketer to a group of his friends (Wheatley 1973). The guests in this 'study' were invited to dine on a meal of what looked like steak, chips and peas (it was the 1970s after all). The only thing that may have struck any of these dinner guests as unusual was how dim the lighting was. However, this aspect of the atmosphere was designed to hide the food's true colour. When the lighting was turned back up to normal part of the way through the meal, imagine the guests' horror when they saw that what they were actually eating was a blue steak, green chips and red peas. A number of the guests suddenly started to feel decidedly ill. Wheatley comments that several of them headed straight for the bathroom. In an even earlier study, Moir (1936) also played around with incongruently coloured foods and observed what his work colleagues made of them:

> *"Moir prepared a buffet of foods for a dinner with scientific colleagues of the Flavor Group of the Society of Chemistry and Industry in London. Many of the foods were inappropriately colored, and during the dinner several individuals complained about the off-flavor of many of the foods served. Several of the individuals reported feeling ill after eating some of the foods, despite the fact that only the color was varied. The rest of the food was perfectly wholesome, with the requisite taste, smell and texture."* (reported in Moskowitz 1978, p. 163)

Of course, an argument could be made here that blue colour in a red meat is unlikely to work in even the best chef's hands, given the very strong associations that this colour has with meat that has 'gone off' (not forgetting the brouhaha that surrounded blue Smarties, and the oft-made claim that there are virtually no foods in nature that are blue). Indeed, one of the reasons why the chef (or home cook) has to be especially careful when using sensory incongruity in the design of food and beverage products relates to the fact that we, as humans, are very sensitive to the potential risk of poisoning (Koza *et al.* 2005). First and foremost, the consumer/diner needs to be reassured that what they are eating is 'safe'. Only then can they relax and start to enjoy the surprise of the multisensorially incongruent dining experience that is before them. That is, it would appear as though sensory incongruity has been used traditionally as a means by which to *shock/challenge* (this can easily misfire; remember the blue steak)[6] rather than to *delight/please* (obviously more likely to succeed) diners. Certainly in the above-mentioned cases, the response to sensory incongruity was unequivocally negative. What seems common to the above examples (Moir 1936; Wheatley 1973; Hitchcock and Gottlieb 2003) is the unpleasantness/unease that was associated with the introduction of this sensory incongruity. This, or so we would wish to argue here, is something that has changed in recent years. Whether it is people's attitudes that have changed, or merely the environment in which such disconfirmation of expectation normally takes place, isn't clear from what we have seen so far.

In the last few years, the ability to change the colour and other visual attributes of a food has been made easier by the development of a variety of augmented and virtual reality technologies (see Chapter 10 for some ideas concerning the future of technology at the dinner table). Some of our Japanese colleagues have already been experimenting with blue sushi, created with the aid of virtual reality technology (e.g. Sakai 2011). Remember the inimitable Fanny Cradock; she was perhaps the first of the TV celebrity chefs in the UK (see Ellis 2007). It was Fanny who made the artificial colouring of food (in unexpected colours) something that was perceived as sophisticated/pleasant (at least given the tastes of the time). Fanny was particularly famous for serving her mashed potatoes in a variety of vivid colours. Nevertheless, this can perhaps be seen as one of the first times in popular (culinary) culture when unusual/incongruent food colouring was used with the intent of delighting/surprising diners, rather than simply just to shock those who were being fed or as a means of masking the true colour of the food (Wilson 2009). The conclusion here would seem to be that sensory incongruity, what once upon a time may have been rather unpalatable, is nowadays seen as a much more positively valenced experience/sensation. It is savoured in its own right because, at least in the setting of the modernist restaurant, its use often highlights the skills of the chef or the cooking/research team who has managed to work successfully with this challenging medium (Lauden 2001).

That said, the modern-day use of sensory incongruity in a gastronomic setting could only really have come about as a result of the development of techniques to extract flavours from natural products. This aspect of the whole modernist approach to cuisine would never have taken off had it relied on synthetic/artificial flavours and colourings that were so common

---

[6] The audience's reaction wasn't too different when, together with the Spanish chef Maria Jose San Roman (http://www.monastrell.com/restaurantei.php), we served blue pizza to the members of the audience at Spain's largest gastronomy event a few years ago!

previously and used by the likes of Moir (1936) and Alfred Hitchcock, not to mention the majority of laboratory studies of the visual contribution to flavour perception.[7]

7.6   Colour–flavour incongruity

As noted already, introducing a mismatch between the colour and flavour of one element in a dish constitutes perhaps the most common kind of sensory incongruity. One can see both successful and unsuccessful examples of this in the literature, in the marketplace and in the setting of the modernist restaurant. For example, a lot has been written about the failure of Clear Tab cola (e.g. Triplett 1994). Other examples, such as blue raspberry drinks (e.g. Cool Blue Gatorade) and Heinz green/blue/purple ketchup have however proved themselves to be surprisingly successful marketing propositions (e.g. Farrell 2000; see also Wilson 2009 on the early failure of pink margarine in the late 1800s).

It could perhaps be argued that colour–flavour incongruity may be more likely to work well if the associations are artificial (or arbitrary) in new products (or dishes) or when there are no already established colour–flavour associations. For instance, the crossmodal association between green packaging and peppermint-flavoured products, likewise between a brownish-red colour, carbonation and cola-flavoured drinks/products, are really little more than extremely well-entrenched conventions (Garber *et al.* 2008; Spence 2011; Piqueras-Fiszman *et al.* 2012; Spence and Piqueras-Fiszman 2013). For many years, cola has been firmly associated with a specific *colour* as much as with a particular flavour. That is, cola-flavoured products have in some sense now become indelibly linked with a specific hue (a visual flavour if you will) in the mind of the consumer. (By contrast, no such clear flavour expectation holds in the case of the colour of a mixed fruit smoothie, for example.) Likewise, when separated from its packaging, a clear carbonated beverage will likely set up flavour expectations associated with lemonade, Sprite or carbonated water (cola flavour would likely be the last thing on anyone's mind).

The concept of a 'Cool Blue raspberry' fruit drink may have worked so well precisely because the consumer had no specific expectations associated with that particular colour, at least not when the drink was situated in the soft drinks aisle. Shankar *et al.* (2010) demonstrated that a blue drink gives rise to different flavour associations as a function of the country in which it is presented (i.e. among different groups of consumers; see also Chapter 6). Once

---

[7] The widespread (i.e. commercialized) opportunity to elicit sensory incongruity can largely be traced back to the industrial development of synthetic colorants and artificial flavourings. That said, people have been colouring their food for thousands of years (e.g. red wine; see Tannahill 1973; Wilson 2009).

again, then, we return to the cultural differences in crossmodal associations and in crossmodal perception that are present in this area. In fact, it could be argued that incongruent colouring has a much longer history in the beverage sector than anywhere elsewhere. One need only think of Blue Curaçao, the exotic blue mixer for cocktails that imparts an orange flavour to a cocktail (while the liqueur itself is naturally colourless, it is often given an artificial blue or orange colour).

In the field of molecular gastronomy (or 'scientific cooking'), the distillation of aromas has in recent years enabled chefs to apply the aromas of certain ingredients to uncoloured foods in order to play extensively with the uncertainty factor (e.g. Martin 2007; this can be contrasted with the more traditional use of artificial flavourings described by Rosenbaum 1979; Classen *et al.* 2005; Wilson 2009).

7.7   Format–flavour incongruity

In many cases, we associate certain formats in foods with specific tastes/flavours. For instance, jellies are commonly associated with sweetness. But just how pleasant do you imagine it would be to taste olive-oil-flavoured jelly (see Figure 7.3)? Similarly, *savoury* ice-creams are nowadays considered by most people (at least in the UK) to constitute a conceptually incongruent format–flavour combination.[8] Why? Well, most of us intuitively believe that certain savoury ingredients feature only in main courses and are usually served warm (or at the very least, cooked). Seeing (not to mention tasting) these ingredients in the ice-cream format is therefore, at the very least, striking. That said, it should be noted that savoury ices were actually a reasonably common feature of fashionable meals a century ago in the UK (e.g. Marshall 1888; Colquhoun 2007). In other words, there is often an important historical (not to mention cultural) component to what counts as sensory incongruity at the dinner table.

When sensory incongruity is unexpected (e.g. when the incongruity is hidden from the diner), it is likely to be perceived as even more incongruent, hence giving rise to a more pronounced negative *disconfirmation of expectation*. In this case however, even when the fact that the dish has a savoury flavour is known in advance (that is, prior to tasting), some level of incongruity will still remain due to the diner's more general unfamiliarity with products of this kind. For example, Sagartoki's Porcini Mushroom Lasagna provides an example of *visible* incongruity since the diner will presumably see from the menu that the ice-cream that accompanies the lasagna is mushroom flavoured (and will remember this, since it's one of the star *tapas* served there). By

---

[8] Although savoury iced concoctions are far more common in some other countries; the ice kerchang found in street stalls all over the Malaysian peninsula (shaved ice, fruit concentrate, firm jelly, kidney beans and sweetcorn) is yummy!

Figure 7.3   Olive oil jelly candy. If you'd expect this green jelly to taste like lime, you're in for a surprise! *See colour plate section*

contrast, the smoked-salmon-flavoured ice-cream that was given to the participants in a psychology experiment (without telling most of them its flavour in advance) can be seen as providing an informative example of *hidden* incongruity (see Yeomans *et al*. 2008). As already described in Chapter 3, the bright red colour together with the format (and the misleading label of the ice-cream) likely set up, in the minds of the participants, a visual flavour expectation of a red-berry-flavoured (e.g. strawberry) ice. This belief was totally disconfirmed when the participants subsequently tasted the salmon flavour (cf. Chapter 3).

Another familiar example of colour incongruity found in the marketplace is a light-green ice-cream. While most Westerners would think mint or perhaps pistachio, green-tea-flavoured ice-cream is very popular in Japan; this is therefore a natural cross-culturally incongruent example of colour–flavour incongruity. The truth is that more and more importance is nowadays being given to a new 'cold cuisine', particularly in the creation of the frozen savoury world. Indeed, the concept is already starting to spread to parlours, kiosks and beach huts across Europe (Schlack 2011). For example, the Humphrey Slocombe company now delivers savoury ice-creams to the top restaurants in the San Francisco area (http://www.humphryslocombe.com). Away from the culinary context, another example of this type of format incongruity that we have all come across at some point is related to the flavour of strawberry syrup. Why is it that when a certain candy tastes like strawberry syrup we say "*Yuck, it tastes like children's medicine!*" but when the medicine has the same taste, we kind of like it? If the flavour is the same, and actually the medicine has that flavour for that specific reason, then why the negative connotation of that flavour when it comes to real candies?

Denis Martin, the Swiss experimental gastronomist working out of Vevey, Switzerland (http://www.denismartin.ch/) has developed some intriguing examples of incongruent colour/format–flavour combinations that he labels '*Contraires*' (see Martin 2007). In his '*pan con tomate*' what appears visually to be bread is actually tomato (see Figure 7.4). Martin plays with misleading appearances, ensuring that his diners have different (and hopefully more memorable) experiences as a result of the surprising flavours that they first anticipate with their eyes (based here as much on their form or shape as on their colour) and what they experience in their mouth. Both shape and colour cues exert a significant effect on our ability to identify flavours orthonasally (Demattè *et al.* 2009). Other chefs such as Homaro Cantu also love to play

Figure 7.4 '*Pan con tomate*' (developed by Denis Martin). *Source*: Reproduced with permission of Denis Martin. *See colour plate section*

these tricks on their diners, although unfortunately we don't have enough space here to describe them all.

Even the order of presentation of foods (or dishes) within a meal can determine what is considered (or perceived) as being sensorially incongruent. Chefs are increasingly playing with the order of presentation of flavours in their menus, for example by including sweet tastes in their starters and savoury tastes/flavours in their desserts (Nolan 2013). In this way, the classical structure of the meal is starting to break down; an unquestionable revolution is currently taking place in first courses and desserts, closely connected to the concept of this symbiosis between the sweet and savoury worlds. We invite the reader to guess whether the flavours of the dishes presented in Figure 7.5 are savoury or sweet and whether they are appetisers or desserts (without reading the caption first)!

7.8   Smell–flavour incongruity

One commercial example of unintended sensory incongruity is that of some herbal/fruit teas which often deliver an intense orthonasal aroma. However, the flavour *expectation* that is elicited by orthonasal olfactory cues and, to a lesser extent, by visual cues leads to the prediction of a stronger flavour that is simply not delivered. Consumers often report being disappointed with the actual flavour of the hot beverage, since the in-mouth experience does not meet their prior (orthonasal) olfactory expectations. This is a case of incongruity (a taste or flavour that is *less* intense than expected) which consumers would likely never want to pay for! For some people, an example of a natural product that falls into this smell–flavour incongruity category is ginger. The characteristic spicy, sweet aroma of this root originates from a heady mixture of volatile oils. Its pungent taste comes from non-volatile phenylpropanoid-derived compounds, particularly the gingerols and shogaols. The latter compounds are derived from the former when ginger is either dried

(a)   (b)   (c)

Figure 7.5 Appetisers served at elBulli: (a) Parmigiano marshmallows; (b) spring rolls made out of lightly sugared cotton; and (c) black olive 'Oreos'. *Source*: Reproduced with permission of Blake Jones. *See colour plate section*

or cooked (see McGee 2004, pp. 425–426). Many people love the sweet smell of freshly baked ginger cookies, but not their pungent spicy taste. See Rozin (1983) for other examples of foods that smell great but which can taste disappointing (e.g. freshly ground coffee) and vice versa (e.g. Limberger cheese).

What is interesting here is that while everyone is aware at some level that certain odours smell very differently retronasally from how they present orthonasally, as yet no one knows precisely why this should be so (but see Ge 2010). If we did, then perhaps more novel foods could be designed to smell of one thing orthonasally (i.e. when we sniff the food) and to taste of something else (i.e. when the volatile compounds hit the back of the nose when we swallow; known as retronasal smell). Just imagine a diner's surprise if dramatic examples of smell–flavour incongruity were to be delivered at the dining table, such as a liquid that smelled (orthonasally) of strawberries until you swallowed a mouthful, whereupon it suddenly, magically, transformed into the retronasal flavour of pineapple. One might even feel as though one was in Willy Wonka's Chocolate factory, having one of those everlasting gobstoppers which flavour changes constantly (Dahl 1964).

> *"'Everlasting Gobstoppers!' cried Mr Wonka proudly. 'They're completely new! I am inventing them for children who are given very little pocket money. You can put an Everlasting Gobstopper in your mouth and you can suck it and suck it and suck it and suck it and it will never get any smaller!'*
>
> *'It's like gum!' cried Violet Beauregarde.*
>
> *'It is not like gum,' Mr Wonka said. 'Gum is for chewing, and if you tried chewing one of these Gobstoppers here you'd break your teeth off! And they never get any smaller! They never disappear! NEVER! At least I don't think they do. There's one of them being tested this very moment in the Testing Room next door. An Oompa-Loompa is sucking it. He's been sucking it for very nearly a year now without stopping, and it's still just as good as ever!'"* (Dahl 1964, p. 116)

7.9 Interim summary

Thus far in this chapter we have seen how chefs are increasingly seeking to deliver, and diners are increasingly coming to expect, surprise in the context of a meal at a modernist restaurant. One of the most popular means of delivering on that promise of surprise is through the use of sensory incongruity. We have explained how that is typically delivered by foods that taste quite different from how their appearance would imply. We have also highlighted some of the problems associated both with ensuring that diners are aware of the incongruity (due to the phenomenon of sensory dominance) and how the experience can be managed to increase the likelihood that it will be perceived by the diner as a pleasant (rather than unpleasant) surprise. We now look a little more closely at the diner's response to such dishes, and what role individual differences may play in mediating this response.

7.10   The diner's response to sensory incongruity

Having taken the decision to introduce sensory incongruity into the design of one's food, one might ask what the likely reactions of the diner are going to be (Cardello 2003). Likely reactions include attentional capture, surprise and memorability, all described in the following sections.

7.10.1   Attentional capture

As an example of the unusual, sensorially incongruent dishes (at least when the incongruity is visible) are more likely to catch a diner's eye. It is to be expected that a diner will unconsciously fix their attention on the dish in order to examine its 'wrongness' and localize the source of their surprise. Sensory incongruity therefore forces the diner into a more elaborate, or attentive, form of information processing (Garber *et al.* 2008).

7.10.2   Surprise

Ludden *et al.* (2009) have put forward a model of surprise that comprises a series of processes and actions. When people are surprised by something, they usually interrupt any other ongoing activities that they may happen to be engaged in in order to focus on, explore and gain more information about their unexpected experience and its causes (and meaning). These reactions can be accompanied by facial and sometimes also verbal expressions. Surprise is certainly what one is likely to feel on first tasting Ricard Camarena's *bizcocho aireado de avellana, gianduja, limón y chocolate* (which consists of an airy sponge of hazelnut, gianduja, lemon and chocolate). Given the name of the dish, the diner expects a sponge cake. The dish's ethereal appearance undoubtedly supports such an assumption. But once the diner puts a spoonful of this food into his/her mouth, it melts like only the very best ice-cream can. "Mmm … what is this?" is the first thing that comes to mind. The diner then tries another spoonful and the wonder still doesn't go away. Even when the dish is finished, the diner may still be somewhat uncertain about how to define it (see www.arrop.com).

7.10.3   Memorability

All of the physiological and behavioural responses that result from a surprising sensory experience help to increase its memorability in the mind of the diner. The fact that consumers/diners may halt their conversation in order to fix their attention on the sensory incongruity that is present on the plate is likely to result in their better remembering both the food itself and their overall dining experience (and they are also likely to remember it for longer; see Köster 2006). To give but one personal example, the first author still remembers the occasion on which he first felt the envelope in which the menu was presented at Heston Blumenthal's The Fat Duck restaurant (even though it happened a decade or so prior to the writing of this book). The envelope's visual texture gave the impression of normal vellum, but to the touch it felt remarkably like human skin! How often does one remember the menu from a restaurant, one might ask? Sensory incongruity was used to create a long-lasting memorable impression, yet another example of sensory marketing in the dining room.

7.11 Molecular gastronomy and surprise

The majority of examples that have been discussed so far in this chapter have come from the sphere of molecular gastronomy (or modernist cuisine). The creative cuisine, or *cuisine d'auteur*, unites cooks, food scientists, designers and psychologists in order to share their expertise and innovate at a more rapid pace than ever before. Previously unimaginable textures and unexpected flavours and aromas are now being combined in order to create unique multisensory culinary experiences that, in some sense, *provoke* the diner. One certainly does not go to one of these restaurants merely in order to eat (in the sense of filling oneself up). Rather, one goes there to enjoy the theatre and spectacle (cf. Pine and Gilmore 1998, 1999; Hanefoors and Mossberg 2003; see also Chapter 11), the multisensory experience (hopefully) in every single morsel of food (see Figure 7.6 for an example of sensory incongruity operating at many different levels).

   A fundamental question to ask here is *why* it is that so many chefs want to use incongruity (sensory or otherwise) in the first place. There are, or so we would like to argue here, several answers to this question. It may be partly because the consequent decontextualization, irony, spectacle, innovation and novelty have all become valued attributes within certain styles of cooking, especially (or perhaps only) in the area of modernist cuisine. Of course, the problem with surprise is that it is hard to pull it off more than once with a given dish (unless, that is, diners happen to have a very poor memory). In other words, surprise in a particular dish is only ever going to be a one-off thing. While there can be some enjoyment in taking a friend or partner to see their response on experiencing a new dish, the bottom line here is that the experimental gastronomist always needs to be on the look-out for new ways in which to deliver something new to their regular diners (Svejenova *et al.* 2007), should they be lucky enough to have any. That is, it requires a constant striving for the next 'new thing', a non-stop process of inspiration, innovation and

Figure 7.6 Tartufo al cerdo Ibérico y aceituna verde (Truffles with Iberian pork aroma, and green olive). Photography taken by Francesc Guillamet. *Source*: Reproduced with permission of Francesc Guillamet. *See colour plate section*

research. Little wonder then that Ferran Adrià, one of the world's most inventive chefs, closed his elBulli restaurant after two (most probably) exhausting decades and is currently investing his time in research. Jean François Revel (1985) captured the quandary beautifully when he wrote: "*Great cuisine is by definition open cuisine, the opposite of cooking blocked by a regional attitude. The first is condemned to invent, to search for the new, the second is condemned to conserve, for better and for worse, what has been put out over the centuries.*"

Does this mean that in the longer term we will see more innovation in the field of molecular gastronomy than in other areas of cooking? The developments in food technology and agrochemistry that have taken place in recent decades certainly mean that cooks have increasingly been able to dissociate tastes/flavours from their 'natural' (traditional) colours, oral-somatosensory textures and contexts in ways that people (e.g. the Italian Futurists; see Berghaus 2001) could only have dreamed of previously. Techniques such as distillation, spherification and thermogelification, together with the introduction of novel ingredients such as xantana, are now allowing a growing number of chefs (not to mention those at home, given the recent release of the molecular gastronomy kit designed for home use; http://www.molecule-r.com/; Youssef 2013) to play with a very wide range of contrasting textures, temperatures, flavours and aromas, applied in combination or even in separate layers

(Adrià *et al.* 2007; Martin 2007; Blumenthal 2008). Far from the caricature of foams, smoke, warm gelatines, meat that is not meat, caviar that isn't caviar, savoury sweets, sweet savouries, desserts that go first and starters that come last (Berghaus 2001; Nolan 2013), this cuisine makes every effort to create novel eating experiences and stimulate all of their diners' senses.

One can however ask whether sensory incongruity is really something that is specifically tied to this style of gastronomy. The reality is that this gastronomic movement could certainly be regarded as a declaration of intent (see Alícia and elBullitaller 2006). The chefs who subscribe to this approach to cooking and food preparation have undoubtedly pioneered the widespread introduction of intentional sensory incongruity in the culinary setting (at least within a certain kind of dining context). Of course, the relevance of such a *manifesto* is reflected in the fact that at least some of what happens behind the kitchen doors of these restaurants can be expected to percolate down to the high street in a few years' time (see Chapter 11 for further discussion on this theme).

The key question here though is whether sensory incongruity is really something that can succeed only within the confines of a Michelin-starred restaurant environment. There is certainly a danger that it is not something that one can easily make work in the home (and hence supermarket) setting.[9] It could be argued that these chefs do actually have the convenient setting, tools, expertise and imagination to play around, explore, surprise and, what is more, challenge the senses of their diners. But what happens when that food (think ice-cream) is served in a shopper's home to a dinner party of expectant friends?[10] In addition, the cookbooks that these chefs launch for those kitchen enthusiasts aren't always exactly the most user-friendly. The recipes typically require a range of expensive kitchen equipment to execute. Hence, at one level it seems as though such recipes are designed to keep molecular gastronomy for the culinary elite (and for visual entertainment for enchanted TV audiences worldwide; but see Youssef 2013 for a welcome exception). This discussion then raises the more general question of when exactly sensory incongruity is acceptable (or rather, likely to succeed)? The atmosphere in which the food is served probably plays a huge role here (e.g. King *et al.* 2004, 2007). Indeed, there is lots of evidence to demonstrate that our evaluation of food and beverage items is critically dependent on the environment in which they are served (see Chapter 9).

In addition, an incongruent sensory experience will undoubtedly be affected by its repetition; comfort food it most certainly is not! Having the formerly surprising molecular gastronomy dish a second time around is always going to be less exciting, diminished somehow. How should the chef/restaurateur deal

---

[9] As A. A. Gill once put it in his Sunday Times column: "*You don't want to find Blumenthal spin-offs in every best-kept village, you don't want sardines on toast sorbet in the freezer section at Tesco*" (Gill 2005).

[10] No home chef, after all, wants to end up giving their guests an experience like Wheatley's (1973) when he presented his dinner guests with the blue steak.

with this? If the underlying flavour/complexity of the dish is worth savouring in its own right, then the loss of surprise might not be expected to detract too much from the overall experience. However, no matter how good the dish is, the diner is always going to feel a sense of loss that they are missing out on something of the spectacle, and the dish will struggle to live up to their memory of the first occasion on which they came across it. One might also consider whether reading about restaurants, and watching endless cookery programs on TV that show the surprise of other diners eating at these restaurants, takes away at least part of the pleasure or surprise when the diner finally gets the chance to try it for him/herself. Or perhaps it is precisely because it is a multisensory experience that the sensorially incongruent dish should not be affected too much by being overexposed on people's TV sets at home. Sensory incongruity is, after all, something that is experienced differently by different individuals; hence there is perhaps a sense in which it is something that one has to experience and assess for oneself.

7.12 Sensory incongruity and the concept of 'naturalness'

As has already been mentioned in Section 7.5, sensory incongruity has traditionally been perceived as unnatural. In the area of food, the research suggests that people nowadays generally tend to prefer foodstuffs that are (or at least are perceived as) being 'natural' (Rozin *et al.* 2004).[11] However, what exactly counts as congruent or, for that matter, 'natural' when it comes to food turns out to be an intriguing question. Take once again the 'Beetroot and Orange Jelly' served at The Fat Duck restaurant as an example (see Figure 7.2). When this dish is brought to the table, the waiter explains: "*This is the beetroot and orange jelly. We recommend that you start with the beetroot.*" Most people who are paying attention will start by sampling the purple-coloured jelly, shown on the left of the plate. However, the secret is that golden beetroots, which are orange, are used, while blood-red oranges are a deep vivid purple to begin with. In this dish, the culinary team has managed to play with the diners' colour–flavour mappings but do so in a manner that is entirely natural.

Similarly, oyster leaves (also known as vegetal oysters) are completely natural leaves that just so happen to taste like the self-same mollusc. When such leaves are presented in the context of so much sensory incongruity and scientific cooking in a molecular gastronomy restaurant, the natural assumption (if you will excuse the pun) is that this flavour in this format must somehow be the result of scientific preparation involving some sort of 'smoke and mirrors'. But this categorically is not the case and, in fact, they are usually presented alone in their natural form (sometimes even a single leaf on a plate;

---

[11] This debate is also linked to the organic and slow food movements (Petrini 2001, see also Chapter 3).

Figure 7.7   An oyster leaf as served at Alinea. A bit incongruent? Yes, but incongruency of the most natural sort. *See colour plate section*

see Figure 7.7) with just a touch of mignonette (vinegar and shallots) as, for example, served by Grant Achatz in his Alinea restaurant in Chicago.[12] These selected examples will hopefully be sufficient to highlight the fact that the multisensory incongruity that is increasingly being used in gastronomic settings need not be achieved by means of unnatural techniques or ingredients (see also Wilson 2009).[13]

Recently, supermarkets in Britain (and presumably in many other parts of the world) have introduced the pine berry. Such peculiarly coloured fruit initially struck many consumers as rather 'unnatural'. Strawberries should be red, right? It turns out that these 'white strawberries' are simply the result of cross-breeding the South American strawberry *Fragaria chiloensis*, which grows wild in some parts of Chile, and the North American strawberry

---

[12] For anyone wanting to try oyster leaves at home, they can be purchased from Koppert Cress BV (http://greatbritain.koppertcress.com/).

[13] The home chef might want to consider whether they really want to master a recipe that creates surprise in their dinner guests' minds, given that they will not be able to serve it to them a second time around.

Figure 7.8   These natural pine berries (hint: look at the right side of the image) are often served in restaurants as a means to amaze the diner, who very likely does not come across such fruits very often. For example, this fruit is found in the dessert Sotobosque served at 41°, the bar founded by Adrià in Barcelona. *Source*: Reproduced with permission of Miguel Angel Castillo & Rocío García Martin, http://la-cocina-creativa.blogspot.com.es. *See colour plate section*

Fragaria virginiana (see Figure 7.8). The key point here is that sensory incongruity need not be unnatural. An increasing number of chefs are now starting to capitalize on the most unusual appearance of this and other fruits in order to surprise their diners. Another similar case, but this time more related to shape incongruity, is that of the finger lemon (or Buddha's hand, a citrus yellow fruit segmented into finger-like sections; look it up on the web). This fragrant fruit (although it may not look like it) can be found growing in parts of China and Japan.

One can see the introduction of foods such as the pine berry and the finger lemon, the purple potato (the Purple Majesty variety; Poulter 2011), not to mention the red banana and the yellow raspberry (the latter products were launched by Marks and Spencer, the golden yellow colour is a completely natural mutation; see Anonymous 2008) as yet more examples of the rise of sensory incongruity in the marketplace. Presumably, no one would have thought of introducing these strangely coloured/shaped fruits in a western supermarket setting a decade or two ago, say.[14] One can therefore see the supermarket chains riding on the growing wave of interest/popularity around

---

[14] After all, isn't this an improvement on selling us visually perfect, but tasteless, fruit-and-veg?

sensory incongruity but one which pairs it with a format that is seen as inherently natural, that is, fruit-and-veg. In conclusion, in/congruity may now be something that consumers are increasingly coming to dissociate from the concept of un/naturalness. Such a change, or so we would like to argue here, has in part been facilitated by the modernist chef, often working together with the big food/flavour houses.

7.13   Individual differences in the response of diners to sensory incongruity

Sensory incongruity elicits diverse responses from different people. For example, it appears to work especially well in those food and beverage products that are designed for (or targeted at) children (e.g. Garber *et al.* 2008; Spence 2012a). Here, one can think of Skittles Confused, Fruity Smarties, green, blue and purple Heinz tomato Ketchup (Farrell 2000) and pink and blue Parkay margarine (Anonymous 2007; Garber *et al.* 2008). It is, however, an open question as to whether children like these products *because* they are sensorially incongruent (and they are aware of such sensory incongruity), or whether instead they like them just because they happen to be brightly coloured, regardless of any sensory congruity/incongruity that this gives rise to. Ideally, children may be drawn to such foods precisely because of the disgusted responses that adults make to them. It is interesting to note the suggestion of Garber *et al.* (2008, p. 590) that children's colour–flavour associations are not as well formed as those of adults, and hence are unlikely to be violated (but see also Spence 2012a). The Jelly Belly Bean Boozled collection certainly dares the consumer to compare some of their most popular flavours with the grossest flavours one can imagine.

> *"You order pigeon because you like pigeon. It arrives at the table in a banana fancy-dress costume and tastes like rabbit. And I want to grab the chef by his swarthy Latin muttonchops and ask him why he has ruined my dinner.*
> *Now I just order something from the menu that I don't like, knowing that there is a good chance that it'll taste like something I do."* (Clarkson 2012)

One might ask whether there are also 'fun' food product categories for adults where sensory incongruity has been introduced in order to enhance that entertaining experience, as in the examples given above for children. One example here is provided by Blue Curaçao. Indeed, as mentioned earlier in this chapter, the domain of cocktails seems to be one where people have for many years now expected to be entertained (Spence 2013a). After all, people are certainly not drinking these multisensorially stimulating drinks for their nutritional content! In fact, one can now see the molecular gastronomy approach being slowly transferred from the kitchen to the cocktail bar (see Uyehara 2011; Spence 2013a). Hence, it may be that sensory incongruity is more acceptable in those areas where the value of the eating and drinking experience is more closely related to that of its entertainment (as opposed to its nutritional content; see also Kass 1994), even in the adult market.

> *"But here's the catch – you won't know which ones are which! The black Licorice bean looks exactly like the Skunk Spray bean! Sweet, luscious Caramel Corn might also be Moldy Cheese. You may think you're tasting our world-famous Buttered Popcorn bean, but what you'll be biting into could actually be Rotten Egg."*
>
> (http://www.jellybelly.com/our_candy/beanboozled.aspx; see also http://itthing .com/world-of-candies for more 'appetizing' treats)

Another cohort of consumers for whom sensory incongruity might well be expected to work especially well is those who are neophilic (Rozin 1999; see also Zuckerman 1979). In this context, this is the name given to those who have a predisposition to seek out novelty and variety in food, trusting (and actively seeking out) the unfamiliar. They tend to associate the trying-out of new foods with excitement, diversity and sophistication. In fact, many look upon food experimentation as a means by which to enrich their lives (see Veeck 2010). Our prediction is that sensory incongruity will evoke a more positive reaction in this group of diners than in neophobes[15] (Raudenbush and Frank 1999). Neophobes also tend to rate new foods less favourably than do neophiles (Logue 2004, p. 90; but see Hobden and Pliner 1995).[16]

Counter to what one might have expected, the willingness to try out new foods is not necessarily correlated with a person's age. It also depends on many other personal factors, for example nationality, education, cultural background, etc. (Otis 1984; Henriques *et al.* 2009). How, then, to categorize oneself? There is actually a questionnaire that anyone can try to assess how neophilic/neophobic they are. The Food Neophobia Scale (Pliner and Hobden 1992) is a widely used tool among sensory and consumer researchers. It consists of 10 statements such as: *"I am constantly sampling new and different foods"*; *"I don't trust new foods"*; *"If I don't know what a food is, I won't try it"*; *"I like foods from different cultures"*; *"Ethnic food looks too weird to eat"*. By indicating one's level of agreement with each of the statements on a 7-point scale, one can soon find out!

---

[15] Unfortunately for your first author, who certainly likes to experiment with unusual taste sensations wherever possible, he happens to be married to one of the most neophobic women on the planet.

[16] According to MacClancey (1992, p. 43) the philosopher Wittgenstein hated any change to his diet. He apparently once said to his host that he didn't care what he ate, just as long as it was always the same.

7.14 Conclusions

According to the literature reviewed in this chapter, there are actually a number of reasons as to why sensory incongruity, and the associated surprise that it may engender, might be an end in-and-of-itself in the context of the perfect meal. To summarize, the key goals include:

1. challenging a diner's senses;
2. provoking the diner;
3. creating more entertaining and memorable eating/dining experiences (it could be said that making a dish sensorially incongruent will result in the diner paying more attention to the food, thus experiencing it and reflecting on it more fully; this might be harder to achieve in the presence of sensory congruity); and
4. highlighting the technical skills of the dish's creator.

The degree to which these goals may be achieved is ultimately going to depend on a diner's willingness to try new foods and/or experiences (i.e. it may depend, at least in part, on whether they are neophobic or neophilic). A person's cultural background may also play a role here, but to date there has been less research on the question of cross-cultural differences. The environment in which a diner comes across such examples of incongruity undoubtedly contribute in determining their response too.

As we have tried to make clear throughout this chapter, one should not necessarily confuse sensory incongruity with notions of 'unnatural'. For while sensory incongruity is often associated with surprise and 'unnaturalness', it is not actually always that easy to determine what is natural and what is artificial (Downham and Collins 2000). This is true both in processed foods, but also in the case of unprocessed foods (remember the oyster leaf? see also Rozin 2005; Wilson 2009). Furthermore, it is also worth remembering that what we consider as natural today has not always been so (think only of the humble orange carrot)[17] and, what is more, will not always continue being so. In some cases then, engaging with sensory congruity requires the chef to track something of a moving target.

The fact that a food or beverage does not match our initial expectations at a multisensory level does not necessarily imply that the combination of sensory inputs will not ultimately be perceived as harmonious and pleasant. As we

---

[17] Once upon a time, carrots were purple. According to popular legend, they were selectively bred in order to have the orange colour of the Dutch royal family in the seventeenth century (Dalby 2003; Macrae 2011; Greene 2012, p. 81). Another (perhaps more plausible) reason for why orange carrots were favoured over the purple variety is because the latter would colour the soups, stews, etc. in which they were placed.

have attempted to show in this chapter, sensory incongruity is perceived at a number of different levels which can be combined in order to enhance a diner's food experiences. While we have chosen to focus on *sensory* incongruity, it is worth noting that incongruity can also be conceptual, verbal (i.e. elicited by the descriptive verbal label; see Chapter 3) or perhaps even seasonal (consider your response should someone serve you iced coffee in the middle of a European winter or a Hot Cross bun in July).

In conclusion, the research that has been reviewed in this chapter has focused on sensory incongruity at the dining table. We have attempted to contextualize the idea of surprising the diner by means of the use of sensory incongruity in terms of the chef's search for novelty, something that chefs have been striving towards for well over a century (see Beaugé 2012). In the next chapter, we will go on to look at another kind of surprise. Chapter 8 discusses the removal of what, as we have already seen, might be the most important sense when it comes to our appreciation of food and drink: vision. Read on as we cast a critical eye over the rise of the dine-in-the-dark trend.

References

Adrià, F., Soler, J. and Adrià, A. (2007) *Un Día en elBulli (A Day at elBulli)*. elBulli-books, Barcelona.

Alícia and elBullitaller (2006) *Léxico Científico Gastronómico (A Scientific Gastronomic Lexicon)*. Editorial Planeta, Barcelona.

Anonymous (2007) 'Anything' and 'Whatever' beverages promise a surprise, every time. *Press release*, 17th May.

Anonymous (2008) *Food newsflash. Kitchen Angels*, **Winter**, 19.

Auvray, M. and Spence, C. (2008) The multisensory perception of flavor. *Consciousness and Cognition*, **17**, 1016–1031.

Beaugé, B. (2012) On the idea of novelty in cuisine: A brief historical insight. *International Journal of Gastronomy and Food Science*, **1**, 5–14.

Berghaus, G. (2001) The futurist banquet: Nouvelle Cuisine or performance art? *New Theatre Quarterly*, **17(1)**, 3–17.

Blumenthal, H. (2008) *The Big Fat Duck Cookbook*. Bloomsbury, London.

Blumenthal, H. (2013) *Historic Heston Blumenthal*. Bloomsbury, London.

Cardello, A. V. (1996) The role of the human senses in food acceptance. In: *Food Choice, Acceptance and Consumption* (eds H. L. Meiselman and H. J. H. McFie), pp. 1–82. Blackie Academic and Professional, New York.

Cardello, A. V. (2003) Consumer concerns and expectations about novel food processing technologies: Effects on product liking. *Appetite*, **40**, 217–233.

Cardello, A. V., Maller, O., Masor, H. B., Dubose, C. and Edelman, B. (1985) Role of consumer expectancies in the acceptance of novel foods. *Journal of Food Science*, **50**, 1707–1714.

Carlsmith, J. M. and Aronson, E. (1963) Some hedonic consequences of the confirmation and disconfirmation of expectancies. *The Journal of Abnormal and Social Psychology*, **66**, 151–156.

Clarkson, J. (2012) Heston's grub is great – but so what if your date is ugly? *The Sunday Times*, 6 May. Available at http://www.thesundaytimes.co.uk/sto/comment/columns/jeremyclarkson/article1031654.ece (accessed January 2014).

Classen, C., Howes, D. and Synnott, A. (2005) Artificial flavours. In *The Taste Culture Reader: Experiencing Food and Drink* (ed. C. Korsmeyer), pp. 337–342. Berg, Oxford.

Colquhoun, K. (2007) *Taste: The Story of Britain Through its Cooking*. Bloomsbury, London.

Dahl, R. (1964) *Charlie and the Chocolate Factory*. Alfred A. Knopf, New York.

Dalby, A. (2003) *Food in the Ancient World from A to Z*. Routledge, London.

Delwiche, J. F. (2012) You eat with your eyes first. *Physiology and Behavior*, **107**, 502–504.

Demattè, M. L., Sanabria, D. and Spence, C. (2006) Cross-modal associations between odors and colors. *Chemical Senses*, **31**, 531–538.

Demattè, M. L., Sanabria, D. and Spence, C. (2009) Olfactory identification: When vision matters? *Chemical Senses*, **34**, 103–109.

Desmet, P. M. A. and Schifferstein, H. N. J. (2008) Sources of positive and negative emotions in food experience. *Appetite*, **50**, 290–301.

Downham, A. and Collins, P. (2000) Colouring our foods in the last and next millennium. *International Journal of Food Science & Technology*, **35**, 5–22.

Duncker, K. (1939) The influence of past experience upon perceptual properties. *American Journal of Psychology*, **52**, 255–265.

Ellis, C. (2007) *Fabulous Fanny Cradock*. Sutton Publishing Limited, Stroud.

Farrell, G. (2000) What's green. Easy to squirt? Ketchup! *USA Today*, 10 July, 2b.

Garber, L. L. Jr., Hyatt, E. M. and Boya, Ü. Ö. (2008) The mediating effects of the appearance of nondurable consumer goods and their packaging on consumer behavior. In: *Product Experience* (eds H. N. J. Schifferstein and P. Hekkert), pp. 581–602. Elsevier, London.

Ge, L. (2012) Why coffee can be bittersweet. *FT Weekend Magazine*, 13/14 October, 50.

Gill, A. A. (2005) Table talk. *The Sunday Times*, 17 April.

Graves, P. (2010) *Consumer.ology: The Market Research Myth, the Truth about Consumers and the Psychology of Shopping*. Nicholas Brearly Publishing, London.

Green, D. M. and Butts, J. S. (1945) Factors affecting acceptability of meals served in the air. *Journal of the American Dietetic Association*, **21**, 415–419.

Greene, W. (2012) *Vegetable Gardening the Colonial Williamsburg Way: 18th Century Methods for Today's Organic Gardeners*. Rodale, New York.

Hanefors, M. and Mossberg, L. (2003) Searching for the extraordinary meal experience. *Journal of Business and Management*, **9**, 249–270.

Hekkert, P. (2006) Design aesthetics: Principles of pleasure in design. *Psychology Science*, **48**, 157–172.

Henriques, A. S., King, S. C. and Meiselman, H. L. (2009) Consumer segmentation based on food neophobia and its application to product development. *Food Quality and Preference*, **20**, 83–91.

Hitchcock, A. and Gottlieb, S. (eds) (2003) *Alfred Hitchcock: Interviews*. University of Mississippi Press, Jackson, MI.

Hobden, K. and Pliner, P. (1995) Effects of a model on food neophobia in humans. *Appetite*, **25**, 101–114.

Hultén, B. (2011) Sensory marketing: The multi-sensory brand-experience concept. *European Business Review*, **23**, 256–273.

Hultén, B., Broweus, N. and van Dijk, M. (2009) *Sensory Marketing*. Palgrave Macmillan, Basingstoke.

Kass, L. (1994) *The Hungry Soul: Eating and the Perfecting of Human Nature*. The Free Press, New York.

King, S. C., Weber, A. J., Meiselman, H. L. and Lv, N. (2004) The effect of meal situation, social interaction, physical environment and choice on food acceptability. *Food Quality and Preference*, **15**, 645–653.

King, S. C., Meiselman, H. L., Hottenstein, A. W., Work, T. M. and Cronk, V. (2007) The effects of contextual variables on food acceptability: A confirmatory study. *Food Quality and Preference*, **18**, 58–65.

Köster, E. P. (2006) Memory for food and food expectations: A special case? *Food Quality and Preference*, **17**, 3–5.

Koza, B. J., Cilmi, A., Dolese, M. and Zellner, D. A. (2005) Color enhances orthonasal olfactory intensity and reduces retronasal olfactory intensity. *Chemical Senses*, **30**, 643–649.

Krishna, A. (ed) (2010) *Sensory Marketing: Research on the Sensuality of Products*. Routledge, London.

Krishna, A. (2013) *Customer Sense: How the 5 Senses Influence Buying Behaviour*. Palgrave Macmillan, New York.

Lathigra, K. (2013) History man. *The Financial Times*, 5/6 October, 11–21.

Lauden, R. (2001) A plea for culinary modernism: Why we should love new, fast, processed food. *Gastronomica*, **1(1)**, February.

Lindstrom, M. (2005) *Brand Sense: How to Build Brands through Touch, Taste, Smell, Sight and Sound*. Kogan Page, London.

Logue, A. W. (2004) *The Psychology of Eating and Drinking*. Brunner-Routledge, New York.

Loken, B. and Ward, J. (1990) Alternative approaches to understanding the determinants of typicality. *Journal of Consumer Research*, **17 (September)**, 111–126.

Ludden, G. D. S. and Schifferstein, H. N. J. (2007) Effects of visual-auditory incongruity on product expression and surprise. *International Journal of Design*, **1**, 29–39.

Ludden, G. D. S., Schifferstein, H. N. J. and Hekkert, P. (2009) Visual-tactual incongruities in products as sources of surprise. *Empirical Studies of the Arts*, **27**, 63–89.

MacClancy, J. (1992) *Consuming Culture: Why You Eat What You Eat*. Henry Holt, New York.

Macrae, F. (2011) What's for dinner? Rainbow coloured carrots and super broccoli that's healthier and sweeter. *DailyMail Online*, 15 October. Available at http://www.dailymail.co.uk/health/article-2044695/Purple-carrots-sale-Tesco-supermarket-Orange-year.html (accessed January 2014).

Maric, Y. and Jacquot, M. (2013) Contribution to understanding odour-colour associations. *Food Quality and Preference*, **27**, 191–195.

Marshall, A. B. (1888) *Mrs A. B. Marshall's Cookery Book*. Robert Hayes, London.

Martin, D. (2007) *Evolution*. Editions Favre, Lausanne.

Masurovsky, B. I. (1939) How to obtain the right food color. *Food Industries*, **11(13)**, 55–56.

McGee, H. (2004) *On Food and Cooking: The Science and Lore of the Kitchen*. Scribner, New York.

Mielby, L. H. and Bom Frøst, M. (2010) Expectations and surprise in a molecular gastronomic meal. *Food Quality and Preference*, **21**, 213–224.

Moir, H. C. (1936) Some observations on the appreciation of flavour in foodstuffs. *Journal of the Society of Chemical Industry: Chemistry and Industry Review*, **14**, 145–148.

Moskowitz, H. R. (1978) Taste and food technology: Acceptability, aesthetics, and preference. In: *Handbook of Perception*, volume VIA, (eds E. C. Carterette and M. P. Friedman), pp. 157–194. Academic Press, San Diego, CA.

Nolan, S. (2013) The Heston effect? List of food trends moves away from the traditional and embraces the unusual. Available at http://www.dailymail.co.uk/news/article-2302999/The-Heston-effect-List-food-trends-moves-away-traditional-embraces-unusual.html (accessed January 2014).

Oliver, R. L. and Winer, R. S. (1987) A framework for formation and structure consumer expectations: Review and propositions. *Economic Psychology*, **8**, 469–499.

Olson, J. C. and Dover, P. A. (1979) Disconfirmation consumer expectations through product trial. *Journal of Applied Psychology*, **64**, 179–189.

Otis, L. P. (1984) Factors influencing the willingness to taste unusual foods. *Psychological Reports*, **54**, 739–745.

Pelchat, M. L., Johnson, A., Chan, R., Valdez, J. and Ragland, J. D. (2004) Images of desire: Food-craving activation during fMRI. *NeuroImage*, **23**, 1486–1493.

Petrini, C. (2001) *Le Ragioni del Gusto* (*Slow Food: The Case for Taste*, translated W. McCuaig). Columbia University Press, New York.

Pine, II, B. J. and Gilmore, J. H. (1998) Welcome to the experience economy. *Harvard Business Review*, **76(4)**, 97–105.

Pine, II, B. J. and Gimore, J. H. (1999) *The Experience Economy: Work is Theatre and Every Business is a Stage*. Harvard Business Review Press, Boston, MA.

Piqueras-Fiszman, B., Velasco, C. and Spence, C. (2012) Exploring implicit and explicit crossmodal colour-flavour correspondences in product packaging. *Food Quality and Preference*, **25**, 148–155.

Pliner, P. and Hobden, K. (1992) Development of a scale to measure the trait of food neophobia in humans. *Appetite*, **19**, 105–120.

Poulter, S. (2011) The purple potato that lowers blood pressure. *Daily Mail*, 15 October. Available at http://www.dailymail.co.uk/news/article-2049294/Purple-potato-helps-lower-blood-pressure-doesnt-make-weight.html (accessed January 2014).

Raudenbush, B. and Frank, R. A. (1999) Assessing food neophobia: The role of stimulus familiarity. *Appetite*, **32**, 261–271.

Revel, J.-F. (1985) *Un Festin en Paroles/Histoire Littéraire de la Sensibilité Gastronomiuque de l'Antiquité à Nos Jours (A Feast in Words/Literary History of Culinary Sensibility from Antiquity to Today)*. Éditions Suger, Paris.

Rosenbaum, R. (1979) Today the strawberry, tomorrow… In: *Culture, Curers and Contagion* (ed. N. Klein), pp. 80–93. Chandler and Sharp, Novato, CA.

Rozin, E. (1983) *Ethnic Cuisine: The Flavor-Principle Cookbook*. The Stephen Greene Press, Brattleboro, VT.

Rozin, P. (1999) Food is fundamental, fun, frightening, and far-reaching. *Social Research*, **66**, 9–30.

Rozin, P. (2005) The meaning of "natural": Process more important than content. *Psychological Science*, **16**, 652–658.

Rozin, P., Spranca, M., Krieger, Z., Neuhaus, R., Surillo, D., Swerdlin, A. and Wood, K. (2004) Preference for natural: Instrumental and ideational/moral motivations, and the contrast between foods and medicines. *Appetite*, **43**, 147–154.

Sakai, N. (2011) Tasting with eyes. *i-Perception*, **2(8)**, http://i-perception .perceptionweb.com/journal/I/article/ic945 (accessed January 2014).

Schifferstein, H. N. J. (2001) Effects of product beliefs on product perception and liking. In: *Food, People and Society: A European Perspective of Consumers' Food Choices* (eds L. Frewer, E. Risvik and H. Schifferstein), pp. 73–96. Springer Verlag, Berlin.

Schifferstein, H. N. J. and Hekkert, P. (eds) (2008) *Product Experience*. Elsevier, Amsterdam.

Schifferstein, H. N. J. and Spence, C. (2008) Multisensory product experience. In: *Product Experience* (eds H. N. J. Schifferstein and P. Hekkert), pp. 133–161. Elsevier, London.

Schlack, L. (2011) Catch cold: London's Chin Chin Laboratorists puts the cool factor back into ice cream. *Hemispheres Magazine*, **July**, 29–30.

Shankar, M. U., Levitan, C. and Spence, C. (2010) Grape expectations: The role of cognitive influences in color-flavor interactions *Consciousness and Cognition*, **19**, 380–390.

Sifton, S. (2011) In London, stalking the new. *The New York Times*, 24 May. Available at   http://www.nytimes.com/2011/05/25/dining/new-restaurants-in-london.html? pagewanted=all&_r=0 (accessed January 2014).

Spence, C. (2002) *The ICI Report on the Secret of the Senses*. The Communication Group, London.

Spence, C. (2007) Audiovisual multisensory integration. *Acoustical Science & Technology*, **28**, 61–70.

Spence, C. (2011) Mouth-watering: The influence of environmental and cognitive factors on salivation and gustatory/flavour perception. *Journal of Texture Studies*, **42**, 157–171.

Spence, C. (2012a) The development and decline of multisensory flavour perception. In: *Multisensory development* (eds A. J. Bremner, D. Lewkowicz and C. Spence), pp. 63–87. Oxford University Press, Oxford.

Spence, C. (2013a) Sound advice. *The Cocktail Lovers*, **Winter (6)**, 18–19.

Spence, C. (2013b) Multisensory flavour perception. *Current Biology*, **23**, R365–R369.

Spence, C. and Piqueras-Fiszman, B. (2013) The multisensory packaging of beverages. In: *Food Packaging: Procedures, Management and Trends* (ed. M. G. Kontominas), pp. 187–233. Nova Publishers, New York.

Spence, C., Levitan, C., Shankar, M. U. and Zampini, M. (2010) Does food color influence taste and flavor perception in humans? *Chemosensory Perception*, **3**, 68–84.

Spence, C., Puccinelli, N., Grewal, D. and Roggeveen, A. L. (in press) Store atmospherics: A multisensory perspective. *Psychology & Marketing*.

Stevenson, R. J. (2009) *The Psychology of Flavour*. Oxford University Press, Oxford.

Svejenova, S., Mazza, C. and Planellas, M. (2007) Cooking up change in haute cuisine. *Journal of Organizational Behaviour*, **38**, 539–561.

Tannahill, R. (1973) *Food in History*. Stein & Day, New York.

Triplett, T. (1994) Consumers show little taste for clear beverages. *Marketing News*, **28(11)**, 2–11.

Uyehara, M. (2011) Cutting edge cocktail trends. *Time Out (New York)*, **816**, 28.

Van der Laan, L. N., de Ridder, D. T. D., Viergever, M. A. and Smeets, P. A. M. (2011) The first taste is always with the eyes: A meta-analysis on the neural correlates of processing visual food cues. *NeuroImage*, **55**, 296–303.

Veeck, A. (2010) Encounters with extreme foods: Neophilic/neophobic tendencies and novel foods. *Journal of Food Products Marketing*, **16**, 246–260.

Wang, G.-J., Volkow, N. D., Telang, F., Jayne, M., Ma, J., Rao, M., Zhu, W., Wong, C. T., Pappas, N. R., Geliebter, A. *et al.* (2004) Exposure to appetitive food stimuli markedly activates the human brain. *NeuroImage*, **212**, 1790–1797.

Wansink, B. (2002) Changing eating habits on the home front: Lost lessons from World War II research. *Journal of Public Policy & Marketing*, **21**, 90–99.

Wheatley, J. (1973) Putting colour into marketing. *Marketing*, **October**, 24–29, 67.

Wheeler, E. (1938) *Tested Sentences that Sell*. Prentice & Co, New York.

Wilson, B. (2009) *Swindled: From Poison Sweets to Counterfeit Coffee: The Dark History of the Food Cheats*. John Murray, London.

Yeomans, M., Chambers, L., Blumenthal, H. and Blake, A. (2008) The role of expectancy in sensory and hedonic evaluation: The case of smoked salmon ice-cream. *Food Quality and Preference*, **19**, 565–573.

Youssef, J. (2013) *Molecular Cooking at Home: Taking Culinary Physics out of the Lab and into your Kitchen*. Quintet, London.

Zuckerman, M. (1979) *Sensation Seeking: Beyond the Optimal Level of Arousal*. Erlbaum Associates, Hillsdale, NJ.

8
Looking for Your Perfect Meal in the Dark

8.1   Introduction

Since the opening of the Blindekuh (Blind Cow) restaurant in Zurich in 1999 and the Unsicht (which means invisible) Bar in Cologne, Germany in 2001, the trend towards dining in the dark (and here we are not talking about romantic candlelight but rather utter darkness, where it's impossible even to see the hand in front of your face), has flourished across Europe, North America and parts of Asia (Long 2010).[1] The idea for Blindekuh apparently originated in 1998 during the 'Dialog im Dunkeln' (Dialogue in the dark) exhibition at the Zurich Museum of Design. The enthusiasm shown by the visitors and the opportunity to create jobs for visually impaired people led the Blind-Liecht foundation to open a restaurant, bar and platform for education and culture in the dark in a former Methodist chapel in Zurich (http://www.blindekuh.ch/).

Pioneered by the likes of Axel Rudolph, psychologist, and owner of the Unsicht-Bar, the concept was developed with the idea of 'shedding some light' on the sensory world of the blind. This empathic approach is meant – or better said, was originally meant – to place the blind at something of an advantage relative to their normally sighted counterparts. Nowadays, though, we would argue that the dining experience of such a restaurant is actually very different

[1] Dans Le Noir, a restaurant chain started by Edouard de Broglie in Paris in 2004 (http://www.danslenoir.com/), currently provides this kind of dining experience in London (Robathan and Whyman 2006).

The Perfect Meal: The Multisensory Science of Food and Dining, First Edition.
Charles Spence and Betina Piqueras-Fiszman.
© 2014 John Wiley & Sons, Ltd. Published 2014 by John Wiley & Sons, Ltd.

from that of a blind person eating and drinking at a conventionally lit establishment. The central question that we wish to address in this chapter, then, is what exactly makes a visit to one of these restaurants so appealing (if anything). And can turning off the lights really make the food and drink taste better?

It is often said that food tastes better in the dark. If that were true then it would imply that we could all improve the quality of the dining experience simply by turning the lights off, or by giving our diners a blindfold to wear. Surely this is something that every chef would want to know about, if it were true. In the following, we investigate whether such claims really are to be believed and, in so doing, address the question of whether we are all more likely to experience the perfect meal in the dark. Distinct from the question of whether food tastes better in the dark, is what food tastes like to the roughly 6% of the population who are colour blind. As we will see below, such individuals see (and hence experience) food very differently from the rest of us, and yet few chefs seem to take this segment of the population into consideration when preparing their meals. The food scientists also seem to have neglected this section of the population when it comes to thinking about how they experience food, surely a worrying omission given how important visual cues appear to be to our appreciation of food.

One of the first things that the diner notices is that the food in dine-in-the-dark restaurants is normally served in bite-sized pieces and without bones. You are far more likely to find yourself eating cubes of meat than trying to manhandle a T-bone steak, and you are far more likely to be given a side-serving of mashed potatoes, say, than a helping of fresh garden peas (Yang 2007). Your soup bowl will probably also have 'ears' so that you can drink the contents rather than having to navigate spoonfuls of hot liquid towards your mouth without spilling them down your front. It is however not only how the food is served that differs when compared to a normally illuminated restaurant. Complex combinations of flavours are also notable by their absence. It turns out that diners often find it difficult to distinguish between flavours in the absence of any visual cues (not all that surprising given what we saw in Chapter 6). Nor is one offered a full menu; in fact, the only decision that the diner typically has to make in such a restaurant is whether to go for the meat, fish or vegetarian option. Sometimes there may also be a 'surprise' option as well (Robathan and Whyman 2006; Mielby and Bom Frøst 2012). Furthermore, the names of the dishes often do not describe the food, or the way in which it has been prepared (see Chapter 3 on the crucial importance of naming to our enjoyment of food). In many cases in fact, the descriptions of the dishes are quite simply mystifying. Take for example the main course from the beef menu at Unsicht-Bar. Here, the intention would seem to be to deliver a novel and surprising multisensory dining experience, especially since few of us will have eaten in complete darkness before.

8.2 The social aspects of dining in the dark

It's often claimed that darkness alters the way in which we relate to others, even those sitting next to us. It's certainly true that the selective sensory deprivation served up by the dine-in-the-dark experience challenges the diner's everyday notions of intimacy and enjoyment when it comes to the social consumption of food. The utter darkness affects how we interact with those around us. Indeed, the guests at such restaurants are sometimes intentionally placed at long benches, and hence in close proximity with strangers. Consequently, there is often no place for intimate conversation; the talk is more likely to revolve around issues such as one's inability to find the tableware or what to do about a spilled glass of wine. Indeed, that's one plausible explanation for why drink-in-the-dark bars do not appear to have caught on thus far (if in fact they exist at all) and why tablecloths are often noticeable by their absence (Mielby and Bom Frøst 2012). That said, this might also be a good way to help break the ice if you find yourself on a 'blind date'. Chen Long, the owner of the Whale Inside Dark restaurant in Beijing, says that the Chinese tend to be shy and that dining in the dark helps them to lower their social inhibitions. His restaurant is apparently particularly popular among Internet daters (Yang 2007).

Speaking also becomes fraught with uncertainty. How do you know whether those sitting at your table are really paying attention to your witty repartee if you can't see them (Newman 2003)? And how can you be sure that prying ears aren't listening in? As a listener, you find yourself making audible grunts at the appropriate points in the conversation (at least in the conversation that you are supposed to be participating in) in order to indicate that you are tuned in to whoever happens to be speaking at the time. This may be one of the reasons why diners don't generally choose to repeat the dark dining experience at home, no matter how memorable it may have been when experienced in the setting of the restaurant.[2]

In addition, we could say that when finding oneself in utter darkness at these types of restaurants one very likely feels, more than ever, at the mercy of the chef/owner of the restaurant. Indeed, diners sometimes specifically mention their feeling of vulnerability (Robathan and Whyman 2006). The diner really has to put their trust in the hands of the chef and serving staff. Every meal, embedded as it is in ritual, has a host who has a major influence on the diner's overall experience. The nature of the interaction is also related to the trust that builds up between the host and his/her guests (Mielby 2007). It can be argued that this feeling of trust is an essential element of the experience of dining in the dark, and is certainly at least as important as it is in a normal restaurant (i.e. one that is conventionally illuminated).

---

[2] The only exception to this presumably being for those wanting to recreate the erotic scene from the 1986 movie $9\frac{1}{2}$ *Weeks*, in which Mickey Rourke blindfolds Kim Basinger and feeds her various tasty titbits.

8.3   Why are dining in the dark restaurants so popular nowadays?

One powerful driver underlying the growth of dining-in-the-dark restaurants may be the influential idea of the 'experience economy'. The key notion here – one that has been around at least since Philip Kotler's classic paper on store atmospherics was published back in 1974, re-popularized by Pine and Gilmore (1998, 1999) in their book and *Harvard Business Review* article of the same name – is that consumers should be encouraged to pay for (and apparently seek out) 'experiences'. This is not only true when it comes to products and services, but also for culinary experiences (see also Hanefors and Mossberg 2003). According to Mielby (2007), the exquisite eating practice is a dialectic episode produced by skilled chefs and diners who are both qualified to enjoy the experience. Indeed, in recent years, many of the most successful companies have managed to differentiate themselves from their competitors in the marketplace precisely by selling engaging multisensory experiences. Many consumers seem to love the experience even when the product offering itself sometimes isn't necessarily '*la crème de la crème*'.

> "*It was an extraordinary experience to eat a whole meal in the dark at Unsicht-Bar.*" (Mielby and Bom Frøst 2012, p. 235)

The dark dining concept fits right in here (Mielby and Bom Frøst 2012). What these restaurants are selling is very much the 'experience'. But this time, rather than delivering something that stimulates more of a diner's senses than the competition, the perhaps counterintuitive proposition is that *less* (senses intervening) can sometimes deliver *more* in terms of the overall experience. It could perhaps be argued that the dark dining concept also plays to the growing concern among some diners (or at least some vocal critics) that delivering the 'experience' has actually become more important than the food itself, at least in certain eating establishments and restaurant chains (see Goldstein 2005; Gill 2007 for some critical commentaries on this theme).

While the notion of dining-in-the-dark started out as a whole meal eaten in darkness, some chefs interested in highlighting the role of each of the diner's senses in the multisensory perception of flavour have taken to serving meals where each course emphasizes a different sense. For the sight course, this is often done by serving/eating the food in the absence of vision. Take for example the Sensory menu served by Caroline Hobkinson in 2012 at The House of Wolf restaurant in London (see below). At least one course was devoted to each sense, and when it was time for the Sight course the diners were instructed to put on a blindfold.[3]

---

[3] One of the more interesting dishes here is the sound course. The diners were encouraged to insert earplugs and 'see' (or should that be 'hear') how their dining experiences changed as a result (see also Spence 2012 on this topic).

Look. Listen. Smell. Touch. Eat!

Amuse bouche

Insert your earplugs

Devour the freshly baked Bread roll without the use of your hands

Neuroscience has revealed a deep 'cross modal' connection, sounds can actually change how we perceive food experiences.

Can you hear the taste?

***

Sight

Blindfold yourself

Your waiter will describe the dish to you

A Cracker bread is placed in front of you

The Smell of Roast Peppers and Fresh Rosemary is distributed

Remove your blindfold

Can you see the taste?

***

Smell

Salmon Sashimi accompanied by a Syringe filled with Ardbeg Ten Years Old.

Revered as the peatiest and smokiest Single Malt.

Inject the Salmon with the Whisky and eat it

Reconstruct the taste of Smoked Salmon with the Smokey Scent

Taste sensations are picked up chemically by our tongue.

Most of flavour is smell.

Can you smell the taste?

***

Touch

Palate cleanser

HENDRICK'S Gin infused Cucumber Granita

Slurp with texture treated spoons with Rose Water Crystals and Maldon Sea Salt

Main

Saddle of Venison with foraged Prunes, Chanterelles and Wild Cherries.

Grab the hand carved long tree branch and spear it.

Can you feel the taste?

***

Sound

Sonic cake pop

Please take your phone

Dial 0845 680 2419

A low note brings out the Bitter, a high pitched sound brings out the Sweet flavour, research at Oxford University has proved.

Can you dial a taste?

It is interesting to note here that what may have started out as dining in dark is now expanding into other activities also enjoyed in the dark. For example, John Metcalfe's 'A Darker Sunset', a contemporary musical performance, was delivered in total darkness at Kings Place in London in 2013. According to Metcalfe, the reason for this was to remove all 'unnecessary' stimulation (by switching the lights off) and allowing the audience to focus their attention squarely on the music, while at the same time exploring the relationship between the performer and audience (but see Ozbaydar 1961). Elsewhere in London, the Dark collective arranges talks in which, you guessed it, the audience listen to the speaker while seated in complete darkness. Meanwhile, a series of Blackout talks and music events were organized in the UK, garnering praise from the critics and public alike in terms of delivering "*an experience not easily forgotten*" and "*a unique and all-immersive sensory trip*". Other entrepreneurial souls have started 'dating in the dark' events, which combine all the pleasures of dining in the dark with a novel approach to dating (Newman 2003; cf. Yang 2007). The multisensory blindfolded dinners organized in New York by artist Dana Salisbury (owner of Dark Dining, a performance/catering company) can be seen to merge a number of these elements into a single event. She combines blindfolded dining with dance, music and other more theatrical elements. As she puts it: "*Why in the dark? Because it awakens the senses and presents new pleasures*" (Allen 2012, p. 86; Gleason 2006; http://www.darkdiningprojects.com/). At certain of her events Salisbury also has her guests caress each other's hands (in the dark), asking them to "feel below the skin, below muscle and bone" (Guterman 2006, p. 481). Another of her intriguing offerings is the Hand-Fed Dark Dining Parties that she organizes in diners' own homes.

> "*I have just begun offering another program, private Hand-Fed Dark Dining Parties held in diners' homes. Clients prepare their favourite menu and invite up to three friends to join them. Shortly before they eat, I arrive and pass around the blindfolds. The diners sit back and are hand-fed. The art is in the feeding itself and the experience of being fed; think of it as performative feeding, a sensory and certainly psychological voyage. Through shifting my body language, I influence the affect of each bite and expose the intimate yet simple act of feeding as something loaded with meaning.*" (Dana Salisbury, personal communication, 2006)

Taken together, a sound business case can certainly be made for offering the 'dine in the dark' experience. What is so striking in hindsight is why no one seems to have thought of doing this prior to 1990 (Allen 2012). But what is in it for the diners themselves? In the remainder of the chapter, we critically evaluate the various arguments that have been put forward over the years in support of the concept.

8.4   Seeing or not seeing (correctly) the food

If *"Eye appeal is half the meal"* as gourmands often claim, what happens if you can no longer see the food that you are eating? According to folk intuition, the result ought to be a heightening of the other senses: *"You smell better, you are more receptive to differences in texture, consistency and temperature … it's a holistic experience"* (Rudolph; cited in Read *et al.* 2011, p. 16; see also Sautter 2002). But is that claim really true? Does it stand up to empirical investigation? The key question here concerns how the absence of one sense (in particular vision, what many consider the most important or dominant sense) affects the perception of food via the others, and how our overall multisensory eating experiences are impacted as a result?

Note here that there are at least two competing influences on people's perception of food when the lights go out. On the one hand, visual cues influence our sensory expectations regarding the taste and flavour of foods (e.g. Deliza and MacFie 1996; Simmons *et al.* 2005; Spence 2010a, b). This is referred to by some as 'visual flavour' (see Spence *et al.* 2010). Our hedonic expectations (Hurling and Shepherd 2003), our taste evaluations (Wilson and Gregson 1967; Spence 2011) and even our total food intake are all determined, at least in part, by whatever it is that we happen to see (Linné *et al.* 2002; Wansink *et al.* 2005). We therefore argue that removing such salient visual cues, which are normally available to us both prior to and during the consumption of food, is likely to diminish the overall dining experience (Spence 2010a). No matter whether we realize it or not, sensory expectation and anticipation constitute a good part of the pleasure of a meal.[4]

8.4.1   The importance of colour to food

While most of us are used to seeing a range of colours on our plate from the various ingredients being served (at least that's what is recommended), being invited to eat a menu entirely composed of food of a particular colour might seem rather unnatural. We would very likely miss the additional colour, not to mention the colour contrast (even if the food presented has not been coloured artificially) on the plate. Take for example the extreme example of this approach, the entirely black meal served by the fictional character De Essientes in Huysmans' (1884, p. 13) novel *Against Nature*.[5] What is more,

---

[4] In his *The Omnivorous Mind* (2012, pp. 86–87) John Allen puts forward a speculative explanation for the dramatic effects of the removal of visual cues in terms of the cross-sensory connections in the gustatory and orbitofrontal cortices. It isn't altogether clear just how much weight such a post-hoc neuroscience-based explanation really carries, however.

[5] This meal was apparently based on Grimod de la Reynière's (1783) funeral supper served in Paris (Weiss 2002, pp. 103–105).

this fictional dinner was served on a black table cloth in a dining room draped in black fabric, while listening to a hidden orchestra playing funeral marches.

> "*Dining off black-bordered plates, the company had enjoyed turtle soup, Russian rye bread, ripe olives from Turkey, caviar, mullet botargo, black puddings from Frankfurt, game served in sauces the colour of liquorice and boot-polish, truffle jellies, chocolate creams, plum puddings, nectarines, pears in grapefruit syrup, mulberries and black-heart cherries. From dark-tinted glasses they had drunk the wines of Limagne and Roussillon, of Tenedos, Valdepeñas, and Oporto. And after coffee and walnut cordial, they had rounded off the evening with kvass, porter and stout.*" (Huysmans 1884, p. 13)

However, examining black food provides a means of studying colour concepts that are sometimes used to enhance the presentation of food. Improved economic circumstances permit a growing number of diners to seek out novel gastronomic experiences and, accordingly, a restaurant with a certain reputation for culinary creativity and innovation may have a competitive advantage. Let's not forget that black foods contribute to culinary diversity and usually have a valuable place on the menu. In Peter Greenaway's film 'The Cook, the Thief, his Wife and her Lover', when asked about menu pricing at *Le Hollandais* restaurant the Cook responds: "*I charge a lot for anything black. Grapes, olives, blackcurrants … Black truffles are the most expensive. And caviar … Don't you think it's appropriate that the most expensive items are black?*" (Greenaway 1989).[6] Chef Grant Achatz, whose dishes have been mentioned several times in this book already, had a dish called 'Halibut. Black pepper, coffee, lemon' that is served in a white monochromatic presentation. The surprising element (see the previous chapter) is that, in fact, most of the elements in this dish are naturally black (coffee, black pepper, black liquorice, vanilla) but are served in a white form (see Figure 8.1a).[7]

Meanwhile, the 'White funeral' meals organized by chef-artist Marije Vogelzang (http://www.marijevogelzang.nl; Vogelzang 2010) consist entirely of white food and especially designed white crockery. The idea behind these meals is to let the diner focus on all the other sensory properties of the food (the aroma, the taste and, most importantly, the tactile and oral texture of the foods that are served; see Figure 8.1b). It's certainly a real-life example in which limiting a meal's colour palette (in this case, 'removing' it completely) can be used to encourage a diner to discover more of the flavour of the foods

---

[6] It is interesting to note here how many of the foods we describe as having a particular colour do not actually have the ascribed hue. When, after all, was the last time you saw a truly 'white' wine (Spence 2010b)?

[7] When something similar (e.g. serving a totally white dinner) was tried at a luncheon organized by the Colour Society of Australia in 1997, the results were pretty disappointing; 'boring' was apparently the term that was most frequently heard (Hutchings 2003, p. 143).

(a)   (b)

Figure 8.1 (a) 'Halibut black pepper, coffee, lemon' served by Grant Achatz in his restaurant Alinea. Note that all of the black food elements are presented in white form. *Source*: Reproduced with permission of Ron Kaplan/LTHForum.com. (b) Detail of one of Marije Vogelzang's White funeral meals. *Source*: www.marijevogelzang.nl. Reproduced with permission of Marije Vogelzang. *See colour plate section*

via their oral-somatosensory textures perceived through their hands (since no cutlery is provided), and use their mouth and nose instead (see Chapter 5).[8]

> "*I eat only white foodstuffs: eggs, sugar, scraped bones; fat from dead animals; veal, salt coconuts, chicken cooked in white water; mouldy fruit, rice, turnips; camphorated sausages, things like spaghetti, cheese (white), cotton salad and certain fish (minus their skins).*" (Eric Satie on food; from Orledge 1995)

There have been occasional reports of individuals who, out of choice, would only eat foods that were white. For example, the designer Yves Saint Laurent had a fondness for all things white when it came to food (Crumpacker 2006, p. 113). Meanwhile, the Hollywood producer Leland Hayward never ate anything but 'white' food', for example scrambled eggs, custard, vanilla ice-cream and chicken hash from the Beverly Hills Hotel (see Hayward and Henry 1977/2011, pp. 28–29). Digressing a little, the Swiss sculptor Alberto Giacometti also falls into the same category given that, in his later years, he would only eat boiled eggs, and those just before midnight (see http://www.giacometti-stiftung.ch/index.php?sec=alberto_giacometti&page =biografie&language=en). In fact, only eating white foods appears to be more common among artists than one might imagine. Eric Satie's favourite colour was white, given its association with purity (Orledge 1995; Strunk 1998).

While none of us may realize just how important colour is to our perception of food, the profound effect it has is nicely illustrated by the unfortunate

---

[8] Marilyn Monroe was also once served a white meal consisting of oysters, soufflé, champagne and grapes, all served off of a white marble table, on which she danced after the meal (Crumpacker 2006, p. 94).

case of an individual (a painter as it turns out) that was vividly described by Oliver Sacks in his best-selling 1995 book *An Anthropologist on Mars* who, after a car crash, suddenly lost his ability to see colour. Everything appeared to him in shades of black and white. The experience may have been something like looking at the plant photographs taken by the English photographer Charles Jones (c. 1895–1910; see Figure 8.2). Sacks describes how unpleasant it suddenly became for the artist to eat, so much so that the brain-damaged painter started to choose foods that naturally had a black or white coloration (think black olives and mozzarella, coffee and yoghurt) or else dined with his eyes closed. One therefore presumes that he would have been a fan of the dine-in-the-dark restaurant concept.

Figure 8.2 In the absence of any colour cues fruits and vegetables start to look strange and rather unappealing, an experience that may have been somewhat similar to that of the colour-blind painter described by Oliver Sacks (1995) in his book *Anthropologist on Mars*

"The 'wrongness' of everything was disturbing, even disgusting, and applied to every circumstance of daily life. He found foods disgusting due to their greyish, dead appearance and had to close his eyes to eat. But this did not help very much, for the mental image of a tomato was as black as its appearance. Thus, unable to rectify even the inner image, the idea of various foods, he turned increasingly to black and white foods – to black olives and white rice, black coffee and yoghurt. These at least appeared relatively normal, whereas most foods, normally coloured, now appeared horribly abnormal." (Sacks 1995, p. 5)

What about all those people who are colour blind? Colour blindness affects around 1 in every 10 men (and a rather smaller number of women). Can you imagine the difficulties that this population face with food (while buying, preparing and consuming it; see Justin Broackes 2010, a colour-blind philosopher who has written extensively on how those with the condition see the world)? Dichromacy is subdivided into three types: protanopia, deuteranopia and tritanopia. The former type is mainly a loss of red sensitivity, known as red/green colour blindness; the second is loss of green sensitivity; and the latter is the absence of blue sensitivity, producing a defect in blue/yellow vision (http://facweb.cs.depaul.edu/sgrais/colorvisiondefi.htm). Most red/green colour-blind individuals will not be able to tell by eye when served a steak whether it has been cooked rare or well done (see Figure 8.3a). They are also unlikely to be able to tell the difference between green and red peppers or between tomato ketchup and chocolate sauce. Meanwhile, for those diners with deuteranopia, a plate of cooked spinach might well remind them of beef stew (see Figure 8.3b).

In addition, let's not forget that a food's colour often provides a reliable indicator as to its quality and to the likely off-taint in meats and fish. Hopefully the latter is not something that restaurant diners normally need to worry about; the more serious concern here is the danger that for those with a food allergy the blind waiter may bring them the wrong dish (Robathan and Whyman 2006). Nevertheless, being presented with a green-greyish piece of meat, regardless of how well it has been cooked, can be very off-putting. Colour also serves as an indicator to the ripeness of the fruit, say (Schaefer and Schmidt 2013). Many colour-blind individuals presumably have to rely on their sense of touch or smell when trying to pick a ripe banana.

In brief, being colour blind is certainly a problem when it comes to trying to enjoy a meal. That said, there is surprisingly little published research on this topic and there are not many chefs who seem to take such a common visual deficit into consideration when designing their dishes. Let's admit it however: it's already challenging enough trying to make sure that one's dishes are equally appealing to everyone, even when they have normal colour vision.

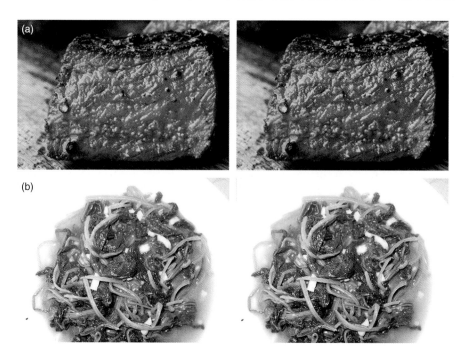

Figure 8.3 (a) A rare steak for a normal-sighted person on the left, compared to how a red/green colour-blinded person would see the same steak (i.e. as well cooked) on the right. (b) A plate of cooked spinach for a normal-sighted person (on the left) compared to how a green colour-blind diner might see it (as a meat stew, perhaps?). *See colour plate section*

8.4.2   Do our other senses really become more acute in the dark?

Removing vision certainly does allow one to concentrate more on the taste and aroma of food and drink (Marx *et al.* 2003; Wiesman *et al.* 2006). It is important to note that we humans have only limited attentional capacity, and vision tends to capitalize on the available neural resources. In fact, it has been estimated that more than half of our brain's processing resources are in some way involved in processing what we see (Gallace *et al.* 2012). As a result, we often don't pay as much attention to the other senses as perhaps we should. Indeed, more often than not, it is what we see that ultimately determines what we perceive. This turns out to be true even when the other senses may be sending our brain a different message (this ubiquitous phenomenon is known by psychologists as visual dominance). As we saw in Chapter 6, colouring certain white wines red can, for example, fool both experts and novices alike into thinking that they are actually drinking a red wine.

The key question here then is whether the tastes, aromas, textures and flavours associated with the consumption of food and drink[9] really do become *more* intense in the absence of sight (when our attention can be directed squarely to all of the food cues). Thus far, the limited scientific evidence that is available on this point argues against this intuitive claim (Jamieson 1981; Scheibehenne *et al.* 2010; but see also Dobrjakova 1939)[10]. For example, in one study no evidence was found to support the claim that taste thresholds were lower (i.e. as would indicate an increased sensitivity to taste) when the participants in the laboratory tasted a variety of solutions with their eyes closed as compared to when they were open (Jamieson 1981).

"My tastebuds were tricked on a couple of occasions – for example, I was abso-lutely convinced I was eating chicken for my main until I found a small bone and realised it was fish." (Robathan and Whyman 2006, p. 40)

More recently, participants in another study also gave similar liking ratings to real foods no matter whether they ate in the darkness or not (Scheibehenne *et al.* 2010), despite the fact that the participants claimed to have paid more attention to the taste of the food in the former case. Unfortunately however, in this experiment no assessment was made of whether or not the participants' ratings of flavour intensity were affected when the lights were turned off. That said, the diners did say that they found it significantly more difficult to eat and that they paid significantly less attention to how much they ate under cover of darkness. This could be the reason why, when served a supersized portion, those who found themselves dining in the dark ate almost 20% more than those who could see the supersized portion in front of them. What is more, in the dark participants tended to underestimate the amount of food that they had eaten, while the reverse was true for those who ate under normal lighting conditions. The latter results might then tie in with the surprise that many of us have experienced at the cinema when, after purchasing a (usually oversized) tub of popcorn, we suddenly realize, as the final credits start to roll, that we have only a few unpopped kernels left (cf. Wansink and Kim 2005; Neal *et al.* 2011).

So, you think you know what you are eating? Consider then the meal experience in the Dans Le Noir restaurant in Camberwell London, as described by The Daily Telegraph's Richard Alleyne (2006, p. 32):

---

[9] Not to mention the oral burn associated with capsaicin in chillies, etc. (cf. Vernon and McGill 1961).
[10] By and large, the journalists would appear to be equally unconvinced (see Alleyne 2006; Liddle 2006; Robathan and Whyman 2006; Lane 2010).

The menu

Starter

What it tasted like: Strips of smoked salmon and battered prawns

What it really was: Salmon sashimi with pesto oil, crab spring roll

Main

What it tasted like: Lamb or beef mince moussaka

What it really was: Roasted fillet of barracuda rosemary dauphinois, grilled aubergine, roasted peppers and Asian butter sauce, sweet potato crisps

Dessert

What it tasted like: Pears in raspberry sauce

What it really was: Caramelised apples, Campari syrup, lavender ice cream and coconut biscuit

---

Source: Richard Alleyne, Telegraph, 2 March 2006. © Telegraph Media Group Limited 2006.

Another factor that makes the experience of dining in the dark unique, and for some rather unpleasant, is the uncertainty that is associated with not recognizing what it is that they happen to be taking into their mouths and ultimately swallowing. This uncertainty may well lead to decreased food acceptability ratings and to a decrease in people's willingness to try the food again subsequently (cf. Yeomans *et al.* 2008).

Wansink *et al.* (2012) investigated whether eating in the dark would differentially influence people's acceptance of foods that were either easy to recognize or not. The foods consisted of a beef *enchilada* consumed with a spoon (apparently not that easy to identify in the dark) and a cracker eaten by hand (easily identifiable even in the darkness). Two groups of participants evaluated the beef *enchilada*: one without any information about the product and the other after having been given the product's name. Each group rated the foods twice, once in the dark and the other time under normal indoor lighting conditions. Another two groups of participants carried out the same procedure but with the cracker instead. In the absence of vision, and in the absence of any information regarding the identity of the food, the beef *enchilada* was judged to be around 8% less acceptable and, what is more, less likely to be consumed again (by around 15%) than when eaten under normal lighting conditions. However, for those foods where the initial uncertainty was low (e.g. crackers), no such reduction in food acceptance or intent to consume was reported.

Could this be part of the reason why popcorn has become such a staple for those wishing to snack at the cinema?[11] Returning to our earlier discussion,

---

[11] Note that popcorn also has a number of other benefits when it comes to snacking in the dark: (1) it doesn't stain your clothes if you happen to miss your mouth; and (2) the shape of the kernels and sound of the crunch in the dark provide reassurance that what you are about to swallow is really what you thought it was and not something altogether different.

this might also be one of the main reasons why those chefs working at dine-in-the-dark restaurants tend to deliver flavours and dishes that are relatively easy for people to recognize. One can only wonder what diners choosing the 'surprise menu' have to say on this topic, given that they don't even know whether they are going to be eating fish or meat.

> *"Q: What sorts of foods do we experience differently when we can't see them?*
>
> *Dana (Salisbury): Meat seems to be the biggest challenge; people have a lot of trouble distinguishing between different kinds. That one is always a shocker, to hand someone fish and have them interpret it as meat. But you must understand that this is not only about food. This is about all the senses. So there are performances by artists throughout the meal between courses."* (Gleason 2006)

A lack of sensory expectations can even lead to confusion and to the illusory identification of flavours that aren't actually present (Piqueras-Fiszman and Spence 2011). Here, one might want to know what happens if we discover that what we have eaten wasn't actually what we originally thought it was?[12] Whenever we consume a food that we cannot recognize, we nevertheless still tend to create beliefs about what the food actually was (cf. Cohen 2011). What happens then if those beliefs don't match up to the reality of the situation, should we eventually find out what was really on our plates? What exactly are the consequences likely to be?

> *"Soup, roasted potatoes and meat. Veal? Chicken? Bread, and butter, which we spread messily. Some pudding for dessert. Vanilla? I can't remember. It tasted like vanilla but it might have been chocolate. Maybe it wasn't pudding but it seemed that way. None of the food tasted very good. Bland, bad texture. Indiscernible tastes and textures."* (Lane 2010)

Surprising as it might seem, the experience of eating in a 'dine in the dark' restaurant may also differ depending on whether a diner has his or her eyes open or not. The reason for this being that closing one's eyes can lead to our adopting a more interoceptive state of awareness (i.e. focusing on one's internal state; Marx *et al.* 2003; Wiesmann *et al.* 2006). By contrast, with our eyes open (even in complete darkness) the brain's attentional and eye-movement systems are more active (they are in what Marx *et al.* describe as a more exteroceptive or 'outward looking' state of awareness). It could therefore be argued that, if anything, any beneficial effects on the multisensory flavour perception of dining in the dark are more likely to be felt by those diners who choose to keep their eyes closed.[13] That said, at present there would appear to be little empirical support for the popular and widespread claim that dining in the

---

[12] Hopefully not anything like what Carlson (1930, p. 89) reports on first tasting a tomato that he thought was an apple: "*the disgusting, disagreeable effect on me of that fluid, insipid, warm mass that filled my mouth was something very striking, and I haven't forgotten it in forty years.*"

[13] Here, one could even imagine playing the sound of someone's heartbeat in the background in the restaurant as this is also likely to increase a diner's state of interoceptive awareness.

dark actually makes food and drink taste better or that it increases a diner's sensitivity to tastes and/or flavours.

Another reason that has been put forward to explain why it is that people want to try the 'dine in the dark' experience is the suggestion that it can give the diner a feeling for how the blind experience food; this is discussed in the following section.

8.5   Does dining in the dark really capture how the blind experience food?

This is the empathic claim. However, we would argue that it is unlikely to be true. Why? Normally sighted individuals typically have a great deal of stored knowledge concerning the appearance properties of foods. This means that once they have recognized the food via their other senses, they often can't help but create in their minds a potentially vivid mental image of what the food actually looks like (Lacey and Lawson 2013). They may even retrieve information concerning how it has been cooked and how much they like it (Simmons *et al.* 2005). This multisensory mental image might well then serve as an input and in some sense feed the cognitive process of eating (see Wolpin and Weinstein 1983; Spence 2011; Spence and Deroy 2013).

One other important question here is whether the blind taste and/or smell better than the sighted? Despite the fact that a number of researchers have addressed this question over the years, the available evidence on this point is still rather mixed. While some researchers have indeed documented superior olfactory and/or gustatory abilities in the blind (Cuevas *et al.* 2009), others have failed to observe any such differences (e.g. Rosenbluth *et al.* 2000). That said, something of a consensus now appears to be forming among researchers that the blind (and this includes the congenitally blind) don't perceive tastes, smells and/or flavours that the normally sighted cannot (see also Smith *et al.* 1993; Wakefield *et al.* 2004). That is, their detection thresholds are no different from those of age-matched sighted control participants. Where the blind do sometimes excel is in their ability to put a name to (i.e. to label) the smell, something that the rest of us usually find very difficult. In this regard then, the blind are much like other sensory experts in the food and beverage sector (such as wine tasters; Hughson and Boakes 2001; Lehrer 2009).

For all of the above reasons, we argue that dining in the dark does not really provide the sighted diner with an experience that is equivalent to that of a blind person dining. Furthermore, we have yet to meet a single diner who has had their perfect meal experience while sitting in complete darkness. Should any reader disagree, we would love to hear from you.

8.6 Cooking in the dark

Shifting topics briefly from the diner to the cook, preparing a meal without vision is certainly much more of a challenge than eating without being able to see the food. Several cookbooks have been written specifically for the blind chef however, in which the key guidelines are as follows: avoid instructions that refer to visual cues (e.g. "stir until the sauce clarifies"), instead referring to other senses such as texture or aroma (or even time); refer to measurements such as spoonfuls or cups and not necessarily grams or litres; and, of course, have Braille labels or recorded measurement equipment wherever possible. The cookbooks (in large print for the partially blind or in Braille) normally come with recorded audio. Some blind cooks also scan recipe books and convert them to electronic format to be able to 'read' them via their computer (screen reader software can provide the output either as a Braille display or as a speech output). Despite these possibilities however, some dishes such as the humble *omelette* will more likely end up as scrambled eggs when vision is denied.

Cooking in the dark might then be a very challenging task. There is actually a TV show in the US called 'Cooking without Looking' (http://www.pbs.org/food/shows/cooking-without-looking/), precisely to raise awareness about the challenges facing those who cook blind. It is interesting to note that some culinary programs actually include blindfolded tasting of foods as a standard part of a chef's training. The idea here is that a blindfolded trainee should learn to rely more on their nose and mouth and hence develop their discriminative abilities based on their other senses, rather than always simply relying on vision (see Gopnik 2011). Indeed, the best way in which to assess the doneness of a steak may well be with the fingers and we have heard of chefs assessing the status of their sautéing onions in the pan by the sound that they make (see also McGee 2004, p. 153; Stuckey 2012, pp. 128–129). Blindfolded tasting allows one to try out new combinations of flavours without any preconceptions and, hopefully, build up a library of flavours that may come in handy later.

8.7 Conclusions

Although it is undoubtedly the case that dining in the dark can make for a memorable multisensory experience, the available evidence suggests that you shouldn't go to such a restaurant if you are hoping that the absence of vision will make the food taste any better (Rosenblum 2010). For humans, as for many other creatures, visual cues play a crucial role in our perception of food and drink, as well as in the control of our appetitive behaviours. Furthermore, our sensory expectations regarding food or drink play a surprisingly large part in how we actually experience them (see Chapter 6). Hence, the removal of this (most) important source of sensory information (or part of it, such as any colour cues) will likely cause a detrimental effect in terms of the correct identification, and hence enjoyment, of our dining experience. It may also make the diner a little apprehensive. Nor, as argued in Section 8.5, does the dine-in-the-dark experience really give you an impression of what it's like for the blind to eat.

Returning to the question in the introductory section: what exactly is the appeal of visiting one of these restaurants? Rather than making the food and drink taste better, a claim that has yet to be substantiated empirically, or even giving one the sense of how the blind experience food, we would like to suggest that it is the feeling of constant unexpectedness (and surprise) that is what makes the experience so interesting for diners. As Mielby and Bom Frøst (2012, p. 234) put it: "*The second course was a salad, and almost every bite was a surprise.*" In this regard, and in this regard alone, dining-in-the-dark shares something with the experience of diners at a typical molecular gastronomy restaurant (see the previous chapter).

Finally, it is perhaps worth noting that although statistics regarding repeat custom at such restaurants are hard to come by, a straw poll of our friends and colleagues suggests that while many enjoyed the unusual sensory experience on offer at one of the growing number of dine-in-the-dark restaurants, few expressed any desire to repeat their experience (Spence and Piqueras-Fiszman 2012; see also Alleyne 2006; Liddle 2006). In conclusion, although it is an interesting experience, given what we have seen in this chapter it would seem very unlikely that diners will ever find their perfect meal in the dark.

References

Allen, J. S. (2012) *The Omnivorous Mind: Our Evolving Relationship with Food.* Harvard University Press, London.

Alleyne, R. (2006) Dinner is served … shame you can't see it. *The Daily Telegraph*, 2 March, p. 3.

Broackes, J. (2010) What do the colour-blind see? In *Color Ontology and Color Science* (eds J. Cohen and M. Matthen), pp. 291–405. MIT Press, Cambridge, MA.

Carlson, A. J. (1930) Physiology of hunger and appetite in relation to the emotional life of the child. In *The Child's Emotions: Proceedings of the Mid-West Conference on Character Development* (pp. 81–90). University of Chicago Press, Chicago.

Cohen, J. (2011) Wine expertise and the palate. Presentation given at the *Wine and Expertise Conference*, University of London in Paris. Paris, France. Oct 13th.

Crumpacker, B. (2006) *The Sex Life of Food: When Body and Soul Meet to Eat.* Thomas Dunne Books, New York.

Cuevas, I., Plaza, P., Rombaux, P., de Volder, A. and Renier, L. (2009) Odour discrimination and identification are improved in early blindness. *Neuropsychologia*, **47**, 3079–3083.

Deliza, R. and MacFie, H. J. H. (1996) The generation of sensory expectation by external cues and its effect on sensory perception and hedonic ratings: A review. *Journal of Sensory Studies*, **11**, 103–128.

Dobrjakova, O. A. (1939) Parallelism in the changes of electrical sensitivity in the organs of vision and taste under the conditions of optical and taste stimulation (in Russian). *Journal of Physiology of the USSR*, **26**, 192–199.

Gallace, A., Ngo, M. K., Sulaitis, J. and Spence, C. (2012) Multisensory presence in virtual reality: Possibilities and limitations. In: *Multiple Sensorial Media Advances and Applications: New Developments in MulSeMedia* (eds G. Ghinea, F. Andres and S. Gulliver), pp. 1–40. IGI Global, Hershey, PA.

Gill, A. A. (2007) *Table Talk: Sweet and Sour, Salt and Bitter*. Weidenfeld & Nicolson, London.

Gleason, P. (2006) No peeking before eating: Dana Salisbury's sightless diners. *Topic Magazine*.

Goldstein, D. (2005) The play's the thing: Dining out in the new Russia. In: *The Taste Culture Reader: Experiencing Food and Drink* (ed. C. Korsmeyer), pp. 359–371. Berg, Oxford.

Gopnik, A. (2011) *Sweet Revolution. The New Yorker*, 3 January. Available at http://www.newyorker.com/reporting/2011/01/03/110103fa_fact_gopnik (accessed January 2014).

Greenaway, P. (1989) *The Cook, the Thief, his Wife and her Lover*. Dis Voir, Paris.

Guterman, G. (2006) Valentine's Day dark dining. Concept and artistic direction by Dana Salisbury. CamaJe Bistro, New York City. 14 February 2006. *Theatre Journal*, **58**, 480–481.

Hanefors, M. and Mossberg, L. (2003) Searching for the extraordinary meal experience. *Journal of Business and Management*, **9**, 249–270.

Hayward, B. and Henry, B. (1977/2011) *Haywire*. Vintage Books, New York.

Hughson, A. L. and Boakes, R. A. (2001) Perceptual and cognitive aspects of wine expertise. *Australian Journal of Psychology*, **53**, 103–108.

Hurling, R. and Shepherd, R. (2003) Eating with your eyes: Effect of appearance on expectations of liking. *Appetite*, **41**, 167–174.

Hutchings, J. B. (2003) *Expectations and the Food Industry: The Impact of Color and Appearance*. Plenum Publishers, New York.

Huysmans, J.-K. (1884) *À Rebours* (Against Nature). Charpentier, Paris.

Jamieson, D. (1981) Visual influence on taste sensitivity. *Perception and Psychophysics*, **29**, 11–14.

Kotler, P. (1974) Atmospherics as a marketing tool. *Journal of Retailing*, **49**(Winter), 48–64.

Lacey, S. and Lawson, R. (eds) (2013) *Multisensory Imagery*. Springer, New York.

Lane, L. (2010) Eating blind. Available at http://www.huffingtonpost.com/lea-lane/eating-blind_b_701736.html (accessed January 2014).

Lehrer, A. (2009) *Wine and Conversation* (2nd edition). Oxford University Press, Oxford.

Liddle, R. (2006) Table talk: Dans Le Noir. *The Sunday Times*, 6 May. Available at http://www.thesundaytimes.co.uk/sto/style/food/article200348.ece (accessed January 2014).

Linné, Y., Barkeling, B., Rossner, S. and Rooth, P. (2002) Vision and eating behavior. *Obesity Research*, **10 (2)**, 92–95.

Long, R. (2010) Dining in the dark. Available at http://hub.aa.com/en/aw/dana-salisbury-jorge-spielmann-dark-restaurant-director-of-dark-dining-projects (accessed January 2014).

Marx, E., Stephan, T., Nolte, A., Deutschländer, A., Seelos, K. C., Dieterich, M. and Brandt, T. (2003) Eye closure in darkness animates sensory systems. *NeuroImage*, **19**, 924–934.

McGee, H. (2004) *On Food and Cooking: The Science and Lore of the Kitchen* (revised edition). Scribner, New York.

Mielby, L. H. (2007) *Expectations and surprise in a molecular gastronomic meal*. Masters thesis. University of Copenhagen, Faculty of Life Sciences, Department of Food Science, Sensory Science.

Mielby, L. H. and Bom Frøst, M. (2012) Eating is believing. In: *The Kitchen as Laboratory: Reflections on the Science of Food and Cooking* (eds C. Vega, J. Ubbink and E. van der Linden), pp. 233–241. Columbia University Press, New York.

Neal, D. T., Wood, W., Wu, M. and Kurlander, D. (2011) The pull of the past: When do habits persist despite conflict with motives? *Personality and Social Psychology Bulletin*, **37**, 1428–1437.

Newman, K. (2003) Dating in the dark. Available at http://news.bbc.co.uk/1/hi/england/london/3099734.stm (accessed January 2014).

Orledge, R. (1995) *Satie Remembered*. Faber and Faber, London.

Ozbaydar, S. (1961) The effects of darkness and light on auditory sensitivity. *British Journal of Psychology*, **52**, 285–291.

Pine, II,, B. J. and Gilmore, J. H. (1998) Welcome to the experience economy. *Harvard Business Review*, **76 (4)**, 97–105.

Pine, II,, B. J. and Gilmore, J. H. (1999) *The Experience Economy: Work is Theatre and every Business is a Stage*. Harvard Business Review Press, Boston, MA.

Piqueras-Fiszman, B. and Spence, C. (2011) Crossmodal correspondences in product packaging: Assessing color-flavor correspondences for potato chips (crisps). *Appetite*, **57**, 753–757.

Read, S., Sarasvathy, S., Dew, N., Wiltbank, R. and Ohlsson, A.-V. (2011) *Effectual Entrepreneurship*. Routledge, New York.

Robathan, M. and Whyman, K. (2006) Black magic? *Leisure Management*, **26 (2)**, 38–39.

Rosenblum, L. D. (2010) *See What I Am Saying: The Extraordinary Powers of our Five Senses*. W. W. Norton & Company Inc, New York.

Rosenbluth, R., Grossman, E. S. and Kaitz, M. (2000) Performance of early-blind and sighted children on olfactory tasks. *Perception*, **29**, 101–110.

Sacks, O. (1995) *An Anthropologist on Mars*. Picador, London.

Sautter, U. (2002) Dining in the dark. *Time*, 22 July. Available at http://www.time.com/time/magazine/article/0,9171,322741,00.html (accessed January 2014).

Schaefer, H. M. and Schmidt, V. (2013) Detectability and content as opposing signal characteristics in fruits. *Proceedings of the Royal Society London B*, **271 (Suppl)**, S370–S373.

Scheibehenne, B., Todd, P. M. and Wansink, B. (2010) Dining in the dark. The importance of visual cues for food consumption and satiety. *Appetite*, **55**, 710–713.

Simmons, W. K., Martin, A. and Barsalou, L. W. (2005) Pictures of appetizing foods activate gustatory cortices for taste and reward. *Cerebral Cortex*, **15**, 1602–1608.

Smith, R. S., Doty, R. L., Burlingame, G. K. and McKeown, D. A. (1993) Smell and taste function in the visually impaired. *Perception and Psychophysics*, **54**, 649–655.

Spence, C. (2010a) The multisensory perception of flavour. *The Psychologist*, **23**, 720–723.

Spence, C. (2010b) The color of wine – Part 2. *The World of Fine Wine*, **29**, 112–119.

Spence, C. (2011) Mouth-watering: The influence of environmental and cognitive factors on salivation and gustatory/flavour perception. *Journal of Texture Studies*, **42**, 157–171.

Spence, C. (2012) Auditory contributions to flavour perception and feeding behaviour. *Physiology and Behaviour*, **107**, 505–515.

Spence, C. and Piqueras-Fiszman, B. (2012) Dining in the dark. *The Psychologist*, **25**, 888–891.

Spence, C. and Deroy, O. (2013) Crossmodal mental imagery. In: *Multisensory Imagery: Theory and Applications* (eds S. Lacey and R. Lawson), pp. 157–183. Springer, New York.

Spence, C., Levitan, C., Shankar, M. U. and Zampini, M. (2010) Does food color influence taste and flavor perception in humans? *Chemosensory Perception*, **3**, 68–84.

Strunk, O. (1998). *Source Readings in Music History*. W. W. Norton & Company, London.

Stuckey, B. (2012) *Taste What You're Missing: The Passionate Eater's Guide to Why Good Food Tastes Good*. Free Press, London.

Vernon, J. and McGill, T. E. (1961) Sensory deprivation and pain thresholds. *Science*, **133**, 330–331.

Vogelzang, M. (2010). TEDxMunich - Marije Vogelzang - Food Love. Available at http://www.youtube.com/watch?v=1WZv459bOCQ (accessed January 2014).

Wakefield, C. E., Homewood, J. and Taylor, A. J. (2004) Cognitive compensations for blindness in children: An investigation using odour naming. *Perception*, **33**, 429–442.

Wansink, B. and Kim, J. (2005) Bad popcorn in big buckets: Portion size can influence intake as much as taste. *Journal of Nutrition Education and Behaviour*, **37**, 242–245.

Wansink, B., Painter, J. and North, J. (2005) Bottomless bowls: Why visual cues of portion size may influence intake. *Obesity Research*, **13 (1)**, 93–100.

Wansink, B., Shimizu, M., Cardello, A. V. and Wright, A. O. (2012) Dining in the dark: How uncertainty influences food acceptance in the absence of light. *Food Quality and Preference*, **24**, 209–212.

Weiss, A. S. (2002) *Feast and Folly: Cuisine, Intoxication and the Poetics of the Sublime*. State University of New York Press, Albany, NY.

Wiesmann, M., Kopietz, R., Albrecht, J., Linn, J., Reime, U., Kara, E., Pollatos, O., Sakar, V., Anzinger, A., Fest, G., Brückmann, H., Kobal, G. and Stephan, T. (2006)

Eye closure in darkness animates olfactory and gustatory cortical areas. *NeuroImage*, **32**, 293–300.

Wilson, G. D. and Gregson, R. A. M. (1967) Effects of illumination on perceived intensity of acid tastes. *Australian Journal of Psychology*, **19**, 69–73.

Wolpin, M. and Weinstein, C. (1983) Visual imagery and olfactory stimulation. *Journal of Mental Imagery*, **7**, 63–73.

Yang, A. (2007) Dining in the dark; Waiter, I'm at your mercy. *The New York Times*, 22 July. Available at http://query.nytimes.com/gst/fullpage.html?res=9803E5D61531F931A15754C0A9619C8B63 (accessed January 2014).

Yeomans, M., Chambers, L., Blumenthal, H. and Blake, A. (2008) The role of expectancy in sensory and hedonic evaluation: The case of smoked salmon ice-cream. *Food Quality and Preference*, **19**, 565–573.

9
How Important is Atmosphere to the Perfect Meal?

9.1   Introduction

A few years ago Carme Ruscalleda, chef at the Sant Pau restaurant in Catalonia (and back then the only Spanish woman to have been awarded three Michelin stars), had a problem. Her restaurant had a beautiful cliff-top location and all the Mediterranean ambience that one could wish for. However, her fear was that the guests were not appreciating the amazing food that she was serving quite as much as she had hoped they might. She worried that some of her diners were getting too distracted by the beautiful surroundings, the sounds of nature – the waves crashing on the beach below, the birds singing overhead – and hence were not paying her food the attention that she thought it deserved. Carme told us in 2009 how she had eventually decided to bring her diners back inside when it came time to eat, specifically so that she could eliminate the distracting atmosphere of nature (in all its glory). The only thing that guests can do outside is drink their aperitifs before the meal, and their coffee after it. The question to be addressed in this chapter is: was she right? Can changing the atmosphere really have such a dramatic effect on our experience of a meal?

At the outset, it is worth bearing in mind that while the question might seem like a simple one, people appear to disagree fervently. Some of those whom we have spoken to are convinced that the environment in the restaurant, say, has no impact whatsoever on the taste of the food. Take for example the following quote from top French chef Alain Senderens talking about the Michelin

---

The Perfect Meal: The Multisensory Science of Food and Dining, First Edition.
Charles Spence and Betina Piqueras-Fiszman.
© 2014 John Wiley & Sons, Ltd. Published 2014 by John Wiley & Sons, Ltd.

Guide: "*I was spending hundreds of thousands of euros a year on the dining room – on flowers, on glasses,*" he said, "*but it didn't make the food taste any better*" (Steinberger 2010, p. 78). These individuals are betting that their guests can focus their attention solely on the flavours and textures of the food, ignoring 'the everything else'. But can they?

Others, such as Carme Ruscalleda, are convinced that the atmosphere exerts a profound influence over what we think about that which we happen to be eating, and how much we enjoy the overall dining experience. One finds this divergence of opinion not only among established chefs but also among the general public. Indeed, we have worked with a number of top chefs who say that great meal experiences are all (and only) about the food. That said, these are sometimes the chefs whose restaurants are located in old knitting museums. We have come across other aspiring chefs who are happy to let the duty manager put the radio or, worse still, their own personal i-pod selection (and let it blast out over the dining room with who knows what playing). At the other extreme, we also see a growing number of restaurateurs going full tilt on the design of immersive and entertaining multisensory environments. Indeed, the emergence of 'the experience economy' (e.g. Hanefors and Mossberg 2003; Jacobsen 2008) has resulted in some in the trade becoming far more concerned about selling the right atmosphere than about providing the best food that they can or, at the very least, giving equal importance to both elements of the dining experience (e.g. Gill 2007; see also Goldstein 2005). As Pine and Gilmore (1998, p. 99) put it: "*At theme restaurants such as the Hard Rock Café, Planet Hollywood, or the House of Blues, the food is just a prop for what's known as 'eatertainment'.*" There is even a vocal and increasingly active group of restaurant critics out there who has started to complain that the background noise in many restaurants is simply too loud nowadays and that they can no longer taste/enjoy their food (see McLaughlin 2010; Sietsema 2008a, b).[1] Indeed, as we will see in Section 9.11, laboratory-based research is starting to provide support for such claims (e.g. Woods *et al.* 2011; Spence 2012).

> "*A loud noise, for instance, may prevent entirely our ability to smell or taste, yet softly played dinner music can create an environment favourable for elegant dining.*" (Crocker 1950, p. 7)

Chefs and restaurateurs alike obviously need to know just how important the atmosphere really is to the delivery of a great meal. If it turns out that it doesn't influence the diner's perception of the food then they can focus all of their energy on perfecting the dishes that they serve and ignore the rest (the 'everything else'). However, if the atmosphere does turn out to exert anything

---

[1] For those who suffer from misophonia and cannot stand the sound of others eating, loud background music might be just the right thing (Edelstein *et al.* 2013).

like as much influence as some people say, then one is going to need to pay far more attention to the décor and background music than many of those working in the business currently do. In fact, after reading about some of the research that we'll describe in this chapter, one might even start to wonder just how much of the restaurant critics' (or restaurant guides') ratings depend on the soft furnishings and bathroom fixtures and how much on the actual quality of the food being served. In recent years, many of the top restaurants have started to collaborate with well-known architects and interior designers with the clearly stated aim of trying to create a truly unique dining atmosphere (see Horwitz and Singley 2004; Sheraton 2004, p. 144), capturing the public's (not to mention the media's) attention in the process (Kotler 1974; Sharp 2013).

"The Michelin Man, they came to believe, had a yen for luxury and wanted his surroundings to be as sumptuous as his food." (Steinberger 2010, p. 75)

It is our belief that, no matter how good the food is, the diners' impression of a meal is always going to be influenced by the atmosphere of the place in which they are eating. That said, we are also of the opinion that even the most fabulous of multisensory atmospheres can never make up for a fundamentally poor product offering. However, given that people's intuitions regarding the relative importance of getting the atmosphere right when it comes to the delivery of the perfect meal differ so markedly, we are going to need to refer to the latest findings from the new science of gastrophysics in order to help resolve this issue once and for all. Ultimately, we argue that the perfect meal requires both great-tasting food and that the dish is consumed in an atmosphere that has been optimized to bring out the best in both the food and in the diner.

"Atmosphere is every bit as important as food. You can eat great food, but if it's in a tedious place, so what?" (the late, great, Michael Winner 2008, p. 14)

"It is possible that the food alone may have less influence on perceived quality than the environmental factors that come into play." (Mielby and Bom Frøst 2012, p. 238)

Note that the atmosphere in a restaurant is determined by more than simply what it looks or sounds like. The social dynamics and company are obviously very important too, as we saw in Chapter 2 (Allen 2012, p. 264). While the majority of research and money is spent on getting the visual appearance or décor just right, there is now a growing interest in making sure that the background noise levels are not too loud and that any music that happens to be playing over the speaker system is appropriate to the kind of food (and drink) being served. Some researchers are also starting to think about how to harness the evocative power of smell in order to enhance the experience of their diners (Spence 2002). As we will see in Section 9.10 some practitioners have quite literally started to work on the feel of their restaurants, whether in terms of the coverings of the chairs on which their diner sits or else playing with the ambient temperature in the dining room during the course of a meal (Anderson 2004; Spence *et al.* 2013a).

> "*Most of us respond favourably to food only in clean, attractive, or even beautiful surroundings, and appetite for its reception varies accordingly.*" (Crocker 1950, p. 7)

Of course, even before making up one's mind about where to eat, a restaurant's atmosphere can influence us. Just think about walking down a street full of restaurants while on holiday in some unfamiliar city. What are the key factors that should, and actually do, influence our choices of where to eat and on our experience of the food thereafter?[2] Clearly the prices on the menu and the overall cleanliness of the restaurant are going to be high up on everyone's list. We also respond positively to those venues that look popular – a paucity of diners rarely bodes well. It has even been suggested that candlelit restaurants appeal to so many of us because, at some primeval level, they remind us of the caves where our ancestors once lived (and presumably also cooked and ate their dinner). In his best-selling book *The Omnivorous Ape*, Watson (1977, p. 151) describes how the dark interior of the cave was the place where, for a few hours at least, our ancestors would have been relatively safe from danger (see also Lyman 1989).[3]

Before we have decided where to eat, we may feel that we have some clear idea about the type of cuisine that we are looking for. But do the tablecloths in the restaurant really matter? Can they influence the food and restaurant choices that we make? And how influential are the smells percolating out of the front door, or for that matter the cosy candlelit atmosphere seen through the restaurant windows from the street outside, on determining where we ultimately choose to eat? And beyond all of these factors, what other cues influence our decision as to where we spend our hard-earned cash?[4] Is it the

---

[2] The recently deceased North American chef Charlie Trotter took things even further. The staff working under him would apparently sometimes be tasked with picking up the cigarette butts from off the sidewalk outside his namesake Lincoln Park restaurant in Chicago. Trotter didn't want the litter to detract from the experience of approaching his restaurant (Anonymous 2013a)! On our last visit to see Heston Blumenthal in Bray, we also came across someone scrubbing the pavement clean outside the entrance to his restaurant!

[3] However, Watson (1977) would appear to be out of touch with contemporary trends in restaurant design when he goes on to state that diners don't like restaurants with big windows. Obviously, no one told those who worked on Heston Blumenthal's Dinner in London, for example. This restaurant, which was recently awarded its second Michelin star, boasts floor to ceiling windows along its Hyde Park side.

[4] Even the behavioural economists have recently started to recognize this as a problem that merits some serious academic consideration. In his book *An Economist gets Lunch*, Tyler Cowen (2012) recently offered advice on analysing the make-up of the clientele in order to determine which restaurant offers the best option.

Greek taverna or that French bistro, the Italian trattoria or the new burger joint no doubt selling 'proper' burgers?

Given the various different ways in which the environment affects us, it should come as little surprise that over the years many different theories have been put forward in order to explain what may be going on in the mind of the diner. Here, we review the evidence relevant to understanding how the multisensory atmospherics can alter a diner's mood, as well as their level of arousal (Kupferman 1964; Smith and Curnow 1966; Ferber and Cabanac 1987; Konečni, 2008; see also Chapter 2). We also highlight claims suggesting that changes to the background music can affect a diner's experience of the passage of time. We revisit the notion of sensation transference (which first came up in Chapter 4) by showing how what a diner feels about the environment can often carry over to influence what they think about the food itself (North and Hargreaves 1996; Velasco *et al.* 2013). Finally, we highlight the importance of the expectations that certain environments create in the mind of the diner, and how they can be transferred to influence what we feel about the quality and acceptability of the food. By the end of the chapter, we hope to have convinced you that getting the atmosphere just right really does constitute a key factor contributing to the delivery of the perfect meal.

"*People don't come to our restaurants because they're hungry.*" (Joseph H. Baum, cited in Sheraton 2004, p. 138)

9.2 Atmospherics and the experience economy

The concern with getting the atmosphere just right certainly isn't an entirely new phenomenon. Back in the mid-1960s for example, the owner of the highly successful Pier Four restaurant in Boston said: "*If it weren't for the atmosphere, I couldn't do nearly the business I do*" (Anonymous 1965).[5] Nor is it restricted to the restaurant context. Indeed, particularly influential in this regard over the last few decades has been Philip Kotler's work on atmospherics. As the great marketer himself once put it (Kotler 1974, p. 48; see also Wysocki 1979): "... *one of the most significant features of the total product is the place where it is bought or consumed. In some cases, the place, more specifically the atmosphere of the place, is more influential than the product itself in the purchase decision. In some cases, the atmosphere is the primary product.*"

"*There is nothing new about themed restaurants. In fact, show me a restaurant that isn't. Ever since the first public dining-room, restaurateurs have been asking,*

---

[5] Here one might also be reminded of the world-famous Tonga Room & Hurricane Bar that opened at the Fairmont Hotel in San Francisco in 1945 (see Lanza 2004; http://en.wikipedia.org/wiki/Tonga_Room). Until recently, it recreated a spectacular tropical thunderstorm every hour or so.

'What shall we make it look like?' The difference recently has been that, whereas the set used to complement the food, now the food is an adjunct to the marketing – or the 'experience' as it is invariably called." (Gill 2007, p. 102)

More recently, Pine and Gilmore published a *Harvard Business Review* article and book both entitled *The Experience Economy* (Pine and Gilmore 1998, 1999). The authors' central argument in these works is that the path to success for many companies, especially those operating in the service sector (and that includes restaurants and bars) is that the public want to consume experiences rather than simply to purchase products. Take for example the Starbucks coffee chain: much of the phenomenal success of this brand has been put down to the welcoming atmosphere of the stores and the subdued public space (at least when compared to the colours used in fast food restaurants) with couches and bookshelves (Luttinger and Dicum 2006, p. 159). Elsewhere, when it comes to thinking about restaurants and how the senses can potentially be harnessed to enhance the multisensory dining experience, many people mention London's very own Rainforest Café (http://www.therainforestcafe.co.uk/). This restaurant certainly delivers a truly multisensory experience, one that stimulates all five of a customer's senses. On entering the restaurant, diners hear the "Sss-sss-zzz" sound of the rainforest. This, together with the other congruent environmental sensory cues, is designed to make the diners' experience more 'authentic'.

"The mist at the Rainforest Café appeals serially to all five senses. It is first apparent as a sound: Sss-sss-zzz. Then you see the mist rising from the rocks and feel it soft and cool against your skin. Finally, you smell its tropical essence, and you taste (or imagine that you do) its freshness. What you can't be is unaffected by the mist." (Pine and Gilmore 1998, p. 104)

While some critics have baulked at the growing popularity of such experiences (see Heldke 2005; Gill 2007)[6], no one can doubt the phenomenal growth in sales that such experience providers have seen in recent years (see Pine and Gilmore 1998, 1999). In fact, nowadays, one finds the experience economy everywhere – indeed, it can sometimes seem as though everyone on the high street is trying to sell it to us, no matter whether one is talking about Nespresso, Samsung or Nike (Spence *et al.* in press).

---

[6] While it is certainly true that *"Customers seek a dining experience totally different from home, and the atmosphere probably does more to attract them than the food itself"* (Anonymous 1965), it is hard not to be sceptical of venues like the Cage Ke'ilu (Café Make Believe). Pine and Gilmore (1998, p. 101) report that *"The establishment serves its customers plates and mugs that are empty and charges guests $3 during the week and $6 on weekends for the social experience."* Is this really the epitome of the experience economy approach to the marketing of dining? Let's hope not (although it has to be said that Pine and Gilmore appear quite taken with the idea).

Quite understandably, some chefs and restaurateurs have started to rebel against this focus on the atmosphere rather than the food. Indeed, a number of the world's top chefs have actually started to worry that the atmosphere of the surroundings, whether the art on the walls or the music playing over the Tannoy, may be distracting their diners from the great-tasting food that they happen to be serving. At top-end restaurants such as The Fat Duck in Bray, Per Se in New York and Alinea in Chicago, the atmosphere is deliberately kept minimalistic. The idea here is to keep the diners' attention squarely focused on the food itself. As Damrosch (2008, p. 22) puts it when describing the ethos at Per Se, "*there is no art on the walls or music in the room: the focus was on the food and the experience of dining.*"

At Chicago's Alinea, one finds oneself in a similar setting. The dining room lacks any music (but many diners probably don't even realize) and the walls are a monotone grey without any potentially distracting ornamentation. One certainly wonders whether this constitutes a better idea than putting some soft music on in the background? Perhaps the idea is to focus the diner's attention on the different sounds that the various dishes themselves might make (or on the waiters' descriptions of the dishes). That said, back in 2011 chef Grant Achatz announced his own plans to revamp his Michelin-starred restaurant by adding different layers of sensation that would include music as one of the 'ingredients' in the whole multisensory dining experience. At the same time, Achatz hoped to move diners from one space to another during the course of their meal. He even went so far as to suggest that the overhaul of the restaurant might include a cellist or musician emerging on scene to play a single note or a full piece of music (Ulla 2011). Although Achatz was quoted as saying that this idea would soon be realized, we are not sure that it has as yet.

To summarize what we have seen in this section, the last couple of decades has seen a widespread growth of interest in the topic of atmospherics, from those restaurateurs who have read the management books and come to the realization that they can significantly increase their sales simply by focusing on selling multisensory experiences through, at the other extreme, to the top chefs who are increasingly worrying about whether the environmental stimulation may be distracting diners from truly appreciating their food. We would like to argue that both approaches – the minimalistic and the experiential – have a role in today's vibrant marketplace for fine dining. That said, the former approach can easily go wrong if not handled with care.[7] By contrast, no one can argue with the phenomenal success that the experiential approach to marketing has had across a very wide range of sectors in the marketplace (e.g. Pine and Gilmore 1998, 1999; Hanefors and Mossberg 2003; Jacobsen 2008; Homburg *et al.* 2012).

---

[7] Indeed, many a restaurant has ended up closing its doors for the last time because the diners found the floor to ceiling white walls and minimalistic décor too austere. Perhaps such an approach only works if combined with the very best in haute cuisine.

9.3   The Provencal Rose paradox

"The experience is familiar to many. There you are sitting in the sun, au bord de la mer *on the* Côte d'Azur, *eating delicious sea food with a loved one, and drinking Provencal rose. The glass frosted with condensation contains a wine of pale salmon colour; the bottle rests in the ice bucket. The experience is intensely pleasurable and at that moment you can come to believe that this is one of the most enjoyable wines you have ever had. Later you buy a case of the wine and when you open it back home on a cold grey day, it has lost all its savour. This is not an exceptional wine. It is not even that enjoyable. What has brought about this decline in the value of the wine from Provence? Is it that it doesn't travel? We know that wine making techniques and preservation have dramatically increased and there is no reason to suppose – if the wine was properly transported and stored – that the wine has suffered any more than any wine that comes from foreign shores. Its fate cannot be due to a dramatic change in the wine. So what explanation can we give of this paradox of Provencal rose?"* (Smith 2009)[8]

As self-styled philosopher of wine (there are worse jobs!) Professor Barry Smith of the University of London put it in the above quote, we have all had the experience of foods and wines that taste great when we are on holiday in the Mediterranean, but which never quite live up to the expectation when we bring them back home to share with friends on a cold, dark winter's night. Obviously not everyone gets to relax on the shores of the Mediterranean but wherever one grew up, whether South Africa or South America, you are likely to have had a similar experience; this is referred to as the Provencal Rose paradox. While most people are familiar with the experience, there has been surprisingly little scientific research on the topic to date.

"I also recognized that I was perhaps prone to a certain psychophysical phenomenon, common among France lovers, whereby the mere act of dining on French soil seemed to enhance the flavour of things." (Steinberger 2010, p. 5)

All this discussion can make one wonder: if the atmosphere does turn out to have such a profound effect, could it perhaps be that the restaurant critics are actually rating the quality of the atmosphere and environment of the restaurant as much as the food itself? Certainly, according to some writers, the Michelin guide *"overcompensates chefs who invest heavily in*

---

[8] The behavioural economists who worry about questions such as *"Why does Mexican food taste different in Mexico?"* (Cowen 2012, p. 187) typically look for answers in terms of the law of supply and demand (the sourcing of the raw ingredients, etc). However, we argue that the difference in atmosphere from one eating location to another is just as, if often not more, important; *that* is a large part of why the food tastes different in one environment relative to another.

*their setting (and location) and undercompensates those who strictly focus
on cuisine quality*" (Steinberger 2010, p. 75). Chossat and Gergaud (2003)
have gone some way towards answering this very question. They analysed
the reviews of the near-200 restaurants run by the Master Chefs of France
that appeared in the popular Gault-Millau restaurant guide. While the
appearance of the restaurant was found to exert a significant impact on
whether the critics/reviewers classified it as top, reassuringly (for all those
'foodies' out there – or are they all 'gastronomes' now?) this did not turn to
be as important a factor as the quality and creativity of the food itself.

> *"'Aren't your reviews very subjective?' I have often been asked in a somewhat
> accusatory tone. The answer is, 'You betcha … and how!' If I qualify or expand on
> that answer, it is usually with 'they are objectively subjective', meaning I reported
> only what I thought – my opinion – without modifying it to include outside influ-
> ences."* (Sheraton 2004, pp. 98–99; for many years Sheraton was the restaurant
> critic at The New York Times)

The location of the restaurant may also impact on how we rate a dish. Certainly, if the atmosphere or views are great, you will likely be paying for it
(Sheraton 2004; Cowen 2012, p. 66). Gergaud *et al.* (2007) have demonstrated
that, at least for the case of Parisian restaurants, a chef's likelihood of being
awarded Michelin stars is to some degree dependent on the location they
choose for their restaurant. That said, we can all probably think of restaurants
where such a generalization does not hold true.

While it might not be worth the chef/restaurateur paying the amount needed
for the extra Michelin star (Steinberger 2010), the atmosphere can definitely
impact everything from what we think about food to the perceived ethnicity
of a meal, as we will see in the following section. While naturalistic sounds are
sometimes used to enhance the dining experience at The Fat Duck in Bray
(Blumenthal 2007, 2008; Spence *et al.* 2011), it is important to note that environmental atmospherics can also distract the diner. In other words, a Mediterranean atmosphere isn't necessarily a good thing; it really all depends what the
restaurant/chef is trying to achieve in terms of the dining experience. While
listening to the sounds of the sea may be appropriate when a diner happens to
be tucking in to some seafood on the outskirts of Slough, that doesn't mean it
will necessarily be the right thing to listen to while tucking in to the rest of the
sophisticated cuisine being served in a cliff-top restaurant in Catalonia, say.
What is crucial here is that the ambience is *congruent* with the spirit of the
dish and the intentions of the chef.

Having set the scene, it is now time for us to take a look at the evidence
that has been collected by the psychologists and sensory scientists working in
this area.

9.4   Does the atmosphere really influence our appraisal of the meal?

Over the course of a number of studies, Meiselman and his colleagues have systematically assessed just how much the atmosphere influences our appraisal of the meal (Meiselman *et al.* 1988, 2000; Meiselman 2008). In one experiment, these researchers measured the acceptability of pre-prepared meals served in one of several different locations: a grill room, a military refectory, a cafeteria, a science laboratory class and a training restaurant at a university located in the south of England (Meiselman *et al.* 2000). Crucially, the diners who took part in this study rated their meals as significantly more acceptable when served in the restaurant setting, while finding them least acceptable when they were served in the cafeteria.[9] The results suggested that the environment was capable of lifting a diner's ratings of a meal by as much as 10%.

Meanwhile, in another study, Edwards *et al.* (2003) compared ten different physical locations varying on several dimensions including the style of service and dining, the choice of starter, whether the diners had a free choice of food and whether they had to pay when selecting their meal. Once again, the location in which the food was served exerted a significant effect on the overall acceptability of the same meal. When the Chicken à la King and Rice dish was served on a white tablecloth in a 4-star hotel restaurant it received a rating that was almost 20% higher than when the very same dish was served in an army training camp, a university staff refectory or in a private boarding school. These results build on a sizeable body of earlier research showing that a large part of the variance that is seen in terms of people's rating of the acceptability of a given food or dish depends on the location in which it happens to be consumed (see Green and Butts 1945; Belk 1974; Michaels 2010). Taken together, such results clearly suggest that our perception of a meal can be profoundly influenced by the atmosphere in the restaurant. The perfect meal most likely requires the perfect atmosphere if the overall experience is not to be rated as disappointing.

9.5   On the ethnicity of the meal

One of the ways in which the atmosphere influences diners is by modulating the perceived ethnicity of the foods they eat. Evidence concerning the influence of the décor on the diner comes from an influential study conducted by Bell *et al.* (1994). These researchers assessed whether it is possible to alter the ethnicity of a dish without necessarily modifying the food itself. To this end, the Grill Room at Bournemouth University, UK offered up a selection

---

[9] Acceptability in this context refers to a diner's overall liking of the food.

of Italian and British foods over four consecutive days. On the first two days, the food was served with the restaurant decorated as it normally would be (white tablecloths and with the walls and ceiling left unadorned). For the second two days, the food was identical but the restaurant was given an Italian theme: Italian flags and posters were mounted on the walls and ceiling, red and white chequered tablecloths were thrown over the tables and a wine bottle was placed carefully in the middle of each table ... you get the general idea.

After having made their food choices and finished their meals, the customers ($N = 138$) were given a questionnaire in which they had to rate how ethnic their meal had been, as well as how acceptable they found the food. Giving the restaurant an Italian theme was all it took to bias the restaurant's customers toward choosing more pasta and Italian desserts (e.g. ice-cream and zabaglione) and significantly less fish items (such as trout). Furthermore, the Italian feel resulted in the diners' ratings of the perceived ethnicity of the pasta items going up significantly, as did their perception of the Italian identity of the meal as a whole. Such results therefore provide a powerful demonstration of how the diner's perception of the ethnicity of a dish can be influenced simply by changing the visual attributes of the environment in which that dish happens to be served.[10]

While Bell and his colleagues focused on manipulating the décor in the restaurant, Yeoh and North (2010) reported that the ethnicity of the music playing in the background can also influence people's food choices. These researchers served Malay and Indian food while either Malay or Indian music was playing. Once again, people's food choices were biased by the music. That said, Yeoh and North's results also suggested that such effects may be stronger under those conditions in which a diner does not have a strong pre-existing preference to choose a particular kind of cuisine.

When taken together, the results of the studies that have been reported in this section clearly highlight how the atmosphere of the restaurant can influence the perceived ethnicity of the food that we eat. Ford Coppola, a serious gourmand, was right to demand "*musical accompaniments matched to menus – accordion players for an Italian pranzo, mariachi for Mexican comida*" (Sheraton 2004, p. 172). Researchers to date have primarily been interested in investigating the influence of altering just one sensory attribute of the environment at a given time (i.e. either changing only the décor *or* else just manipulating the background music). It would certainly be interesting in future research to assess, for example, the relative importance of visual versus auditory atmospheric cues in determining the food choices and perceived ethnicity of dishes that are chosen by diners (we'll see some examples of holistic approaches in Section 9.12). A second intriguing question for future

---

[10] It should however be borne in mind that this isn't quite the whole story here, since the foods in the study of Bell *et al.* (1994) were also described on the menu using ethnic names (e.g. *maccheroni al formaggio* on the Italian days versus *macaroni and cheese* on the British days). Given what we saw in Chapter 3, this is also likely to have impacted on people's ratings of the foods.

research in this area will be to ascertain whether there are any kinds of superadditive effects here (cf. Spence 2002). For example, will Italian foods be selected even more often (and will estimates of the ethnicity of the dish increase still further) if some Italian opera is played alongside the introduction of the chequered tablecloths and Italian flags and posters. Certainly, according to the fundamental neuroscience, the result of combining different sensory inputs in a congruent manner can sometimes be far bigger than the sum of the effects seen when each of the senses is studied independently (Stein and Meredith 1993; Spence 2002).

> "*Integrated sensory inputs produce far richer experiences than would be predicted from their simple coexistence or the linear sum of their individual products … The integration of inputs from different sensory modalities not only transforms some of their individual characteristics, but does so in ways that can enhance the quality of life.*" (Stein and Meredith 1993)

Elsewhere in the literature, the research shows that people's wine choices can be profoundly influenced (far more than any of us likely realize) by the music that happens to be playing in the background in a store. In one classic study (North *et al.* 1997, 1999), shoppers in a British supermarket situated in the suburbs of a city in the East Midlands bought far more French music when French accordion music was played over the Tannoy (73% of the bottles sold as compared to 27% German), while buying far more German wine when Bierkeller music was played instead (77% of the bottles sold, as compared to 23% French). In this case, the change in people's behaviour was by no means subtle. The other striking result to emerge from the study of North *et al.* (1997, 1999) was that when the customers were questioned after leaving the tills about whether the background music had influenced their purchasing decision, most said that it had not (when the sales figures showed that it very obviously had)! Such results therefore suggest that the restaurateur should sometimes be cautious about taking their diner's verbal comments about why they did, or did not, like a particular dish or meal at face value. All of us are sometimes unaware of the true drivers of our behaviour (Hall *et al.* 2010).

9.6   Tuning up how much money and time we spend at the restaurant

9.6.1   The style and volume of the music

Across a range of contexts, background music has been shown to influence how much people choose to spend and their overall perception of the quality of the service that they receive (see Baker *et al.* 1994; Grewel *et al.* 2003). For example, when music with more 'upmarket' associations (e.g. classical music) is played (rather than, say, Top 40 hits), the shoppers in one study spent significantly more in the setting of a North American wine store (Areni and Kim 1993). *A priori*, there would seem to be little reason to doubt that exactly the same pattern of results would also be obtained should one look at the impact of playing different kinds of music in, say, the context of fine dining (see also Spence *et al.* 2013b).

Early indications that this might indeed be the case come from a study by North and Hargreaves (1998) in which people expressed a willingness to pay more when classical music was played in a student café rather than easy listening, pop or no music. Taking things one step further, Wilson (2003) and North *et al.* (2003) have both demonstrated that diners do indeed spend more when classical music is played in a restaurant. In the latter study, the average spend was 10% higher with classical music than with pop music or else when there was no music (taking the average figure up to £33 from £30 per person; see also Spence *et al.* 2013b).

Of course, one could be forgiven for thinking that such an approach to increasing sales would only work when the restaurant or cafeteria itself has an atmosphere that is appropriate, or congruent, with the associations of classical music. However, Wilson (2003) observed a similar pattern of results in a Sydney restaurant whose name, Out of Africa, would hardly seem congruent with classical music. Nevertheless, the 300 diners in this study spent significantly more when classical, jazz or popular music was played than when there was easy-listening music, or else when the music was turned off. One study whose results can be taken to provide at least partial support for the idea that the congruency between the music and the restaurant environment sometimes does matter comes from Lammers (2003). Here, the diners were observed to spend around 15% more when soft (i.e. quiet) as opposed to loud background classical or soft rock music was played. This result was put down to the fact that the quieter music provided a better match for the 'serene' (the author's description) oceanside restaurant in Ventura, California than the louder music. It should however be borne in mind that only 80 diners were observed in this study, a relatively small number as compared to some of the other restaurant studies discussed in this chapter.

Both the style and loudness of the music influence how long a person chooses to stay in a given venue eating and/or drinking (Jacob 2006). Loud music or noise during a meal will often accelerate the rate at which food is eaten (Stroebele and de Castro 2004). Unsurprisingly, the more discomforting the noise the less time people spend in a restaurant (North and Hargreaves 1996), while the more they like the music, the longer they stay (Wansink 1992). The musical mode (e.g. major or minor; see Knöferle *et al.* 2012) will presumably also have an impact; to date however, this factor has not been studied in the context of the restaurant. Perhaps less obviously, it has also been suggested that changing the type of music playing in the background can influence a diner's (or drinker's) awareness of the passage of time (see

Drews *et al.* 1992; Kellaris and Kent 1992; Kellaris *et al.* 1996; Caldwell and Hibbert 1999). The idea here is that if a certain piece of music happens to make people think that time is passing more slowly, then they might well end up spending longer in whatever environment they are in and hence possibly eating, drinking and ultimately spending more as a result (Roballey *et al.* 1985; McCarron and Tierney 1989; North and Hargreaves 1996, 2008; Yalch and Spangenberg 2000; Caldwell and Hibbert 2002; Sullivan 2002; see also Guéguen *et al.* 2004, 2008; Stafford *et al.* 2012).

9.6.2   The tempo of the music

The atmosphere affects the speed and duration of dining. In one oft-cited early study, Milliman (1986) manipulated the tempo (that is, the number of beats per minute or bpm) of the music playing in a medium-sized restaurant in Texas, and assessed its impact on behaviour. The 1400 diners who unwittingly took part in this study ate significantly more rapidly when fast instrumental music was played than when slower music was played instead. In the latter condition, the diners spent more than 10 minutes longer eating, bringing the total duration of their restaurant stay up to almost an hour. Although there was no effect of musical tempo on how much people spent on their food, there was a marked difference on the final bar bill with those in the slow music condition spending around a third more (equating to each and every patron ordering an average of three extra drinks). To put this into monetary terms, slowing the music down resulted in an increase of nearly 15% in the gross margin at the restaurant; not bad when you consider that this study was conducted on Fridays and Saturdays (i.e. on the restaurant's two busiest days).

In another study, playing fast-tempo instrumental non-classical background music ($M = 122$ bpm) resulted in cafeteria diners taking significantly more bites per minute than when a slow-tempo piece of music ($M = 56$ bpm) was played instead (4.4 versus 3.2 bites per minute, respectively; Roballey *et al.* 1985). Meanwhile, McElrea and Standing (1992) reported that doubling the bpm of the background music increased the speed at which a group of students drank soda in the lab. Taken together, the results from the restaurant and the psychology lab demonstrate the profound effect that the tempo of the background music can have on the behaviour of diners and drinkers. The evidence reported in this section shows that the background music in those places in which we choose to eat can have a major impact on both our eating and drinking behaviours (see Spence 2012 for a review). It can affect everything from how long we choose to stay in a restaurant through to how much we end up eating, from how rapidly we bring the fork or spoon up to our mouths though to how much we end up spending, and from how ethnic we feel the dish to be through to how acceptable we rate the overall multisensory dining experience (see Herrington and Capella 1994, 1996; Wilson 2003; Stroebele and De Castro 2004, 2006; Jacob 2006).

Of course, the fear for many people when they read about such findings is that certain restaurants, and especially fast food chains, might be deliberately changing the sensory qualities of the décor/atmosphere in order to try and make people eat and drink more or, if not more, then at least more quickly (and so free up the table for the next customer). This leads to the inevitable question: could it be the atmosphere (in the restaurant) that is making us fat (see Hill and Peters 1998; Lawton 2004; Wansink 2006)? That is, are certain environments or multisensory atmospheres positively obesogenic? It is undoubtedly true that the latest scientific insights regarding the effect of the atmosphere on dining behaviour provide guidelines that can be used by the restaurateur interested in trying to increase their profit margin. However, it is also worth remembering that there are researchers who are instead trying to increase adaptive eating behaviours in those concerned about their weight (Wansink 2004, 2006) and even in various patient groups (who otherwise might not eat as much as they should), a more intelligent use of the atmospherics (e.g. see Ragneskog *et al.* 1996).

9.6.3   "Pardon?"

Of course, any restaurateur reading about the dramatic effects of the ambient music on the behaviour of their diners and drinkers might well be tempted to simply turn up the volume (not to mention the bpm) of the music playing in their establishments (if they are not of the fine-dining variety). But doing this can also have negative consequences. Have you, for example, ever found yourself in a restaurant that is simply too loud? If so, you would join the growing number of customers who, in recent years, have started to complain about those restaurants and bars where one simply can't hear oneself think, never mind listen to what the person sitting across from you might be saying (see Sietsema 2008a, b; Amis 2009; McLaughlin 2010; Stuckey 2012; Platt 2013).[11] For example, restaurant critic Paul Reidinger was less than complementary when describing the acoustics in the Delfina restaurant in San Francisco (below):

> "*Spare walls, stone floors, and shiny, cold zinc-topped tables amounting to an ideal environment for the propagation of decibels ... a crescendo that's not unlike the approach of a train. All that's missing is a horn and a flashing light.*" (Reidinger 1999)

---

[11] Researchers have recently developed new ways of helping to reduce the din in those restaurants that are too loud (e.g. Acoustiblok sells QuietFiber to counteract just such noise; see Clynes 2012; Stuckey 2012, p. 119).

Both the absence of music and the presence of too much sound (whether music or background noise) can detract from a diner's enjoyment of a meal. There is probably no simple solution here for the restaurateur concerned about getting the ambient sound just right, however. Rather, it is always going to be a matter of playing it by ear (as it were) and using one's judgment (and who knows, the odd experiment) in order to optimize the auditory atmosphere. While this might require some trial and error, the key thing to remember is that as long as you're thinking about how to match the sound to the theme or tone of the restaurant, and whether or not the music is too loud, you will be in a much better position than all of those restaurants where they may care passionately about the food but are let down by their lack of attention to the auditory atmosphere. Without a doubt, anything has to be better than the Christmas carols that were playing while one of your authors was dining in an Indian restaurant recently (and before you ask, it wasn't even Christmas).

"*Inside our dining rooms, one basic way we take care of our guests is by providing an atmosphere of comfort and welcome … I hear noise the way a good chef tastes salt: too much is overbearing; too little can be stifling … With just the right noise level, each table has the luxury of becoming enveloped by its own invisible veil of privacy, allowing animated conversation to flow within that discreet container. Too much noise, on the other hand, aggressively invades the space and interferes with the guests' ability to engage with one another. It's annoying, stressful, and inhospitable.*" (Meyer 2010, p. 246)

9.7   Context and expectation

But why exactly should it be that the background music and décor exert such a dramatic effect not only on how much we eat, drink and ultimately spend, but also on what we think about the food? One idea that has been put forward here, no matter whether we are talking about the perceived ethnicity or price/quality, is that the music and décor may activate superordinate knowledge structures that then prime the selection of certain products or ideas. A second and potentially related explanation is that any eating environment will inevitably set up an expectation in the mind of the diner regarding the likely quality of that which they are served. Time and time again in studies of food acceptability, consumers have been shown to exhibit assimilation behaviours. That is, diners' ratings of a food or meal will often move in the direction of their prior expectations as set by the environment or context in which that meal happens to be served (see Cardello 1994; Cardello *et al.* 1996; Rozin and Tuorila 1993). If the expectation and the experience are not too different from one another then the experience will likely be assimilated to the diner's prior expectation. However, beware the diner whose meal experience falls below the expectations set by the atmosphere – the likely outcome is a negatively valenced disconfirmation of expectation (see Schifferstein 2001). As Danny Meyer, one of the world's most successful restaurateurs and head of the Union Square Hospitality Group, succinctly put it: *"Context is everything"* (Meyer 2010, p. 128).

"One of the simplest yet most powerful influences of eating environment on food acceptance is that consumers are rating their expectation of the food in addition to its actual properties." (Meiselman *et al.* 2000, p. 235)

"What I taste is very much a function of what the context leads me to expect." (Heldke 2005, p. 392)

9.8 The lighting

Beyond the auditory atmosphere (and visual décor), the lighting is probably the next most frequently considered attribute of the dining environment that a restaurateur might want to think about changing (Birren 1963). Obviously, though, doing this is likely going to be rather more expensive, not to mention time consuming, than simply changing the music (unless one simply turns down the house lights and lights up some candles to give a more romantic feel). The lighting (colour, temperature, style, etc.) of the restaurant can be crucial in terms of setting the right atmosphere for the diners. It is commonly asserted that bright or harsh lighting reduces the amount of time that people stay in a restaurant (Sommer 1969). In this section, we will review the literature that has investigated the influence of the different parameters of the environmental lighting on our liking of what we're eating and drinking, as well as on our eating behaviour (and, of course, our spending).

There is some intriguing (at least for the baristas among our readership) laboratory-based evidence suggesting that those who like their coffee strong (i.e. bitter) will drink more of the stuff under brighter ambient lighting, whereas those who like their coffee weaker drink more under dimmer illumination (Gal *et al.* 2007). While this is a book about the perfect meal (not the perfect drink), it turns out that most of the research on lighting has been conducted with drinks (such as coffee or wine) rather than with plates of food (see Gregson 1964; Wilson and Gregson 1967; Sauvageot and Struillou 1997; Oberfeld *et al.* 2009). The reason for this is that changing the lighting can have a dramatic, and occasionally really rather undesirable, effect on the appearance properties of food (Wheatley 1973). Drinks, by contrast, can easily be served from black tasting glasses, and hence one can be certain that any effects result from the impact of the ambient lighting itself rather than from any change in the appearance properties of the food that the different lighting may have caused (see Chapter 6). In one such intriguing study involving the aforementioned black tasting glasses, researchers in Germany reported that tourists on a wine cruise on the Rhine were willing to pay almost 50% more for a bottle of dry Riesling wine when they tasted it in a riverside winery while under red illumination rather than under regular white lighting (Oberfeld *et al*. 2009). The wine was also rated as tasting better under blue or red lighting than under white or green.

There have even been reports that the colour of the environment in which we eat affects our appetite. According to some authors, people eat less under red lighting. Although it is often difficult to figure out what empirical research many such claims are based upon, recent studies have indeed demonstrated that simply serving food from a red plate inhibits people's consumption behaviour (see Chapter 4); all you dieters out there take note. When it comes to painting the entire room red, then the results can be even more dramatic. In his book on emotion, Dylan Evans describes an Italian movie director who once painted the staff canteen a bright red with the idea of getting his actors into the right frame of mind prior to filming some tense scenes. However, this change to the environment was so effective that, within a few weeks, fights had apparently started breaking out among some of the regular diners (Evans 2002, p. 87)![12]

Why should our consumption behaviour and liking of foods be so affected by the lighting? One possibility here is that the atmosphere might *arouse* people and hence make them eat faster (see also Smith and Curnow 1966), or else it might *relax* them and so result in their staying longer and possibly eating (and spending) more (Guéguen and Petr 2006) or sometimes less (Genschow *et al*. 2012). The sensory attributes of the environment may also influence a diner's *mood* (Evans 2002; Schifferstein *et al*. 2011). Potentially relevant to the case of dining, there are models in the environmental psychology literature suggesting that any environment can be categorized in terms of three fundamental dimensions of experience: pleasure, arousal and dominance (Mehrabian and Russell 1974).

9.9   The olfactory atmosphere

Who hasn't been drawn into a restaurant because of the delectable aromas wafting out onto the street? However, while olfactory cues can undoubtedly impact on our mood and well-being (see Spence 2003 for a review), there has been far less consideration of the role that purposefully introduced scents might have in the context of dining. Although restaurants usually do not have an ambient smell, mainly to avoid it interfering with the flavour of diners' orders (at least in the eating space), they may well use it at the entrance with the aim of attracting potential diners in. This is certainly a common strategy for fast food/bakery locals (think Cinnabon or Subway; it's hard to avoid the

---

[12] As discussed in the last chapter, it is worth remembering that removing the lighting altogether (as in a dine-in-the-dark restaurant) has also proved to be a very successful niche concept even if, as we saw, the food does not necessarily taste as good as when it can be seen.

cinnamon or bready aroma, respectively) with their kitchen vents apparently strategically placed in order to spread the smell a good 100 m down the street (though many consumers believe it's some artificial aroma being pumped into the streets).[13] Restaurants might also want to locate the kitchen's vents so that they face the street to attract passing diners.[14] Relevant here, your authors were involved some years ago in a project to evaluate a range of different scents to help freshen up the smell of one large fast food chain. The idea was to find the perfect scent that would not only mask the greasy smell of hot oil, but provide a signature branded pleasant fragrance to roll out across all their franchises.

Meanwhile, entering the front door of Chicago's Alinea restaurant recently, your author found herself walking through a half-lit violet corridor full of glass vases mounted on the walls and hanging from the ceiling, containing flowers that delivered an intense sweet flowery spring smell. Before you have actually entered the main dining space itself, your mind is already being transported somewhere else. This olfactory experience serves as a transition between the hectic world outside and Alinea's carefully curated gastronomic experience inside.[15] But, back to the olfactory environment in the dining area itself: some top restaurateurs take their attempts to control the olfactory environment in their restaurant very seriously. For instance, the wait staff at Per Se in New York are explicitly banned from wearing any scented products for fear that the smell of their perfume or aftershave may overpower, or interfere with, the taste of the dish (Damrosch 2008, p. 22).[16]

But can introducing a scent that is not associated with the food into the restaurant environment exert an influence on diners? Guéguen and Petr (2006) conducted a study in a small pizzeria in Brittany, France to address this very question. These researchers looked at the effects of releasing the scent of lemon and lavender (via an electric fragrance diffuser) versus a no-fragrance baseline condition. The study was conducted over three successive Saturday evenings in a restaurant seating a maximum of 22 diners. Interestingly, the

---

[13] We have even heard of Japanese companies releasing the scent of food into their offices at different times of day on different floors in order to more effectively manage the flow of workers into the staff canteen (Classen *et al.* 1994; Fox 2001).

[14] Another option is to have sweet fragrances (of chocolate, say) in the washrooms in order to tempt people to buy dessert. Indeed, women who have been exposed to a chocolate scent in an unrelated task will end up eating far more chocolate cookies when later asked to evaluate them (Coelho *et al.* 2011). That said, those diners who would like to keep their eating under control might be well advised to keep a small bottle of olive oil in the handbags, since it has been suggested that sniffing the aroma of olive oil results in people feeling fuller after eating (Innes 2013).

[15] "*Adam Tihany explained that when designing an elegant restaurant he likes to plan a separate foyer entrance that hints at the splendour to come, while at convivial, informal brasseries like Artisanal, guests are immediately plunged into the action.*" (Sheraton 2004, pp. 148–149)

[16] The great French oenologist Emile Peynaud (1987, 2005) would undoubtedly have approved: he once said that perfumes and aftershaves should be avoided in the context of tasting fine wine (see also Poupon 1957). Peynaud thought that ambient scents such as perfume would interfere with a wine taster's ability to rate a wine.

diners ($N = 88$) stayed around 15% longer and spent approximately 20% more on the day on which they were exposed to the lavender scent. The authors' suggestion here was that this fragrance may have relaxed the diners (and, as a result, they ended up spending more).[17] Of course, it is worth remembering that neither of the scents used in this study were especially congruent with most people's conception of what a pizzeria should smell like. There may be other fragrances out there that would have had an even more pronounced influence on the behaviour of diners due to their ability to help deliver a multisensory atmosphere that is truly congruent and immersive. In her description of one of the Futurist dinners, David (1987, p. 61) wrote:

> "*Meals were to be eaten to the accompaniment of perfumes … to be sprayed over the diners, who, fork in the right hand, would stroke meanwhile with the left some suitable substance – velvet, silk, or emery paper.*"

What if particular fragrances were to be delivered to compliment a specific dish? The olfactory component, after all, is said by some to contribute as much as 80–95% of our experience of the taste of a food. One of the first people (at least in the modern era) to talk about spraying aromas around while the diners were eating was the famous Italian Futurist Filippo Tomasso Marinetti in *La Cucina Futuristica* (Marinetti and Colombo 1930); by now, this shouldn't come as any surprise to the reader. Such ideas can nowadays be seen echoed in a number of contemporary molecular gastronomy dishes, such as the oyster dish that was served for a time at The Fat Duck restaurant in Bray. In this case, the waiter would spray the scent of lavender over the dish as soon as it had been placed on the table in front of the diner (Blumenthal 2008). As we will see in Chapters 10 and 11, scent-delivery technology is now starting to appear in the experimental workshops – or restaurants – of a number of innovative chefs. Before too long, you may even find scent being released from the flower arrangement in the middle of one's table.

Heston Blumenthal has taken the idea of incorporating scent into the dining experience to a whole new level. At one time, the idea had been to spray a sweet-smelling and vaguely familiar scent onto the doorframe at the entrance to The Fat Duck restaurant so that when the diners walked through the low door on their arrival, they would get a faint whiff. "There is nothing special here", some might say. But that would be to miss the point. Why so? Because the reminder card that arrived a few weeks before your booking would also have been scented. "Why," the diner thinks, "have they sent me this? This, after all, was the one appointment that I was never going to miss."[18] Hence,

---

[17] The concern here, when trying to interpret such results, is that one can never be certain of what effect any differences in the weather may have had on the behaviour of diners, over and above any impact of the ambient scent. This problem is exacerbated by the fact that only one evening was associated with the release of each fragrance.

[18] Especially after having waited so long to get that reservation in the first place (not to mention the fact that the restaurant has already taken your credit card details).

or so the idea goes, when you arrive at the restaurant you are reminded of that familiar smell. If you still have not identified it as the scent of the sweet shop, all should become clear at the end of the meal when the waiter gives the diners a parting gift: a small bag of sweets (as if from the sweet shop). The underlying idea here is that you will take those sweets home and, when you open the bag to taste them, the smell will remind you of your experience at The Fat Duck. The scent has therefore been used to extend your 5-hour dining experience, prolonging it so that it lasts days, weeks or even months. If you doubt that olfactory memories can be that evocative, remember the Proust phenomenon (Chu and Downes 2000, 2002; Herz and Schooler 2002).

9.10   On the feel of the restaurant

One sense that we haven't really looked at in this chapter is touch. There are some chefs out there who are now starting to sit up and pay a lot more attention to the *feel* of their restaurants, and how that can be brought in to compliment the multisensory dining experiences that they offer. Obviously, every restaurateur wants to ensure that their diners are as comfortable as possible. But where to start? With his tactile dinner parties, Filippo Tommaso Marinetti (1876–1944) was perhaps the first to think creatively about the importance of touch and tactile stimulation to the act of eating (see the quote earlier), not to mention its enjoyment by diners. His suggestion was that in order to maximally stimulate the senses, diners should wear pyjamas made of (or covered by) differently textured materials such as cork, sponge, sandpaper and/or felt and eat without the aid of knives and forks to enhance the tactile sensations (see David 1987, p. 61; Harrison 2001, pp. 1–2). Marinetti was, in so many ways, a man ahead of his time. That said, the last few years have seen something of a revolution in terms of our understanding of the role of touch in the experience of eating and drinking and, perhaps more importantly, its exploitation in both everyday eating and drinking as well as in the context of experiential dining. The latest research has for example shown that holding a warm drink in one's hands can improve social interaction (Williams and Bargh 2008). So why not hold the warm soup bowl in one's hands rather than let its heat dissipate into the dining table as we so frequently do?

Taking inspiration from the Italian Futurists, there would certainly seem to be grounds for thinking that simply by enhancing the tactile stimulation delivered by the chair on which a person is sitting at a restaurant, it might actually be possible to enhance (or, at the very least, to alter) a diner's experience. In her recent book *Taste What You're Missing*, Barb Stuckey (2012) reports on a Californian chef who deliberately chooses the throws that he places over the back of the chairs, his aim being to deliver a rich tactile experience to his diners. Meanwhile, others have been experimenting with changing the feel of the table (see Figure 9.1). Even McDonalds have apparently been investing

Figure 9.1   Urban picNYC table, design by Haiko Cornelissen Architecten. *Source*: Reproduced with permission of Haiko Cornelissen Architecten, WY, USA. *See colour plate section*

in the introduction of softer chairs in one of their store rebranding exercises, the idea being to make their stores friendlier (Barden 2013; Hultén *et al.* 2009, pp. 144–145).

One trend that we have noticed recently is the move away from white table-cloths to a much greater emphasis on bare natural wood tables. One might think of this as giving the restaurant a Nordic feel as in Copenhagen's Noma, currently the world's top restaurant (Anonymous 2013b). Such a change can be framed in terms of changing the feel of the restaurant. As Crawford (1997) put it a few years back: "*surfaces made from natural materials are often prefer-able, as irregularity is far more sensual than clinically perfect surfaces.*" Joshua Skenes, chef-owner of Saison in San Francisco, also appreciated the impor-tance of the feel of the restaurant:

> "*You need great food, great service, great wine, great comfort. And comfort means everything. It means the materials you touch, the plates, the whole idea that the silverware was the right weight. We put throws on the back of the chairs.*" (quoted in Stuckey 2012, pp. 85–86)

Presumably the pleasant feelings that are associated with dining from a table covered with a starched tablecloth (as compared to an uncovered plastic table-top, say) might serve much the same purpose, that is, stimulating the diner's sense of touch. In the former case however, or even when thinking about the role that linen napkins might play here, it becomes much harder to separate out any positive effects associated with the sensory properties (e.g. the sight or feel) of the material of the tablecloth (or napkins) from any cultural associations that we may have with such table coverings and fine dining experiences (Visser 1991).

Elsewhere, chefs such as Paco Roncero now have exquisite control over both the temperature and humidity in their experimental workshop/dining spaces (Jakubik 2012). As such, perhaps for the first time in history the chef is in the position of being able to change the feel of the dining space during the course of a meal, potentially matching the temperature of the environment with the dish (see also Chapter 10). Did we mention that the ceramic table where the diners eat in Roncero's Madrid establishment is also heated? It can even vibrate! (Though quite what kind of dish would benefit from this particular feature of the fittings isn't, as yet, altogether clear.) Venues such as The Icebar (now found in several countries) also play with the temperature. They take it to the extreme however: the temperature is permanently set to a rather chilly −5 °C (and so it has to be, since everything is built from ice). By having the temperature so low they effectively differentiate themselves from the competition; an example of tactile sensory marketing, if you will (see Figure 9.2). Should you be thinking about going there yourself, don't worry: every guest is given a cape and gloves (and boots if necessary) to avoid becoming too cold. Potentially relevant here are also the results of a study of New York clothes stores, showing a correlation between the in-store temperature and the price of the merchandise (Fiore 2008). Similarly, note in Figure 9.2 that the bar also has blue lighting to further convey a low temperature (see Sester *et al.* 2013).

Reducing the ambient temperature slightly could make sound financial sense; not only will the restaurateur save on the heating bill in winter, it may also increase sales as people tend to eat more when the environment that they are in is colder (Logue 2004, p. 16; see also Brobeck 1948; Spence *et al.* in press). On the flip side however, they are presumably unlikely to stay as long.[19] Interestingly, Ijzerman and Semin (2009) have also reported that people feel socially closer in a room that is warmer. While the feel of a restaurant, in the very literal sense of the term, is something that restaurateurs have rarely paid all that much attention to previously, there is certainly now a growing opportunity to differentiate oneself on the basis of one's tactile offerings. The restaurateur should be asking him/herself: what is the signature feel of my restaurant? By changing the feel of a restaurant, it may even be

---

[19] Some researchers have now started to investigate whether a person's feeling regarding the environment can be altered, not by changing the actual temperature of the venue but by presenting audiovisual stimuli to promote feelings of warmth or coolness (such as a video of a fireplace or the wintry sound of the wind blowing; Sester *et al.* 2013). Certainly, it felt decidedly chillier in the Wolf's Lair at The House of Wolf restaurant in London recently when chef Josef Youssef (Youssef 2013) played the sound of a decidedly cold wind blowing from the rafters for the first two courses of his eight-course tasting menu. It is presumably no coincidence that the interior of The Icebar (see Figure 9.2) is a cool blue.

Figure 9.2 Image of the interior of the Icebar located in Oslo. *See colour plate section*

possible to enhance the diner's experience of a meal (see Spence *et al.* 2013a for a review). Having the wait staff touch the diners can also be an effective means of engaging with one's customers (Lynn *et al.* 1998; Guéguen and Jacob 2005). When the tactile attributes of the environment are coordinated with the auditory, visual and perhaps even the olfactory atmospherics, who knows what wonderfully immersive multisensory dining experiences we may be enjoying in the years to come.

9.11 Atmospheric contributions to taste and flavour perception

So far in this chapter we have reviewed a number of the key research findings showing how the atmosphere in the places in which we eat and drink can, and frequently does, influence many aspects of our behaviour while dining in a restaurant. These aspects include everything from biasing the diner's perception of the likely price, quality and acceptability of a meal through to subtly altering how much we eat and drink and how much time and money we end up spending on food and drink. But can the atmosphere actually change the very taste and flavour of the food we eat? That would be truly surprising, no? Here, we review the evidence demonstrating that people's ratings of the fundamental sensory-discriminative attributes of the food can be influenced simply by changing the colour of the lighting or turning up the background noise.

The results of early laboratory research suggested that changes in the ambient illumination would sometimes impact on an observer's ability to discriminate the taste (e.g. the sourness) of a drink (e.g. Gregson 1964; Wilson and Gregson 1967; but see Sauvageot and Struillou 1997). One recent example that demonstrates how changing the lighting can, very literally, change the taste (or better said, flavour) comes from research conducted by Oberfeld *et al.* (2009). The attentive reader may remember that this study was also mentioned in Section 9.8 on the impact of illumination on price. The study is also relevant here however, in that the results of Oberfeld *et al.* demonstrate that changing the colour of the ambient lighting (between blue, green, red or white) affects the spiciness of wine. When the wine was sampled in the psychology laboratory, the tasters rated it as being spicier if it was tasted under blue or green lighting rather than under white lighting. The colour of the ambient lighting also had a marginally significant effect on fruitiness, with the ratings in green and white light different from those reported under either the blue or red lighting. The opposite pattern was observed for bitterness.

> " ... *I'm sitting in a restaurant – there's music. You know why they have music in restaurants? Because it changes the taste of everything. If you select the right kind of music, everything tastes good. Surely people who work in restaurants know this* ... " (multiple synaesthete and mnemonist S.V. Shereshevsky, quoted in Luria 1968, pp. 81–82)

When it comes to the effect of music or sounds in taste/flavour perception, researchers have demonstrated that loud music (or noise) can impair a person's ability to detect the sweetness and saltiness in food (Woods *et al.* 2011; see also Peynaud 1987, p. 104). Woods and his colleagues conducted a couple of experiments designed to assess the impact of loud versus quiet background noise (75–85 dB versus 45–55 dB, respectively) on the perception of the sweetness, sourness, crunchiness, liking and overall flavour of a variety of foodstuffs (including Pringles, cheese, rice cakes, biscuits/crackers and flapjack). In one study, rice cakes were rated as being significantly crunchier when loud background noise was played. Meanwhile, salty foods (crisps and cheese) appeared less salty, while sweet-tasting food (biscuits and flapjack) were rated as less sweet. Given what we have seen so far in this section, one might wonder whether crispy or crunchy foods would taste different if eaten at a noisy party, say (Spence *et al.* 2011). In other research, playing loud music has also been shown to suppress the perception of alcohol and to increase the rated sweetness of alcoholic drinks (Stafford *et al.* 2012).

Taken together, the results of the research that has been conducted to date certainly suggests that presenting loud background noise (e.g. as one might perhaps encounter on a plane) can definitely have a negative impact on the taste of food (Masuda *et al.* 2008; Woods *et al.* 2011; Stafford *et al.* 2012; Stuckey 2012, p. 119). Other researchers have suggested that loud noise may simply make it harder for a person to concentrate on the sensory properties of food and drink (Stafford *et al.* 2012; see also Marks and Wheeler 1998). Clearly, when the background music or noise is sufficiently loud, it may simply mask the relevant auditory cues that we normally use to assess/discriminate the textural or oral-somatosensory properties of dry (or other noisy) foods (Spence 2012).

There is a lot more fundamental research that needs to be done in this area. Just take the following: have you ever noticed how many people seem to order tomato juice in the plane in a way that you just don't see them doing on the ground? One intriguing idea here is that umami may be the one basic taste that is unaffected by loud background noise. It turns out that tomato juice is especially high in umami (umami, or MSG, is also found in mushrooms and parmesan cheese). The suggestion that certain tastes are unaffected by background noise would certainly be consistent with early laboratory research showing that background noise didn't impact ratings of tomato juice in the laboratory setting (see Pettit 1958). It would also be consistent with anecdotal reports from those who test airline food for some of the world's biggest airlines (Michaels 2010). Perhaps all those travellers who order a Bloody Mary after the seatbelt sign has been turned off have figured out intuitively what scientists are only now slowly coming to recognize empirically (McCartney 2013; Spence *et al.* 2014).

"The inexplicable blandness of airline food has been pondered at 30,000 feet by generations of travellers." (Connor 2010, p. 13)

The fact that the atmosphere has such a dramatic effect on our experience of a meal can, in part, help to explain why it is that the food served on airplanes is nearly always disappointing, even if it has purportedly been designed by one of the world's top chefs. Think here only of Heston Blumenthal working on the food served on British Airways, Neil Perry working with Qantas and The Rausch Brothers starting to advise Colombia's Avianca. Of course, the cramped conditions associated with the final preparation of the food can't help either (Sheraton 2004).

Do you imagine that anyone has had their perfect meal in the air? Perhaps, but we think it highly unlikely. Various factors impair our perception of flavour while we are flying and, as we have seen already, the very loud background noise or the light plastic cutlery that so many of us are forced to eat with (see Chapter 5) isn't going to help either. Add to that the low cabin air pressure and the dry air, reducing the contribution of those flavourful volatile molecules, while at the same time rapidly evaporating our nasal mucus (Michaels 2010), and it is no wonder that the top chefs seem powerless to do anything to make the food served at 35,000 feet anything like as tasty as they so obviously can on the ground.[20] Some airlines such as Lufthansa and Singapore Airlines have even built specialized testing facilities in order to mimic the conditions of the average passenger without anyone having to leave the ground (Michaels 2010), all with the aim of making their food taste better in the air.

That said, there are a number of creative types who are currently thinking about how to incorporate the senses in order to ramp up the dining experience in the sky. Given what we have seen already in this chapter, noise-cancelling headphones should help for a start. Beyond that, any of the atmospheric cues that we have seen throughout this chapter could improve the experience. Italian music to complement the Italian food? A red and white tablecloth for any Mediterranean cuisine? Heavy cutlery is a must (from what we saw in Chapter 5). Heston's preferred solution is a nasal spray to increase the moistness in your nose (McCartney 2013). Just don't forget to have a tissue handy!

Whether in the air or on the ground, we believe that in the very near future we are going to see music and soundscapes used to enhance rather than mask the taste of flavour of foods. One example of this sort of approach comes from Crisinel *et al.* (2012). These researchers investigated the consequences of manipulating the pitch of a background soundscape on the taste of food. The participants in their study had to evaluate four pieces of cinder toffee while listening to two auditory soundtracks, one designed to be more congruent with a bitter-tasting food and the other to be more congruent with a sweet-tasting food instead. The results revealed that the toffee samples tasted while listening to the presumptively 'bitter' soundtrack were rated as tasting significantly more bitter than when the exact same foodstuff was evaluated while listening to the 'sweet' soundtrack. These results provide some of the first convincing empirical evidence that background sound can be used to modify the taste (and presumably also the flavour) of a foodstuff. As we will see in Chapter 10, these soundtracks have already made it onto the menu at restaurants.

It is also likely that there may be some sort of halo or transfer effect, such that our liking of the background music (or, for that matter, any other background atmospheric cues) may influence our liking of the food (Woods *et al.* 2011). This could provide another route by which getting the music right will enhance the taste of food and drink. For example, in one recent study in New Zealand, people's ratings of the pleasantness of dark chocolate and bittersweet chocolate *gelati* were found to be significantly higher when they listened to preferred music, but significantly less pleasant when the music was changed to something that the participants didn't like (Kantono 2013; see also North 2012; Peralta 2012; Spence and Deroy 2013).

---

[20] The one consolation for those travelling in Economy is that all those bodies help to keep the humidity levels a good 10% higher than those measured in First Class (Michaels 2010).

9.12   Multisensory atmospherics

The majority of the studies that we have looked at so far in this chapter have concentrated on manipulating just a single sense at a time, either changing the décor or lighting or else only manipulating the music or perhaps introducing a scent. However, as was mentioned earlier, the available cognitive neuroscience research suggests that the biggest impact on our experiences and behaviours sometimes occur when several sensory attributes are changed at once and when the (e.g. atmospheric) cues available to each of a diner's senses complement one another. This is precisely the sort of situation in which one might expect to see a superadditive response (both in the brain but also in behaviour), a response that is far bigger than that which can be achieved by manipulating a single sense individually (see Stein and Meredith 1993; Spence 2002; Spence *et al.* in press). Recently, a few researchers have started to look more systematically at the effects of manipulating several senses at the same time in the context of dining and drinking.

Wansink and Van Ittersum (2012) tested the combined impact of changing the lighting and the music on the behaviour of diners in Hardee's, a fast food restaurant in Champaign, Illinois. They found that simply by softening the music and lighting (in order to create a more relaxed atmosphere), diners' ratings of their meals improved significantly (by around 15%). The restaurant in question had two dining areas. In one, the lighting was set at its normal bright level, the colour scheme was also bright and the music playing in the background was loud. Meanwhile, the other 'fine dining' environment had a much more relaxed atmosphere: there were pot plants and paintings, window shades and indirect lighting and don't forget the candles on the tables which themselves were covered with white tablecloths. Soft jazz instrumental ballads could be heard playing in the background. Those diners who were lucky enough to eat in the more relaxed side of the restaurant rated their meal as significantly more enjoyable, while at the same time consuming less (their calorie intake was reduced by an average of more than 150 calories, or 18%). The overall cash spend in this study was unchanged (not necessarily what the avaricious restaurateur wanted to hear). However, Wansink and Van Ittersum managed to finesse their findings by concluding that "*a more relaxed environment increases satisfaction and decreases consumption*" (Wansink and Van Ittersum 2012, p. 1).

Finally in this section, we have to mention the results of a very recent study of multisensory atmospherics conducted by Velasco *et al.* (2013) in Soho, London early in 2013. In order to demonstrate the profound effect of the multisensory environment on flavour perception, these researchers took over a small building and created three very different multisensory environments: one designed to bring out the grassiness on the nose, another to bring out the sweetness of the taste and the third designed to accentuate the woodiness of the aftertaste of the whisky. The 'participants' were brought into the

(a)   (b)   (c)

Figure 9.3 Stills taken from the three environments used demonstrating the impact of the atmosphere on the whisky drinking experience. (a) the 'grassy' room; (b) the 'sweet' room; and (c) the 'woody' room. *Source*: Velasco et al. 2013. *See colour plate section*

building, given a small glass of whisky and a scorecard, and then led through the three rooms over the course of 15 minutes (see Figure 9.3). In each room, the drinkers were asked to rate their experience of the whisky and of the room in which they were tasting it. Over three nights, almost 500 people came through the doors. The results were clear: get the multisensory environment right and it is possible to change the taste of the drink in your hand by 20%, even though people know that it is the same drink. No wonder then that the Provencal Rose paradox phenomenon is so strong when you aren't even sure whether the wine is exactly the same or not.[21] The results of Velasco *et al.*'s study also highlighted a correlation between people's liking of the room that they were in and their liking of the whisky (see also Petit and Sieffermann 2007).

9.13 Conclusions

As the research reviewed in this chapter has so clearly revealed, the multisensory atmospherics of the environments in which we eat and drink can exert a profound effect on many aspects of the meal experience. The reader should hopefully by now be convinced that a key part of the experience of the perfect meal is always going to come from getting each and every one of the individual elements of the atmosphere just right. Everything from the décor through to the background music, even the feel and smell of the restaurant, play a much more important role in our food and drink experiences than many of us realize (and as the evidence reviewed earlier in the chapter showed, it can even affect us when we do not realize that it does). Perhaps it should not come as any

---

[21] Note here that there is an underlying bias to assume that the food and drink stay the same, unless we have some reason to believe that it has changed (Woods *et al.* 2010). Any such bias would have been working against the environment impacting on the experience in this study, since everyone held on to the same glass during the course of the experience and hence knew that the drink did not change.

surprise that a recent report highlighted an increasing trend towards spending on the décor (Mitchell 2013). In fact, curators say that restaurants care more about their décor than ever before (Sharp 2013). That said, it is important to remember that getting the environment right for your perfect meal need not necessitate spending huge amounts of money or finding a restaurant that has been designed by one of the world's top architects. It is more about making sure that the atmosphere matches the food that you are eating and that it sets the most appropriate expectations. Indeed, for some people, the perfect meal may be no more sophisticated than a delicious picnic surrounded by nature on a glorious summer's day (cf. Gill 2007), or eating fish and chips from paper with one's fingers while sitting on the beachfront.

Get the environment right, and one can significantly enhance the perceived acceptability, quality and ethnicity of a dish. The atmosphere can make a meal appear more valuable; it can even change the very taste and flavour of the food itself, especially when we are up in the air. It certainly has an impact on how much we end up consuming and on how much we spend. The environments in which we eat and drink affect our mood: they can arouse us or they can make us more relaxed. Ultimately, one can think of the environment (or atmosphere) as exerting an effect on our dining behaviour because it captures our attention. The atmosphere sends a particular message; it can even serve as an affect-creating medium (from Kotler 1974; see also Schifferstein *et al.* 2011).

Given the evidence showing just how important the atmosphere is to the diner's experience of food and drink, it should come as little surprise that a growing number of restaurants are currently thinking about how best to control the environments in which their diners consume their food, and match it to the food and drink. We will take a look at a range of other futuristic dining scenarios in Chapters 10 and 11. As we started to see at the end of this chapter, technology is increasingly coming to play a role here, offering the restaurateur control over not only the music but also the lighting. Even the ambient temperature and scent may soon be delivered to the diner's table at the press of a button, synchronized with the arrival of the dishes at the table (continue through the next chapter for more). When these toys are put in the hands of the top chefs, they are all used in order to give the entire dining space (or at least the diners' table) a particular feel and/or to target specific diners or group of diners with particular sensory backgrounds to best complement the dish that they are eating. The obvious worry lurking in the background is that, unless one is careful, the restaurateur may end up spending too much time thinking about the décor and entertainment value of the venue and neglect the quality of the food itself.

We believe that gastrophysics, the new science of the table, has been instrumental in helping to demonstrate just how important a role the atmosphere plays in terms of influencing a diner's overall experience of the meal. It is also increasingly starting to provide guidelines to help the restaurateur and aspiring chef to optimize their culinary product offering. What is more, many of the insights that have emerged in this chapter, such as making sure to coordinate the music with the food, are things that any one of us can incorporate into our own everyday dining experiences,[22] no matter whether hosting friends for dinner or thinking what to put on the jukebox in our local gastropub. Returning to the question with which we started this chapter, the Spanish chef Carme Ruscalleda was certainly right to be concerned about the atmospherics of the environment in which her diners enjoy her wonderful cuisine. As we have seen throughout this chapter, getting the multisensory atmospherics just right can make all the difference between the diner having a merely mediocre meal and experiencing what may be their perfect meal.

References

Allen, J. S. (2012) *The Omnivorous Mind: Our Evolving Relationship with Food*. Harvard University Press, London.

Amis, K. (2009) *Everyday Drinking: The Distilled Kingsley Amis*. Bloomsbury, London.

Anderson, A. T. (2004) Table settings: The pleasures of well-situated eating. In *Eating Architecture* (eds J. Horwitz and P. Singley), pp. 247–258. MIT Press, Cambridge, MA.

Anonymous (1965) More restaurants sell an exotic atmosphere as vigorously as food. *The Wall Street Journal*, 4 August, **1**.

Anonymous (2013a) Charlie Trotter: He changed Chicago dining – and America's. *Time*, 18 November, **12**.

Anonymous (2013b) Talking Scandinavian design with Space Copenhagen. *DesignCurial*, 5 August. Available at http://www.designcurial.com/news/talking-scandinavian-design-with-space-copenhagen/ (accessed February 2014).

Areni, C. S. and Kim, D. (1993) The influence of background music on shopping behavior: Classical versus top-forty music in a wine store. *Advances in Consumer Research*, **20**, 336–340.

Baker, J., Grewal, D. and Parasuraman, A. (1994) The influence of store environment on quality inferences and store image. *Journal of the Academy of Marketing Science*, **22**, 328–339.

Barden, P. (2013) *Decoded: The Science Behind Why We Buy*. John Wiley & Sons, Chichester.

Belk, R. W. (1974) An exploratory assessment of situational effects in buyer behaviour. *Journal of Marketing Research*, **11**, 156–163.

Bell, R., Meiselam, H. L., Pierson, B. J. and Reeve, W. G. (1994) Effects of adding an Italian theme to a restaurant on the perceived ethnicity, acceptability, and selection of foods. *Appetite*, **22**, 11–24.

---

[22] For those of you who may be worried about the 'obseogenic' effects of the environment on your own consumption behaviours (Lawton 2004; Wansink 2006; Sobal and Wansink 2007), we would advise that you eat your food from a red plate under red lighting while listening to slow music playing at a quiet volume.

Birren, F. (1963) Color and human appetite. *Food Technology*, **17** (May), 45–47.

Blumenthal, H. (2007) *Further Adventures in Search of Perfection: Reinventing Kitchen Classics*. Bloomsbury, London.

Blumenthal, H. (2008) *The Big Fat Duck Cookbook*. Bloomsbury, London.

Brobeck, J. R. (1948) Food intake as a mechanism of temperature regulation. *Yale Journal of Biological Medicine*, **20**, 545–552.

Caldwell, C. and Hibbert, S. (1999) Play that one again: The effect of music tempo on consumer behaviour in a restaurant. *European Advances in Consumer Research*, **4**, 58–62.

Caldwell, C. and Hibbert, S. A. (2002) The influence of music tempo and musical preference on restaurant patrons' behavior. *Psychology and Marketing*, **19**, 895–917.

Cardello, A. V. (1994) Consumer expectations and their role in food acceptance. In *Measurement of Food Preferences* (eds H. J. H. MacFie and D. M. H. Thomson), pp. 253–297. Blackie Academic & Professional, London.

Cardello, A. V., Bell, R. and Kramer, F. M. (1996) Attitudes of consumers toward military and other institutional foods. *Food Quality and Preference*, **7**, 7–20.

Chossat, V. and Gergaud, O. (2003) Expert opinion and gastronomy: The recipe for success. *Journal of Cultural Economics*, **27**, 127–141.

Chu, S. and Downes, J. J. (2000) Odour-evoked autobiographical memories: Psychological investigations of Proustian phenomena. *Chemical Senses*, **25**, 111–116.

Chu, S. and Downes, J. J. (2002) Proust nose best: Odours are better cues of autobiographical memory. *Memory and Cognition*, **30**, 511–518.

Classen, C., Howes, D. and Synnott, A. (1994) *Aroma: The Cultural History of Smell*. Routledge, London.

Clynes, T. (2012) A restaurant with adjustable acoustics. *Popular Science*. Available at http://www.popsci.com/technology/article/2012-08/restaurant-adjustable-acoustics (accessed February 2014).

Coelho, J. S., Idlera, A., Werle, C. O. C. and Jansen, A. (2011) Sweet temptation: Effects of exposure to chocolate-scented lotion on food intake. *Food Quality and Preference*, **22**, 780–784.

Connor, S. (2010) Science finds the plane truth about in-flight meals. *The Independent*, **15 October**, 13.

Cowen, T. (2012) *An Economist Gets Lunch: New Rules for Everyday Foodies*. Plume, New York.

Crawford, I. (1997) *Sensual Home: Liberate your Senses and Change your Life*. Quadrille Publishing, London.

Crisinel, A. S., Cosser, S., King, S., Jones, R., Petrie, J. and Spence, C. (2012) A bittersweet symphony: Systematically modulating the taste of food by changing the sonic properties of the soundtrack playing in the background. *Food Quality and Preference*, **24**, 201–204.

Crocker, E. C. (1950) The technology of flavors and odors. *Confectioner*, **34** (January), 7–10.

Damrosch, P. (2008) *Service Included: Four-Star Secrets of an Eavesdropping Waiter*. William Morrow, New York.

David, E. (1987). *Italian Food*. Barrie & Jenkins, London.

Drews, D. R., Vaughn, D. B. and Anfiteatro, A. (1992) Beer consumption as a function of music and the presence of others. *Journal of the Pennsylvania Academy of Science*, **65**, 134–136.

Edelstein, M., Brang, D., Rouw, R. and Ramachandran, V. S. (2013) Misophonia: Physiological investigations and case descriptions. *Frontiers in Human Neuroscience*, **7 (296)**, 1–11.

Edwards, J. S. A., Meiselman, H. L., Edwards, A. and Lesher, L. (2003) The influence of eating location on the acceptability of identically prepared foods. *Food Quality and Preference*, **14**, 647–652.

Evans, D. (2002) *Emotion: The Science of Sentiment*. Oxford University Press, Oxford.

Ferber, C. and Cabanac, M. (1987) Influence of noise on gustatory affective ratings and preference for sweet or salt. *Appetite*, **8**, 229–235.

Fiore, A. M. (2008). The shopping experience. In: *Product Experience* (eds H. N. J. Schifferstein and P. Hekkert), pp. 629–648. Elsevier, London.

Fox, K. (2001) The smell report. Social Issues Research Centre. Available at http://www.sirc.org/publik/smell.pdf (accessed February 2014).

Gal, D., Wheeler, S. C. and Shiv, B. (2007) Cross-modal influences on gustatory perception. http://ssrn.com/abstract=1030197 (accessed February 2014).

Genschow, O., Reutner, L. and Wanke, M. (2012) The color red reduces snack food and soft drink intake. *Appetite*, **58**, 699–702.

Gergaud, O., Guzman, L. M. and Verardi, V. (2007) Stardust over Paris gastronomic restaurants. *Journal of Wine Economics*, **2**, 24–39.

Gill, A. A. (2007) *Table Talk: Sweet and Sour, Salt and Bitter*. Weidenfeld & Nicolson, London.

Goldstein, D. (2005). The play's the thing: Dining out in the new Russia. In: *The Taste Culture Reader: Experiencing Food and Drink* (ed. C. Korsmeyer), pp. 359–371. Berg, Oxford.

Green, D. M. and Butts, J. S. (1945) Factors affecting acceptability of meals served in the air. *Journal of the American Dietetic Association*, **21**, 415–419.

Gregson, R. A. M. (1964) Modification of perceived relative intensities of acid tastes by ambient illumination changes. *Australian Journal of Psychology*, **16**, 190–199.

Grewel, D., Baker, J., Levy, M. and Voss, G. B. (2003) The effects of wait expectations and store atmosphere evaluations on patronage intentions in service-intensive retail stores. *Journal of Retailing*, **79**, 259–268.

Guéguen, N. and Jacob, C. (2005) The effect of touch on tipping: An evaluation in a French bar. *Hospitality Management*, **24**, 295–299.

Guéguen, N. and Petr, C. (2006) Odors and consumer behavior in a restaurant. *International Journal of Hospitality Management*, **25**, 335–339.

Guéguen, N., Jacob, C. and Le Guellec, H (2004) Sound level of background music and consumer behavior: An empirical evaluation. *Perceptual and Motor Skills*, **99**, 34–38.

Guéguen, N., Jacob, C., Le Guellec, H., Morineau, T. and Lourel, M. (2008) Sound level of environmental music and drinking behavior: A field experiment with beer drinkers. *Alcoholism, Clinical and Experimental Research*, **32**, 1795–1798.

Hall, L., Johansson, P., Tärning, B., Sikström, S. and Deutgen, T. (2010) Magic at the marketplace: Choice blindness for the taste of jam and the smell of tea. *Cognition*, **117**, 54–61.

Hanefors, M. and Mossberg, L. (2003) Searching for the extraordinary meal experience. *Journal of Business and Management*, **9**, 249–270.

Harrison J. (2001) *Synaesthesia: The Strangest Thing*. Oxford University Press, Oxford.

Heldke, A. (2005) But is it authentic? Culinary travel and the search for the "genuine article". In: *The Taste Culture Reader: Experiencing Food and Drink* (ed C. Korsmeyer), pp. 385–394. Berg, Oxford.

Herrington, J. D. and Capella, L. M. (1994) Practical applications of music in service settings. *Journal of Services Marketing*, **8**, 50–56.

Herrington, J. D. and Capella, L. M. (1996) Effect of music in service environments: A field study. *Journal of Services Marketing*, **10**, 26–41.

Herz, R. S. and Schooler, J. W. (2002) A naturalistic study of autobiographical memories evoked by olfactory and visual cues: Testing the Proustian hypothesis. *American Journal of Psychology*, **115**, 21–32.

Hill, J. O. and Peters, J. C. (1998) Environmental contributions to the obesity epidemic. *Science*, **280**, 1371–1374.

Homburg, C., Imschloss, M. and Kühnl, C. (2012) *Of Dollars and Scents – Does Multisensory Marketing Pay Off?* Institute for Marketing Oriented Management. Available at http://imu2.bwl.uni-mannheim.de/fileadmin/files/imu/files/ap/ri/RI009.pdf (accessed February 2014).

Horwitz, J. and Singley, P. (eds) (2004) *Eating Architecture*. MIT Press, Cambridge, MA.

Hultén, B., Broweus, N. and van Dijk, M. (2009) *Sensory Marketing*. Palgrave Macmillan, Basingstoke, UK.

Ijzerman, H. and Semin, G. R. (2009) The thermometer of social relations mapping social proximity on temperature. *Psychological Science*, **20**, 1214–1220.

Innes, E. (2013) Could olive oil be the key to weight loss? Scientists discover even the SMELL of it can make us feel full. Available at http://www.dailymail.co.uk/health/article-2293948/Could-olive-oil-key-weight-loss-Scientists-discover-SMELL-make-feel-full.html (accessed February 2014).

Jacob, C. (2006) Styles of background music and consumption in a bar: An empirical evaluation. *International Journal of Hospitality Management*, **25**, 716–720.

Jacobsen, J. K. (2008) The food and eating experience. In *Creating Experiences in the Experience Economy* (eds J. Sundbo and P. Darmer), pp. 13–32. Edward Elgar Publishing, Cheltenham.

Jakubik, A. (2012) The workshop of Paco Roncero. Trendland: Fashion Blog & Trend Magazine. Available at http://trendland.com/the-workshop-of-paco-roncero/ (accessed February 2014).

Kantono, K. (2013) Effect of music genre on pleasantness of three types of chocolate gelati. Available at http://prezi.com/em2_up3clq1s/effect-of-music-genre-on-pleasantness-of-three-types-of-chocolate-gelati/ (accessed February 2014).

Kellaris, J. and Kent, R. (1992) The influence of music on consumers' temporal perceptions: Does time fly when you're having fun. *Journal of Consumer Psychology*, **1**, 365–376.

Kellaris, J., Mantel, S. and Altsech, M. B. (1996) Decibels, disposition, and duration: A note on the impact of musical loudness and internal states on time perception. *Advances in Consumer Research*, **23**, 498–503.

Knöferle, K. M., Spangenberg, E., Herrmann, A. and Landwehr, J. R. (2012) It is all in the mix: The interactive effect of music tempo and mode on in-store sales. *Marketing Letters*, **23 (1)**, 325–337.

Konečni, V. J. (2008) Does music induce emotion? A theoretical and methodological analysis. *Psychology of Aesthetics, Creativity, and the Arts*, **2**, 115–129.

Kotler, P. (1974) Atmospherics as a marketing tool. *Journal of Retailing*, **49** (Winter), 48–64.

Kupfermann, I. (1964) Eating behaviour induced by sounds. *Nature*, **201**, 324.

Lammers, H. B. (2003) An oceanside field experiment on background music effects on the restaurant tab. *Perceptual and Motor Skills*, **96**, 1025–1026.

Lanza, J. (2004) *Elevator Music: A Surreal History of Muzak, Easy-listening, and Other Moodsong*. University of Michigan Press, Ann Arbor.

Lawton, G. (2004). Angelic host. *New Scientist*, **184** (December), 68–69.

Logue, A. W. (2004) *The Psychology of Eating and Drinking*, 3rd edition. Brunner-Routledge, Hove, East Sussex.

Luria, A. R. (1968) *The Mind of a Mnemonist*. Harvard University Press, Cambridge, MA.

Luttinger, N. and Dicum, G. (2006) *The Coffee Book: Anatomy of an Industry from Crop to the Last Drop*. The New Press, New York.

Lyman, B. (1989) *A Psychology of Food, More Than a Matter of Taste*. Avi, van Nostrand Reinhold, New York

Lynn, M., Le, J.-M. and Sherwyn, D. (1998) Reach out and touch your customers. *Cornell Hotel and Restaurant Administration Quarterly*, **39**, 60–65.

Marinetti, F. T. and Colombo, L. (1930/1998) *La cucina futurista: Un pranzo che evitò un suicidio (The Futurist kitchen: A meal that prevented suicide)*. Christian Marinotti Edizioni, Milan.

Marks, L. E. and Wheeler, M. E. (1998) Attention and the detectability of weak taste stimuli. *Chemical Senses*, **23**, 19–29.

Masuda, M., Yamaguchi, Y., Arai, K. and Okajima, K. (2008) Effect of auditory information on food recognition. *IEICE Technical Report*, **108 (356)**, 123–126.

McCarron, A. and Tierney, K. J. (1989) The effect of auditory stimulation on the consumption of soft drinks. *Appetite*, **13**, 155–159.

McCartney, S. (2013) The secret to making airline food taste better. *Wall Street Journal*, 13 November. Available at http://live.wsj.com/video/the-secret-to-making-airline-food-taste-better/8367EF44-52DD-41C4-AC4A-FFA6659F3422.html#!8367EF44-52DD-41C4-AC4A-FFA6659F3422 (accessed February 2014).

McElrea, H. and Standing, L. (1992) Fast music causes fast drinking. *Perceptual and Motor Skills*, **75**, 362.

McLaughlin, K. (2010) Pass the salt … and a megaphone. *The Wall Street Journal*, 3 February. Available at http://online.wsj.com/article/SB10001424052748704022804575041060813407740.html (accessed February 2014).

Mehrabian, A. R. and Russell, J. A. (1974) *An Approach to Environmental Psychology*. MIT Press, Cambridge, MA.

Meiselman, H. L. (2008) Experiencing food products within a physical and social context. In *Product Experience* (eds H. N. J. Schifferstein and P. Hekkert), pp. 559–580. Elsevier, Amsterdam.

Meiselman, H., Hirsch, E. S. and Popper, R. D. (1988) Sensory, hedonic and situational factors in food acceptance and consumption. In *Food Acceptability* (ed D. M. H. Thomson), pp. 77–87. Elsevier, London.

Meiselman, H. L., Johnson, J. L., Reeve, W. and Crouch, J. E. (2000) Demonstrations of the influence of the eating environment on food acceptance. *Appetite*, **35**, 231–237.

Meyer, D. (2010) *Setting the Table: Lessons and Inspirations from one of the World's Leading Entrepreneurs*. Marshall Cavendish International, London.

Michaels, D. (2010) Test flight: Lufthansa searches for savor in the sky. *Wall Street Journal*, 27 July. Available at http://online.wsj.com/article /SB10001424052748703294904575384954227906006.html (accessed February 2014).

Mielby, L. H. and Bom Frøst, M. (2012) Eating is believing. In *The Kitchen as Laboratory: Reflections on the Science of Food and Cooking* (eds C. Vega, J. Ubbink and E. van der Linden), pp. 233–241. Columbia University Press, New York.

Milliman, R. E. (1986) The influence of background music on the behavior of restaurant patrons. *Journal of Consumer Research*, **13**, 286–289.

Mitchell, J. (2013) Interior inspiration: Beautiful restaurant design. *DesignCurial*, 23 September. Available at http://www.designcurial.com/news/interior-inspiration-10-of-the-worlds-most-beautiful-restaurant-designs/ (accessed February 2014).

North, A. C. (2012) The effect of background music on the taste of wine. *British Journal of Psychology*, **103**, 293–301.

North, A. C. and Hargreaves, D. J. (1996) The effects of music on responses to a dining area. *Journal of Environmental Psychology*, **16**, 55–64.

North, A. C. and Hargreaves, D. J. (1998) The effects of music on atmosphere and purchase intentions in a cafeteria. *Journal of Applied Social Psychology*, **28**, 2254–2273.

North, A. and Hargreaves, D. (2008) *The Social and Applied Psychology of Music*. Oxford University Press, Oxford.

North, A. C., Hargreaves, D. J. and McKendrick, J. (1997) In-store music affects product choice. *Nature*, **390**, 132.

North, A. C., Hargreaves, D. J. and McKendrick, J. (1999) The influence of in-store music on wine selections. *Journal of Applied Psychology*, **84**, 271–276.

North, A. C., Shilcock, A. and Hargreaves, D. J. (2003) The effect of musical style on restaurant customers' spending. *Environment and Behavior*, **35**, 712–718.

Oberfeld, D., Hecht, H., Allendorf, U. and Wickelmaier, F. (2009) Ambient lighting modifies the flavor of wine. *Journal of Sensory Studies*, **24**, 797–832.

Peralta, E. (2012) The sounds of asparagus, as explored through opera. *The Salt*. Available at http://www.npr.org/blogs/thesalt/2012/05/29/153950254/the-sounds-of-asparagus-as-explored-through-opera (accessed February 2014).

Petit, C. and Sieffermann, J. M. (2007) Testing consumer preferences for iced-coffee: Does the drinking environment have any influence? *Food Quality and Preference*, **18**, 161–172.

Pettit, L. A. (1958) The influence of test location and accompanying sound in flavor preference testing of tomato juice. *Food Technology*, **12**, 55–57.

Peynaud, E. (1987) *The Taste of Wine: The Art and Science of Wine Appreciation* (translated M. Schuster). Macdonald & Co, London.

Peynaud, E. (2005) Tasting problems and errors of perception. In *The taste Culture Reader: Experiencing Food and Drink* (ed C. Korsmeyer), pp. 272–278. Berg, Oxford.

Pine II, B. J. and Gilmore, J. H. (1998) Welcome to the experience economy. *Harvard Business Review*, **76 (4)**, 97–105.

Pine, II, B. J. and Gimore, J. H. (1999) *The Experience Economy: Work is Theatre and Every Business is a Stage*. Harvard Business Review Press, Boston, MA.

Platt, A. (2013). Why restaurants are louder than ever. *Grub Street New York*. Available at http://www.grubstreet.com/2013/07/adam-platt-on-loud-restaurants.html (accessed February 2014).

Poupon, P. (1957) *Pensées d'un dégusteur (1975) Nouvelles pensées d'un dégusteur (New thoughts of a taster)*. Confrérie des Chevaliers du Tastevin, Nuits-Saint-Georges.

Ragneskog, H., Brane, G., Karlsson, I. and Kihlgren, M. (1996) Influence of dinner music on food intake and symptoms common in dementia. *Scandinavian Journal of Caring Science*, **10**, 11–17.

Reidinger, P. (1999) A sound choice. *San Francisco Bay Guardian*, 13–19 January.

Roballey, T. C., McGreevy, C., Rongo, R. R., Schwantes, M. L., Steger, P. J., Wininger, M. A. and Gardner, E. B. (1985) The effect of music on eating behavior. *Bulletin of the Psychonomic Society*, **23**, 221–222.

Rozin, P. and Tuorila, H. (1993) Simultaneous and temporal contextual influences on food acceptance. *Food Quality and Preference*, **4**, 11–20.

Sauvageot, F. and Struillou, A. (1997) Effet d'une modification de la couleur des échantillons et de l'éclairage sur la flaveur de vins évaluée sur une échelle de similarité (Effect of the modification of wine colour and lighting conditions on the perceived flavour of wine, as measured by a similarity scale). *Science des Aliments*, **17**, 45–67.

Schifferstein, H. N. J. (2001). Effects of product beliefs on product perception and liking. In *Food, People and Society: A European Perspective of Consumers' Food Choices* (eds L. Frewer, E. Risvik and H. Schifferstein), pp. 73–96. Springer Verlag, Berlin.

Schifferstein, H. N. J., Talke, K. S. S. and Oudshoorn, D.-J. (2011) Can ambient scent enhance the nightlife experience? *Chemosensory Perception*, **4**, 55–64.

Sester, C., Deroy, O., Sutan, A., Galia, F., Desmarchelier, J.-F., Valentin, D. and Dacremont, C. (2013) "Having a drink in a bar": An immersive approach to explore the effects of context on beverage choice. *Food Quality and Preference*, **28**, 23–31.

Sharp, A. (2013) A feast for the eyes: Inside some of the world's best-designed restaurants where the décor is just as important as the food. *Daily Mail Online*, 24 September. Available at http://www.dailymail.co.uk/news/article-2430031/TOY-New-York-Japans-Hoto-Fudo-Inside-worlds-best-designed-restaurants.html (accessed February 2014).

Sheraton, M. (2004) *Eating my Words: An Appetite for Life*. Harper, New York.

Sietsema, T. (2008a) No appetite for noise. *The Washington Post*, 6 April. Available at http://www.washingtonpost.com/wp-dyn/content/article/2008/04/01/AR2008040102210_pf.html (accessed February 2014).

Sietsema, T. (2008b) Revealing raucous restaurants. *The Washington Post*, 6 April. Available at http://www.washingtonpost.com/wp-dyn/content/article/2008/04/04 /AR2008040402735.html (accessed February 2014).

Smith, B. C. (2009) *The emotional impact of a wine and the Provencal rose paradox.* Unpublished manuscript. Centre for the Study of the Senses, University of London.

Smith, P. C. and Curnow, R. (1966) "Arousal hypothesis" and the effects of music on purchasing behavior. *Journal of Applied Psychology*, **50**, 255–256.

Sobal, J. and Wansink, B. (2007) Kitchenscapes, tablescapes, platescapes, and food-scapes: Influences of microscale built environments on food intake. *Environment and Behavior*, **39**, 124–142.

Sommer, R. (1969) *Personal Space. The Behavioural Basis of Design.* Prentice-Hall, Englewood Cliff, NJ.

Spence, C. (2002) *The ICI Report of the Secret of the Senses.* The Communication Group, London.

Spence, C. (2003) A new multisensory approach to health and well-being. *Essence*, **2**, 16–22.

Spence, C. (2012) Auditory contributions to flavour perception and feeding behaviour. *Physiology and Behaviour*, **107**, 505–515.

Spence, C. and Deroy, O. (2013) On why music changes what (we think) we taste. *i-Perception*, **4**, 137–140.

Spence, C., Shankar, M. U. and Blumenthal, H. (2011) 'Sound bites': Auditory contributions to the perception and consumption of food and drink. In *Art and the Senses* (F. Bacci & D. Melcher, eds), pp. 207–238. Oxford University Press, Oxford.

Spence, C., Hobkinson, C., Gallace, A. and Piqueras-Fiszman, B. (2013a) A touch of gastronomy. *Flavour*, **2**, 14.

Spence, C., Richards, L., Kjellin, E., Huhnt, A.-M., Daskal, V., Scheybeler, A., Velasco, C. and Deroy, O. (2013b) Looking for crossmodal correspondences between classical music and fine wine. *Flavour*, **2**, 29.

Spence, C., Michel, C. and Smith, B. (2014) Airplane noise and the taste of umami. *Flavour*, **3**, 2.

Spence, C., Puccinelli, N. Grewal, D. and Roggeveen, A. L. (in press) Store atmospherics: A multisensory perspective. *Psychology and Marketing*.

Stafford, L. D., Fernandes, M. and Agobiani, E. (2012) Effects of noise and distraction on alcohol perception. *Food Quality and Preference*, **24**, 218–224.

Stein, B. E. and Meredith, M. A. (1993) *The Merging of the Senses.* MIT Press, Cambridge, MA.

Steinberger, M. (2010) *Au Revoir to All That: The Rise and Fall of French Cuisine.* Bloomsbury, London.

Stroebele, N. and de Castro, J. M. (2004) Effects of ambience on food intake and food choice. *Nutrition*, **20**, 821–838.

Stroebele, N. and de Castro, J. M. (2006) Listening to music while eating is related to increases in people's food intake and meal duration. *Appetite*, **47**, 285–289.

Stuckey, B. (2012) *Taste What You're Missing: The Passionate Eater's Guide to Why Good Food Tastes Good.* Free Press, London.

Sullivan, M. (2002) The impact of pitch, volume and tempo on the atmospheric effects of music. *International Journal of Retail and Distribution Management*, **30**, 323–330.

Ulla, G. (2011) Grant Achatz plans to 'overhaul the experience' at Alinea. Available at http://eater.com/archives/2011/11/23/grant-achatz-planning-major-changes-at-alinea.php#more (accessed February 2014).

Velasco, C., Jones, R., King, S. and Spence, C. (2013) Assessing the influence of the multisensory environment on the whisky drinking experience. *Flavour*, **2**, 23.

Visser, M. (1991) *The Rituals of Dinner: The Origins, Evolution, Eccentricities, and Meaning of Table Manners*. Penguin Books, London.

Wansink, B. (1992) Listen to the music: Its impact on affect, perceived time passage and applause. In: *Advances in Consumer Research* (eds J. Sherry and B. Sternthal), pp. 715–718. Association for Consumer Research, Provo, **19**.

Wansink, B. (2004) Environmental factors that increase the food intake and consumption volume of unknowing consumers. *Annual Review of Nutrition*, **24**, 455–479.

Wansink, B. (2006) *Mindless Eating: Why We Eat More Than We Think*. Hay House, London.

Wansink, B. and Van Ittersum, K. (2012) Fast food restaurant lighting and music can reduce calorie intake and increase satisfaction. *Psychological Reports: Human Resources and Marketing*, **111 (1)**, 1–5.

Watson, L. (1977) *The Omnivorous Ape*. Coward, McCann & Geoghegan, Inc., New York.

Wheatley, J. (1973) Putting colour into marketing. *Marketing*, 24–29 October, 67.

Williams, L. E. and Bargh, J. A. (2008) Experiencing physical warmth promotes interpersonal warmth. *Science*, **322**, 606–607.

Wilson, G. D. and Gregson, R. A. M. (1967) Effects of illumination on perceived intensity of acid tastes. *Australian Journal of Psychology*, **19**, 69–72.

Wilson, S. (2003) The effect of music on perceived atmosphere and purchase intentions in a restaurant. *Psychology of Music*, **31**, 93–112.

Winner, M. (2008). A glorious meal – now spot the pantomime dame. *The Sunday Times*, 16 November (News Review), 14.

Woods, A. T., Poliakoff, E., Lloyd, D. M., Dijksterhuis, G. B. and Thomas, A. (2010) Flavor expectation: The effects of assuming homogeneity on drink perception. *Chemosensory Perception*, **3**, 174–181.

Woods, A. T., Poliakoff, E., Lloyd, D. M., Kuenzel, J., Hodson, R., Gonda, H., Batchelor, J., Dijksterhuis, G. B. and Thomas, A. (2011) Effect of background noise on food perception. *Food Quality and Preference*, **22**, 42–47.

Wysocki, B. (1979). Sight, smell, sound: They're all arms in the retailer's arsenal. *The Wall Street Journal*, 17 November, **1**, 35.

Yalch, R. F. and Spangenberg, E. R. (2000) The effects of music in a retail setting on real and perceived shopping times. *Journal of Business Research*, **49**, 139–147.

Yeoh, J. P. S. and North, A. C. (2010) The effects of musical fit on choice between two competing foods. *Musicae Scientiae*, **14**, 127–138.

Youssef, J. (2013) *Molecular Cooking at Home: Taking Culinary Physics out of the Lab and into your Kitchen*. Quarto Books, London.

10
Technology at the Dining Table

10.1 Introduction

The primary question to be addressed in this chapter is how a variety of emerging and rapidly developing digital technologies (see Miller 2012) are increasingly going to play a part in, and hence change (hopefully for the better), our dining experiences in the years to come. Some of us (the lucky few) will initially experience this merging of, or interaction between, technology and gastronomy while dining out at one of the increasing popular restaurants serving modernist cuisine. In the future – and this transition will likely take a couple of years – we will start to find some of the same technologies in more typical restaurants and at home while sitting around the table with friends and family; our dinner guests may either be physically present or else might just be 'virtually' there (e.g. see the fascinating work on 'the telematic dinner party' outlined in Chapter 11). While the tremendous growth of, as well as growing interest in, modernist cuisine in recent years has partly relied on the development and utilization of new technologies in the kitchen (see Chapter 1), we believe that there is tremendous scope here to revolutionize our eating and drinking experiences and behaviours by means of the intelligent marriage of food and drink with the latest in digital technology.

While one sees a number of such technological developments emerging from restaurants (and often reads about them in press releases and news stories), it is worth bearing in mind that various technologies have already made their way, more or less unannounced, into many of our everyday restaurant environments. One already sees, for example, the increasing use of technology at the dining table. Think only of the waiters whose orders are transmitted electronically to the kitchen direct from the tableside (rather than relying on the traditional paper-and-pencil notepad or, worse still, the waiter's memory). However, as well as restaurants starting to provide digital technology at the table, many diners are themselves starting to use their own portable electronic technologies while dining. This might be anything from the diners distractedly fiddling with their smartphones during the meal through to the increasingly common trend for diners to document their meal at a fancy restaurant by using the self-same devices to photograph the dishes and then blog/tweet about the experience (sometimes while eating; O'Hara *et al.* 2012; Anonymous 2013a).[1] Another growing trend is for solo diners to use their mobiles to make up for a lack of company at the dinner table:

"They're attractive, they don't moan about the menu and they barely register when you've had too much to drink. But until now the iPad has hardly been seen as an ideal dining companion." (Anonymous 2013b)

10.2   Technology on the dining table

While digital technologies may initially assist the waiter in transferring the diner's order straight to the kitchen, it may not be all that long before there is no longer any need for a waiter in the first place. At least not if Inamo, a recently opened London restaurant is anything to go by. The diners in this futuristic venue place their orders from an illustrated food and drinks menu that is projected directly onto the table.[2] A chef colleague remembers that Alain Senderens (http://www.senderens.fr/) may have been projecting images onto the tabletop a decade ago (something like art nouveau patterns and fairies according to our source). Philippe Starck has also started to bring technology into his latest Paris dining venture; Asian TV channels are presented continuously on the tabletop in his Asian-themed bar and restaurant Miss Kō.[3] Several restaurants (e.g. Shoyu at Minneapolis-St Paul airport) already use an iPad at each table so that travellers can place their orders and navigate online while regaining their composure between flights. Together with the big TV screens and sleek styling, a friend who recently dined there while waiting

---

[1] Bearing in mind the dramatic growth in mobile devices, twinned with the profound development of technologies increasingly found in such devices (Miller 2012), it's likely that the most prevalent – not to mention sophisticated – technology at the dining table will soon mostly be embedded in such handheld devices.

[2] According to their website (http://www.e-table-interactive.com/): *"At the core of Inamo is our interactive ordering system … You'll set the mood, discover the local neighbourhood, and even order a taxi home."*

[3] For those who like their science fiction, this is somewhat reminiscent of Hugo Gernsback's prediction of a little over a century ago that the restaurant of the future would be projecting daily magazines onto the walls, with the pages changing periodically (see Gernsback 1911, p. 42).

for her delayed flight described it as a "futuristic experience at the airport". Or, as the Yo Sushi chain put it in a recent window display, "Taste the future".

Even more futuristically, at the Robot Restaurant in Harbin, Heilongjiang province, China diners have been cooked for, not to mention served, by 20 robots since the restaurant first opened its doors in June of 2012 (Ward 2013). This is all well and good, but it is important to remember here the importance of the social interaction between diner and restaurant staff in ensuring a truly great meal (see Chapter 2). Indeed, as Danny Meyer, President of the Union Square Hospitality Group in New York, notes:

> "Despite high-tech enhancements, restaurants will always remain a hands-on, high-touch, people-oriented business. Nothing will ever replace shaking people's hands, smiling, and looking them in the eye as a genuine means of welcoming them. And that is why hospitality – unlike widgets – is not something you can stamp out on an assembly line." (Meyer 2012, p. 93)

From the diner's side, though, there certainly ought to be options here to use the increasingly ubiquitous handheld technologies at mealtimes (think smartphones, tablet computers, etc.; Miller 2012). It can't be long now before diners start to use their portable electronic devices in order to help navigate through the menu and hence make better-informed food choices (see Choi *et al.* in press). Technologies such as 'the SatNav of food and menu selection' might help the diner to spot the bargains on the wine list, or else perhaps to translate the menu items while dining abroad (see also Chapter 3). They could also provide helpful information about any of the obscure ingredients that might appear on their menu (see the quote below). Indeed, many more of us might soon find ourselves needing such technological assistance once more restaurants dispense with the need for the waiter to visit the table prior to ordering! Given all of the information that is now at our disposal over the web, one might ask whether it isn't somewhat strange that we mostly still leave the decision about what to order from the menu until we actually arrive at the restaurant itself – a time that most of us would surely rather spend chatting with our dining companions or else savouring an aperitif.

> "What, I asked nicely, is zhug? The quail came spiced with it. 'Zhug,' the waitress smiled sweetly, 'is what the quail comes spiced with.' And? I prompted. 'It's a spice,' she grinned helpfully. And finally: 'Would you like me to ask the chef?' That would be kind. Now, hands up all of you who know what zhug is? Exactly. What is the point of having a menu if you need the Rosetta stone to translate it?" (Gill 2007, p. 113, on Jabberwocky food)

Talking of technology at the dinner table, a number of experimental kitchens, and even a few restaurants, have recently started to experiment

Figure 10.1 At the El Celler de Can Roca restaurant in Girona, Spain the diner can interact with the food, the plate and the surrounding elements thanks to the projection systems on the top of the table. *Source*: Reproduced with permission of Visual 13, Vilajuiga, Spain

with the possibilities associated with projecting images directly onto the food that is placed in front of the diner. For example, at the El Celler de Can Roca restaurant in Spain (rated as the best restaurant in the world according to the 2013 San Pellegrino ranking), the images projected on the food are designed to bring the food to life (e.g. http://vimeopro.com/user10658925/el-celler-de-can-roca/video/40919096). For instance, one projection makes a dish look like the surface of an egg that dramatically cracks open to reveal the food within (see Figure 10.1). Another kind of entertainment that is now being offered by a growing number of restaurants and bars is achieved by incorporating new socializing interactive technologies into their counters, tabletops or even in the walls themselves (e.g. see i-Bar or i-Wall) that produce sound or light up as the diner touches them. Meanwhile, at his restaurant Show Cooking Anteprima in Northern Italy, the Michelin-starred Swiss chef Daniel Facen has mounted screens in the dining area so that his diners can actually see their own dish being prepared in real time from the comfort of their own seat (here taking the idea of the glass-fronted kitchen window to a whole new level; Schira 2011).

10.3   Transforming the dining experience by means of technology

'The sound of the sea' seafood dish, which has been the signature dish on the tasting menu at Heston Blumenthal's The Fat Duck restaurant in Bray for a number of years now, provides an excellent example with which to highlight the way in which digital technologies can be used to deliver a genuinely different kind of multisensory dining experience (see Anonymous 2007; Spence *et al.* 2011; Schöning *et al.* 2012). With this dish, the waiter arrives at the table holding in one hand a plate of seafood that looks very much like an artistically arranged seashore and, in the other, a seashell out of which dangles a pair of iPod earphones (see Figure 10.2). The diners are instructed to insert the earphones before starting to eat, whereupon they hear the sounds of the sea: the waves crashing gently on the beach together with a few seagulls flying around overhead. This soundscape was developed specially for the dish by Condiment Junkie, a London-based sonic design agency.

In the case of 'The sound of the sea' dish, the technology (nothing more than a miniature iPod) completely transforms the dining experience, both by enhancing the taste/flavour of the food itself (see Spence *et al.* 2011 for evidence on this score)[4] and by getting the diner to pay more attention to the gustatory (and auditory) experience. Some diners have even been known to find the multisensory experience so powerful that they have broken into tears while tasting this dish (e.g. de Lange 2012). When your authors dined at Blumenthal's flagship restaurant recently, it was striking how nearby tables of erstwhile talkative diners were suddenly silenced once they had put their earphones in. People are more likely to take notice of the flavours/textures at play in a dish if their attention is squarely focused on the food rather than, say, on the latest gossip being conveyed by one's dining companions (see Spence 2012a).

The idea here is presumably that diners should come away from 'The sound of the sea' dish thinking rather more carefully about the multisensory dining experience and, in particular, the role that sound (and ultimately technology) can play in the experience of what it is that one is eating and drinking. As Blumenthal himself puts it: "*Sound is one of the ingredients that the chef has at his/her disposal*". Here it is interesting to note that back in the 1930s the Italian Futurists were already serving frogs legs while diners listened to the sounds of

---

[4] Here at the Crossmodal Research Laboratory in Oxford, we conducted research together with Heston Blumenthal demonstrating that people rate seafood as tasting significantly better when listening to a soundtrack like the sound of the sea than when listening to another (incongruent) soundtrack, such as the sound of farmyard chickens, or pre-recorded restaurant cutlery noises (Spence *et al.* 2011).

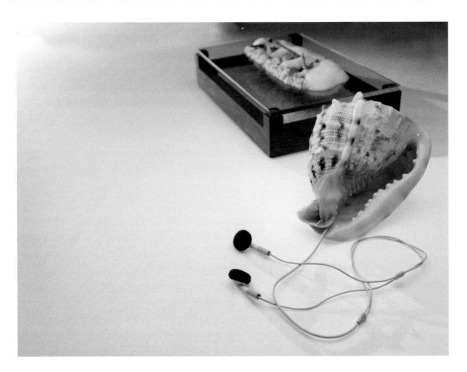

Figure 10.2 'The sound of the sea' seafood dish (for a number of years, the signature dish served on the tasting menu at Heston Blumenthal's The Fat Duck restaurant) provides an excellent example of how digital technologies can be used to enhance the multisensory dining experience. The experience of eating seafood is rated as significantly more pleasant while listening to waves crashing gently on the beach and seagulls flying overhead. Photograph by Ashley Palmer-Watts, *Source*: Photograph by Ashley Palmer-Watts. Reproduced with permission of Lotus PR and The Fat Duck

frogs croaking in the background (see Berghaus 2001). This is just one more example of how the Futurists were, in so many ways, way ahead of their time when it comes to contemporary trends in fine dining (see also Chapter 11).

The 'Messi's Goal' dessert from El Celler de Can Roca provides another example of the innovative use of technology at the table. This dish (a tribute to the famous football player from Barcelona FC) comes to the table as half a football ball covered with artificial grass perfumed with the smell of freshly cut grass (see Gopnik 2011) and three meringues, a white chocolate ball and a small bowl covered with a net on top. The diner is also given a small MP3 player wired up to a speaker. What the diner hears in this dish is the sound of an excited football commentator and the excited roar of the crowd at the Bernabeu stadium in Madrid at the 2012 classic match, from which Barcelona FC emerged victorious. The diner eats a scented meringue as the commentator announces that the world-famous footballer, Lionel Messi, is dodging Real Madrid players (aka 'Meringues') and approaching the goal. When the commentator announces that a goal has been scored, the well-behaved diner is supposed to throws the chocolate ball into the little netted bowl, thus breaking it, and eating the ball with the supposedly yummy taste of glory! We guess that the fans of Real Madrid FC, if served this dessert, would certainly not enjoy it anywhere near as much as a *culé*. (Remember that we saw in Chapter 2 that if a diner is put in a negative mood, their food does not 'taste' as good!)

Much of the current excitement around the merging of digital technology with food at the dining table lies precisely in the fact that it holds the potential to radically change our experience of dining, and to do so in a manner that many diners seem to genuinely appreciate. This will likely happen first at the tables of the Michelin-starred molecular gastronomy restaurants (such as The Fat Duck). However, as we have already mentioned at several points throughout this book, we firmly believe that within a couple of years a number of the more successful of these technological innovations will start appearing in the context of home dining.

10.4 Augmented Reality (AR) food: A case of technology for technology's sake?

In recent years, the proceedings of many international conferences on human–computer interaction (HCI; such as Siggraph, Ubiquitous Computing, etc.) have started to include contributions from those researchers working on the development of a variety of food-related augmented reality (AR) applications. It is certainly true that computer-mediated human–food interactions are now starting to attract a growing amount of research interest from the HCI community (e.g. Grimes and Harper 2008; Choi *et al.* in press; see also http://www.gastrotechdays.com/).

A few years ago, Hashimoto and colleagues (working out of Japan) developed what they described as a straw-like user interface (Hashimoto *et al.* 2007, 2008). This AR device could be used to recreate the sounds and feeling (or vibrations) that one would normally associate with sucking a particular liquidized (or mashed) food up through a straw. To operate, the user simply places the straw-like device over a mat showing the food that one would like to try and then sucks on the straw. When one of your authors tried this cutting-edge technology at a conference in Japan a few years back, he found the audio-tactile (i.e. multisensory) experience to be surprisingly realistic, despite the fact that no actual food passes the user's lips. Such technologies often enable their users to experience food in a completely different way. They can also help to bring out the playful elements in our interaction with food (Wei and Nakatsu 2012).[5]

---

[5] Many of these applications have also garnered more than their fair share of media coverage (e.g. Fellett 2011; Winter 2012; http://video.repubblica.it/tecno-e-scienze/la-dieta-degli-occhiali-ingrandiscono-il-cibo/97531?video=&ref=HRESS-27).

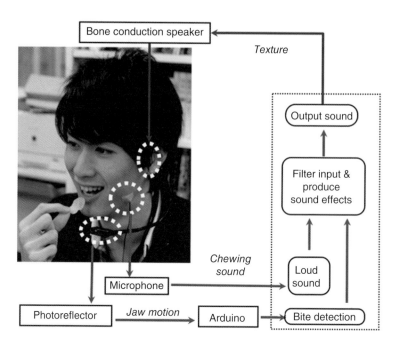

Figure 10.3 The playful 'Chewing Jockey' detects the user's jaw movements and then plays back a specific pre-recorded sound. *Source*: Reproduced with permission of Naoya Koizumi

Some researchers are currently working on technologies that will enable their users to listen to a variety of different sounds whenever they happen to close their jaw while eating (see Koizumi *et al*. 2011; Tanaka *et al*. 2011). The technology shown in Figure 10.3, known as the 'Chewing Jockey', incorporates a light sensor that can detect when a user's jaw moves and then plays back a specific pre-recorded sound. For example, the sound of someone screaming could theoretically be presented while the user munches away a mouthful of Jelly Babies. Alternatively, a microphone taped to the user's jawbone can also be used to amplify the user's own self-produced biting sounds. As yet, though, it's hard to see any practical application for this technology other than simply its entertainment value (but see Masuda and Okajima 2011 for an attempt to enhance the perceived texture and pleasantness of foods by such means). Another related example is the EverCrisp App, developed by Kayac Inc., Japan. The idea here was to develop an app for mobile devices that would enhance the crunch of noisy (e.g. dry) food products simply by changing the sound that people heard as they bit into a particular food (see Chapter 6 for the background to the incorporation of auditory cues to the perception of crispness/crunchiness).

Many other research groups (predominantly, it would seem, those based in Asia) are now developing a veritable array of AR and virtual reality (VR)

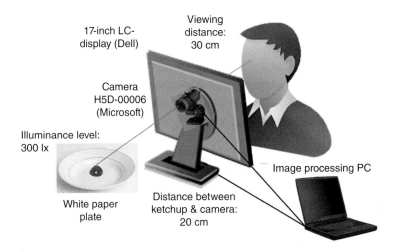

17-inch LC-display (Dell)

Viewing distance: 30 cm

Camera H5D-00006 (Microsoft)

Illuminance level: 300 lx

White paper plate

Distance between ketchup & camera: 20 cm

Image processing PC

Figure 10.4 AR and VR developments that enable their users to change the apparent colour, texture and even the size of the food that they are eating. *Source*: Spence and Piqueras-Fiszman 2013

applications that enable their users to change the apparent colour, texture and even the size of the food that they are eating (e.g. Narumi *et al.* 2010, 2012; Okajima and Spence 2011; Sakai 2011; Figure 10.4). While such technological innovations undoubtedly help to highlight just what is possible through the marriage of digital technology and food at the dinner table, it could be suggested that many of those working in the HCI/Ubiquitous Computing arena have focused a little too much of their creative energies on showcasing what the technology can deliver without necessarily spending enough of their time thinking about the practicalities associated with implementing the technology, no matter whether in the context of the high-end restaurant or the home-dining setting.

10.5   Using QR codes to change our interaction with food

Another potentially interesting technology when it comes to the experience of food results from embedding QR (quick response) tags in/on food (e.g. see Schöning *et al.* 2012).[6] Take for example QR Code Cookies (Qkies; http://qkies.de). Once a food item (e.g. a cookie) incorporates such a tag, diners can then use their mobile device to scan it and may be surprised to see

---

[6] See also http://www.youtube.com/watch?v=cbsKSPdOSX0; http://www.newlaunches.com/archives/one_cookie_seven_flavors_with_meta_cookie_ar_headgear.php.

what the designer/chef had in mind (e.g. see the online links listed above). Elsewhere, Narumi *et al.* (2011) have used this tagging approach in order to develop a multisensory display that, according to the developers at least, could change the perceived flavour of food by means of visual and olfactory AR. The device recognizes the digital tag on an item of food, and then changes the visual appearance of the food accordingly. At the same time, an appropriate aroma may be added to the food.

Just such a futuristic device was trialled at a recent Digital Olfaction Society meeting held in Berlin (April 2013; http://www.digital-olfaction.com/). The experience was certainly intriguing but not, we would argue, something that the modernist chefs need worry about incorporating into their front-of-house service just yet. At present, the QR Code is often burnt into a food (such as a biscuit) and the singed smell that this can give rise to tends to spoil the effect when you actually put the morsel of food into your mouth. While this problem can be avoided by using edible paper QR tags stuck onto the food, this still limits the kinds of foods that this technology can be used for (especially in the context of fine dining).

Elsewhere, at a Futurist participatory dining and performance installation, QR codes were used to link diners to 'food porn' images such as *"crushed up doughnuts that just somehow look dirty"* (Anonymous 2011).

10.6 Fostering healthy eating through the incorporation of technology

Over and above any potential use of digital technology to enhance the experience of food and drink or to provide entertainment for the diner, a number of researchers have now started to turn their attention to the question of whether digital technologies can be used to help people control/modify their eating behaviours. There can be no doubt but that the worldwide obesity crisis represents one of the more serious challenges facing society today (e.g. see the World Health Organization 1998; Caballero 2007; Sheridan 2007). Given the failure of many traditional (e.g. informational) approaches to tackling this crisis, researchers are increasingly coming to consider what alternative strategies can be used to help people modify their food behaviours (e.g. see Marteau *et al.* 2012). Relevant in this regard, Toet *et al.* (in press) have recently been trialling a range of digital cutlery and plateware. For example, these researchers have developed a sensor-rich spoon that vibrates if it detects that the person using it is eating too rapidly. The idea here is that the technology might provide a subtle nudge (see Thaler and Sunstein 2008) that will hopefully encourage the overweight or health-conscious diner to eat more slowly (and hence, ultimately, to eat less).

The EsTheremine talking fork (or, better said, fork-like instrument) developed by Japanese researchers (see Kadomura *et al.* 2011) could also be used to deliver health-related messaging to diners. A similar concept has been presented by HAPILabs. For example, the HAPIfork (and spoon; http://www.hapilabs.com/) is a recently developed eating implement that can measure how long a diner has been eating for, how much time they leave between successive mouthfuls and how many mouthfuls they have taken in total. The fork uses this information to provide feedback on the user's eating habits which can be viewed online via a web interface (similar to those devices one uses to track one's sports performance). There are also a number of mobile apps that allow one to monitor one's eating behaviour on the go (some with a 21-day training plan included) to get you on the right path. In addition, certain of these devices can also vibrate if necessary in order to remind their user to slow down. Why pay for a personal trainer when your own cutlery can help you to eat more healthily? In another project funded by Philips Research, Toet *et al.* (in press) investigated the feasibility of having people eat from plates that had digital scales embedded within them, in order to calculate the total amount of food that they had consumed.

Such health- and well-being-related uses of technology seem more likely to make their first appearance in the home dining environment rather than in the restaurant setting. Who, after all, goes out to eat if they are trying to watch their weight (Howard *et al.* 2012; Scourboutakos *et al.* 2013)? While such research on the use of digital technology to improve our eating behaviours is still very much in its infancy, we would argue that it nevertheless represents a promising, not to mention potentially important, area for future research.

In recent years, Philips Research has also been working on developing a concept that goes by the name of the Diagnostic Kitchen. The idea here is to allow users to take an accurate and personally relevant look at what they happen to be eating (http://www.design.philips.com/philips/sites/philipsdesign/about/design/designportfolio/design_futures/food.page). Rather than relying on general information such as the 'recommended daily intake', the idea is that the technology could scan a diner's food in order to analyse how well it matches their current needs. By using the 'Nutrition monitor', consisting of a scanning 'wand' and a swallow-able sensor, one can determine exactly how much to eat in order to match one's digestive health and nutritional requirements. All of this could obviously be of great benefit for those trying to stick to a healthy diet. In a related vein, Hoonhout *et al.* (in press) have been investigating a number of possibilities (and challenges) associated with the digitization of menu/recipe recommendations, and how digital technology could potentially be used to support healthy food choices (e.g. by providing the diner with the relevant nutritional information at the most appropriate time).

One other intriguing AR application here has emerged from research reported by Narumi *et al.* (2011). These Japanese scientists have developed an AR technology (based on the food tagging system mentioned earlier) that is capable of modifying the visually perceived size of a handheld food item (and the apparent amplitude of the aperture of the hand's grip). The health-related idea here is that people might eat less if it appeared as though they were consuming a larger piece of food than was really the case (cf. Spence *et al.* 2012 for a review of the literature on size perception and food consumption). The results of preliminary research using this device have been encouraging. For example, in one experiment Narumi *et al.* demonstrated that people consumed less when the food that they had been given to eat (a large biscuit in this case) was made to look larger than it actually was. We would however argue that longer-term follow-up studies are needed before we can come to any meaningful conclusions about the impact of such new technologies on people's eating behaviours in more naturalistic environments.

Technology can also be used to help us enjoy eating more healthily, even if we do not find vegetables particularly tasty. The latest research has shown how it is even possible to stimulate the diner's taste buds directly via means of technology (Ranasinghe *et al.* 2011; Marks 2013). In this case, a system was built that was capable of digitally delivering and controlling the sensation of taste on the human tongue by means of combined electrical and thermal stimulation. This system supposedly gives the user the impression that they can perceive certain tastes where there may be none, or else complements (or masks) certain others tastes which are actually present. Of course, in all of the cases just mentioned, the proof will be in the pudding in terms of whether these technologies ultimately turn out to be any more successful than previous attempts to make a significant impact on the obesity crisis. Thinking more strategically, it could be argued that such technologies will only stick in the marketplace if they are capable of providing a demonstrable benefit in terms of enhancing the diner's experience. This is a topic that we discuss in more detail in the following section.

10.7   Technology and distraction

Given the increasing appearance of technology at the tables of a number of the world's most *avant-garde* restaurants (e.g. The Fat Duck, Blumenthal 2008; El Celler de Can Roca), and its likely appearance at the home dining table of the future, one perhaps needs to take a step back and consider whether it is necessarily always a good thing to bring technology to the dining table. While the use of technology in this domain certainly holds the potential to enhance the diner's/drinker's experience, or to allow a restaurant to differentiate itself from the opposition in the challenging world of fine dining, it is important to remember that it can also provide an unwanted form of distraction (not to mention result in the adoption of anti-social habits by diners). Brazilian bar Salve Jorge created something of a stir in 2013 when it introduced the Offline Glass, a beer glass whose base is partly cut away so that it can only stand alone (i.e. without tipping over) if the customers put their mobile phones underneath, thus 'disconnecting' and, hopefully, being more sociable as a result (http://www.youtube.com/watch?v=-c6DNB7zWBA).

Indeed, going back to the question of healthy eating, one worrying finding to note here is that food consumption increases by as much as 15% when people are distracted by the radio/TV while eating (Bellisle and Dalix 2001; Blass et al. 2006; Braude and Stevenson 2014).[7] This is all the more worrying when it is realized that half of all meals are consumed in a room where the TV is on (Gore et al. 2003); this is what is known as secondary eating, that is, eating or drinking while doing something else (see Hamrick et al. 2011; Pollan 2013, p. 190). The relevant percentage when it comes to the change in consumption associated with people being distracted by their mobile devices at the dinner table is currently unknown (but one might hazard a guess that it is likely to be at least as important). It is also worth noting that the latest research has shown that people rate sweet, sour and salty drinks as less intense when they were engaged in a demanding secondary task than when they were concentrating on their food (van der Wal and van Dillen 2013).

10.8   Using technology to control the multisensory atmosphere

Another way in which digital technology is increasingly being put in the hands of the diner is illustrated by those restaurants where the customer can actually change the atmosphere (at present, primarily the colour of the lighting) in their dining space (e.g. the Pod restaurant in Philadelphia). Interestingly, Philips Research has been working on similar technologies for use in the home setting. The idea here is to enable the home owner (and hence potentially also the diner) to control the multisensory atmosphere by choosing from a range of different pre-selected ambient lighting, music and possibly even scent combinations, all designed to convey a particular multisensory mood or ambiance (see also Pelgrim et al. 2006).

Given such technologies, one could for example think about adding real value to the dining experience by marrying the opportunity to control the multisensory atmospherics with research findings showing how the experience might potentially be enhanced (see the previous chapter on this score). Suddenly, the technology starts to convey a meaningful benefit (over and above any entertainment value that it may deliver). For example, could the

---

[7] See also Cowen (2012, Chapter 2) for the longer-term impact of TV on food quality.

technology be used to season the food/drink (i.e. to make it taste sweeter, for example) without necessarily having to reach for the sugar (and all the associated calories; Oberfeld *et al.* 2009; Crisinel *et al.* 2012)? The answer (see below) would appear to be yes, it can. In conclusion, the range of scientific insights that are now available concerning the effects of the multisensory atmosphere on the pleasantness and enjoyment of food could be used in the service of digital technology, potentially giving it a purpose in terms of enhancing the diner's gastronomic experience rather than merely providing an entertaining distraction.

That said, and given what we have seen in the previous chapter, the most pronounced enhancement of the customer's eating and drinking experiences are likely to occur when they are put in charge not only of the ambient lighting, but also the music/sounds that they happen to be hearing. It is to the auditory attributes of the environment that we turn in the following section.

10.9   On the neuroscience of matching sound to food (and how technology might help)

It is not just the visual atmosphere that can be changed in order to enhance the taste/flavour of that which is being consumed. A large body of empirical research now shows the profound effect that what we listen to has on everything from the food and drink choices that we make through to the experience of the very taste of the food and drink itself (see Chapter 9 on this topic). In the future, we anticipate that technology will increasingly be used for the personalized delivery of music and/or soundscapes to individual tables (where, for example, a group of friends may be sharing a bottle of wine, say), or even to an individual diner or drinker. In fact, the last few years have seen something of an explosion of research interest in the matching of music/soundscapes to specific tastes, flavours and food textures (see Hirsch 2011; Mesz et al., 2011, 2012; Pelaccio, 2012; Spence 2011a,b, 2013a, b).

In a recent collaboration between academia and sonic design (Crisinel *et al.* 2012), the Crossmodal Research Laboratory in Oxford conducted a series of studies in which the participants were given pieces of bittersweet toffee to taste while listening to one of two soundscapes developed by Condiment Junkie. The toffee in this case had been prepared by The Fat Duck experimental research kitchen in Bray. The soundscapes themselves were developed on the basis of prior laboratory research by Crisinel and Spence (2010a,b) that had highlighted how people typically pick lower-pitched tones as matching (or corresponding crossmodally with) bitter tastes while reporting that higher-pitched sounds provide a better match for sweet tastes (see Knöferle and Spence 2012 for a review of the literature on matching musical parameters to specific tastes and flavours). The results revealed that people's perception of the intensity of the bitterness and sweetness of the toffee was modified by around 10% simply by varying the pitch of a soundtrack that was playing over headphones.

A variant of 'The bittersweet symphony' dish was featured on the menu at The House of Wolf, an experiential dining restaurant that opened in London late in 2012.[8] In both cases, when the Sonic Cake Pop dish (a ball of bittersweet toffee on a stick covered with chocolate) was brought to the table, the diners were provided with two numbers to call. Depending on which number was dialled, the diner could listen to a soundtrack that had been designed to emphasize either the sweetness of the dessert or instead bring out its bitter notes. We would argue that dishes such as the Sonic Cake Pop highlight the second route by which technology (the technology which most people bring to the dinner table) holds the potential to influence the dining experience.

Exciting ideas and opportunities are now also starting to emerge here around the intelligent pairing of music designed to support/complement specific brand experiences. For example Courvoisier developed a smell kit containing six small bottles, each containing one of the key notes found in their XO Imperial Cognac. The aromas included the distinctive smells of crème brûlée, ginger biscuits, candied orange, coffee and iris flower. Next, a composer was commissioned to generate a series of short musical clips to match each of these aroma notes in the Cognac based on the Oxford Crossmodal Research Laboratory's prior findings. The idea here was that consumers could listen individually to each of these musical pieces. Once they had established (or, better still, correctly guessed) the association between the three pieces of music and the various aromas, they could then listen to another musical composition in which each of these individual musical elements had been skilfully integrated into a single, more complex, arrangement (i.e. to match in some sense the complex interplay of aromas that you might find in a fine cognac). Available as an app (http://courvoisier.com/uk/le-nez-de-courvoisier-app), one can imagine the tech-savvy diners enhancing their experience of their post-dinner digestive in the setting of the restaurant by bringing out their handheld device.[9] While subsequent laboratory-based testing here in Oxford revealed that the composer in this particular case had failed to unambiguously capture the distinctive aromas in his musical compositions, the iterative process of refinement (between the science lab and the creative designer or composer) will likely allow the latter to home in on the optimal musical expression (or at least the most unambiguous one) with which to convey specific tastes/flavours (see Spence 2013a).

---

[8] The dish also featured on the menu at the 2012 annual Experimental Food Society banquet also held in London (http://www.experimentalfoodsociety.com/).

[9] See also Crisinel *et al.* (2013) and Spence (2011b, 2012a) for a similar approach to crossmodally matching the music to the taste of Starbucks coffee.

It is also worth noting that those chefs who recommend that you cook their recipes while listening to a particular piece of music (e.g. Pelaccio 2012) may also be influencing how the dish eventually turns out; the latest research suggests that listening to sweet music makes people mix sweeter drinks than when they listen to other sorts of music.

10.10   On the future of technology at the table: digital artefacts

If such insights regarding the crossmodal matching of music and soundscapes to food and drink were to be delivered by means of musical plateware – that is, cups and plates that made music whenever they were picked up or rotated, or which change the sound they made as the level in your glass slowly goes down – who knows what entertaining and possibly enhanced eating and drinking experiences might be had. See, for example, the musical coffee cup (http://www.youtube.com/user/virsomio?nomobile=1;   http://www.youtube.com/watch?v=MY3NcckUffM). Heineken's recent Ignite bottle, which lights up when people cheer and move it around and which dims when left alone for a while, is just at the prototype stage (http://www.youtube.com/watch?feature=player_embedded&v=fU6SmBPcWFA; see also Beigl *et al.* 1999). Today's technology is enabling more and more innovative companies to embed interactive printed electronic labels, for example, in packaging (Gyekye 2011). The first prototypes have consisted of interactive bottle labels that activate flashing lights when a bottle is held, for example the limited-edition pack of Bombay Sapphire Gin that was launched in 2012, produced with electroluminescent ink (http://www.designweek.co.uk/news/gin-packaging-that-lights-up-in-your-hand/3035380.article). While this is but one example of how active packaging can be used to interact with the costumer, there is more scope for interaction here than ever before. Returning to the topic of how technology might be used to help the consumer to drink in a more responsible manner, MIT graduate Dhairya Dand invented 'Cheers', alcohol-aware ice-cubes that glow and 'groove' in time to the ambient music. These ice-cubes incorporate LEDs and an accelerometer that keeps track of how often the glass is brought up to a drinker's lips, changing colour (from green to red) if one's cubes decide that you have had a few too many sips (López 2013). Other ideas here involve the use of digital artefacts,[10] including responsive placemats or beermats that could potentially be used to enable a consumer to select the music they like and hopefully enhance their experience while using an interface that is as naturalistic as possible (Butz and Schmitz 2005). While we are on the

---

[10] Note that digital technologies will increasingly appear in digital artefacts, such as the Mediacup, where the computer becomes less and less visible to the diner who will increasingly find him/herself interacting with the technology at a more implicit level (see Beigl *et al.* 1999).

topic, why not a digital charger plate that would allow the diner to select their preferred musical accompaniment for a given dish (see also Mogees, http://www.brunozamborlin.com/mogees/ for some ideas of just what is possible on the tabletop these days). Hyperdirectional loudspeakers that can direct sound at an individual or particular table may also provide some intriguing opportunities for the targeted delivery of experiential soundscapes (what are sometimes referred to as sound showers) to diners in the years to come.[11]

At the Crossmodal Research Laboratory here in Oxford, together with Condiment Junkie, we are currently working with one London bar/restaurant on the futuristic-sounding idea of being able to deliver targeted music/soundscapes to diners/drinkers. For example, one idea that we are playing around with is that a person comes up to the bar, orders a bottle of wine for his table, say, and when (s)he returns to his table the background music delivered specifically to that table will have changed to match the wine that has been ordered. Indeed, in a recent cocktail tasting master class that was organized by Condiment Junkie at London's 69 Colbrooke Row, different ambient and headphone-delivered soundscapes were paired with a series of tasty cocktails (Spence 2013a,b). It was amazing to see just how dramatically the multisensory experience changed once the soundscape matched the taste or flavour of the drink (not to mention the drinker's individual tastes).[12]

Spotting a marketing opportunity when they see one, food designers Sam Bompas and Harry Parr made a set of musical spoons, one to go with each of five flavours of baked beans (see Chapter 4; Anonymous 2013c). In this case, each of the spoons had a tiny MP3 player hidden inside. A diner couldn't hear anything until they put the spoon in their mouth; sound vibrations would then travel via the diner's jawbone to their ear (transmitted via bone-conducted sound). The flavour–music combinations included cheddar cheese with a rousing bit of Elgar, fiery chilli with a Latin samba, blues for the BBQ-flavoured beans and Indian music for the curry-flavoured beans (could have been yours for just £57 at Fortnum & Mason)!

---

[11] While headphones obviously allow for the targeted delivery of sound, they provide a less than optimal means of sound delivery in social situations. While wearing the iPod earphones works for one of the many courses on the tasting menu at The Fat Duck, diners would likely not comply if they had to keep the earphones in place throughout the entire course of the meal. Indeed, the very first attempt to introduce sound at the table involved a pair of over-ear headphones. Informal trials in the restaurant soon revealed that people, especially ladies of a certain age, felt too self-conscious and perhaps not quite as alluring as they had hoped to feel with this technology strapped over their stylish new hairdo's. This experimental dish never quite made it onto the menu.

[12] Note here that supertasters (see Chapter 6) will match a louder sound to bitter tastes than medium or non-tasters (Bartoshuk 2000). Linda Bartoshuk uses a technique known as crossmodal matching to illustrate that, on average, supertasters match the sweetness of the taste of Coke to a 90 decibel tone, whereas medium tasters will choose a tone of 80 decibels instead. This might not sound like a lot, but 10 dB difference actually represents a doubling of the sound level (Stuckey 2012, p. 26).

10.11   The SmartPlate

Working at a more conceptual level, Romanian designer Julian Caraulani (finalist in the 2012 Electrolux Design Lab competition) developed a concept going by the name of the SmartPlate (see http://www.electrolux.co.uk/Global-pages/Promotional-pages/Electrolux-Design-Lab/Electrolux-Design-Lab-Finalists-Present-Concepts-that-Stimulate-the-Senses/). The idea was that this intelligent piece of plateware would 'understand' food and somehow transform it into sound. According to the online description (http://www.yankodesign.com/2012/09/27/musical-dinner/), this digital artefact (if it were ever to be built) would complete the circle of senses by which we understand what we eat. The idea is that the plate would connect wirelessly to the user's mobile device and, in theory at least, it could be used to measure the ingredients, identify them and then attach musical notes, harmonies and rhythms to each of them. The user would be able to listen actively, to compose, and to interact with recipes of sound, sharing the experience in the most intimate of ways: that is, by means of music. While this design idea is undoubtedly intriguing (and undoubtedly successfully captures the current buzz around the synaesthetic matching of sensations; see Spence 2012b, 2013a), the practicalities associated with matching tastes/flavours to sound might actually prove somewhat harder to develop than the description cited above might lead one to believe.

10.12   Anyone for a 'Gin & Sonic'?

One of the challenges that we are currently working on here at the Crossmodal Research Laboratory in Oxford goes by the name of the 'Gin & Sonic'. Denis Martin, a 2-Michelin-starred chef who runs the restaurant Denis Martin in Vevey, Switzerland, uses a balloon and a liquid nitrogen bath in order to create a gin and tonic, the likes of which has never been seen before. In this dish, the drink is poured into a balloon which is then inflated and tied up. The balloon is subsequently carefully submerged in a bath of liquid nitrogen (careful now) and turned rapidly until it freezes. Once the skin of the balloon has been peeled off, what is left is a perfectly hollow white sphere of deep-frozen gin and tonic. While the dish itself is visually dramatic (not to mention delicious), the one thing that is missing from this multisensory dining experience is the 'Schh...' of the tonic gently fizzing in the glass (frozen tonic, after all, makes no noise). We are currently collaborating with Condiment Junkie to try to embed an actuator (a device that can transform any rigid surface into a loudspeaker) into the plateware in order to bring back the sound of carbonation, this time through the plateware (see Zampini and Spence 2005 on the importance of sound to the perception of carbonation).

As the above examples have hopefully made clear, digital technology is increasingly being used to help bring sound and food/drink together in new and creative ways, often embedded in digital artefacts. What is particularly exciting in terms of the future of such cross-sensory matching is that experimental dining spaces, such as the workshop of Paco Roncero in Madrid, are now increasingly being fitted out with the requisite technology needed to deliver specific visual, auditory, olfactory and increasingly multisensory experiences to diners (see Jakubik 2012; http://www.estudiocbaselga.co.uk/estudio/news; see also Chapter 11).

10.13 The tablet as twenty-first century plateware?

Another of our current favourite ideas around the theme of bringing digital technology to the dining table relates to the possible use of tablet computers as intelligent twenty-first century plateware (see Figure 10.5). Consider for a moment how the eating experience could be enhanced if people were to stop being so distracted by their tablets (and other handheld mobile devices) while eating. Imagine what possibilities might open up if one were to start serving the very food itself from a tablet?[13] One idea that immediately springs to mind

Figure 10.5 Eating direct from a tablet computer is one of the future plateware possibilities that may be well worth considering. *Source*: Reproduced with permission of Adam Scott. *See colour plate section*

---

[13] Acknowledging, of course, the difficulty of keeping such technology clean. That said, waterproof tablets have recently started appearing on the shelves.

here is that it would be possible to change the screen colour (and hence the plate colour) in order to bring out the sweetness in a dish, say. This suggestion is based on recent findings (as discussed in Chapter 4) showing that strawberry desserts are rated as tasting more than 10% sweeter, not to mention 15% more flavourful, when eaten from a white plate as compared to when exactly the same food is tasted from a black plate instead (Piqueras-Fiszman *et al.* 2012). If matters were that simple, though, there would probably be no need for digital plateware at all (just make sure you have a set of white plates at home). As we saw in Chapter 4 however, it turns out that the optimal plate colour (in term of enhancing the taste/flavour of a dish) actually depends on the particular food that is being served. A tablet computer screen might then be ideal in terms of generating precisely the right colour background (not to mention shape) to bring out the taste of the particular dish being eaten from its surface. Who knows, one might also be able to trigger particular kinds of musical accompaniment (or soundscapes; see previous sections) depending on what exactly the diner chooses to eat from the plate at a given time.

Talking of which, Philips Design has also been exploring a new range of plateware concepts in its latest design probes,[14] in particular, investigating how the integration of light, conductive printing, selective fragrance diffusion, micro-vibration and the integration of other sensory stimuli might affect the eating experience (http://www.design.philips.com/philips/sites/philipsdesign/about/design/designportfolio/design_futures/food.page). For example, researchers engaged in one design probe going by the name of 'Multisensorial Gastronomy' have been exploring how the eating experience can be enhanced or altered by stimulating the senses using the integration of electronics, light and other stimuli. Developed in collaboration with Michelin-starred chef Juan Maria Arzak (http://www.arzak.info/arz_web.php?idioma=En), the four design concepts of interactive tableware – Lunar Eclipse (bowl), Fama (long plate), Bocado de Luz (serving plate) and the Eye of the Beholder (platter) – react to the food that is placed on them.

In some sense, one can think about the increasing use of technology at the table as providing chefs the means with which to deliver on some of the wondrous culinary/theatrical ideas that were first proposed and, on occasion, tried out by the Italian Futurists (see Berghaus 2001). Just take as an example the Ultraviolet restaurant that opened in 2013 in Shanghai (Kessel 2013). Here one sees technology being used to control all aspects of the multisensory dining atmosphere. For example, the restaurant's chef Paul Pairet serves fish and chips while diners listen to the sound of The Beatles, a Union Jack flag is projected onto the tabletop and the image of wet windowpanes (what else?) covers the walls (Gonzalez 2013). Meanwhile, fans are used to disperse the

---

[14] *"The FOOD design probe: A far-flung design concept; A provocative and unconventional look at areas that could have a profound effect on the way we eat and source our food 15-20 years from now"* (http://www.design.philips.com/).

appropriate fragrances through the dining space coordinated with the food currently being served. Of course, one wonders how so much investment in the technology can ever pay off when the dining space can only seat ten at a time.[15]

10.14 Tips from the chef at the tips of your fingers

There are apps for everything today and the gastronomy sector will certainly not remain immune to such developments for much longer, mostly when there are so many people that may not be able to go (or be tempted to go) to fine-dining restaurants. What do they do? The answer is obvious. San Pellegrino, sponsor or the annual competition of the world's 50 best restaurants, launched an app called 'The Perfect Host' (Leahy 2012). This features numerous recipes grouped in different themes, and even includes useful tips on how to set the table! One can safely assume that this app is targeted at the fine-dining end of the market. Anyone who thought that Ferran Adrià was not going to launch his own app certainly doesn't know what the maestro is capable of... his 'Adrià en casa' (Adrià at home) app offers over 30 three-course menus, each simple to prepare and, according to the app, affordable. There are tips on how to 'organize ourselves at home' (in terms of time, buying the ingredients, etc.) and it boasts pre-configured timers and options to email the shopping list, take pictures and then post them on your preferred social network. Note that your final dishes might not look quite as nice as those shown in the app, since some are presented on designer plateware especially created for elBulli, sorry. Other TV chefs such as Nigella Lawson have also thought about developing their own distinctive apps (Poole 2012, p. 32; see also http://www.makemyplate.co/). In brief, it is no longer sufficient to simply take pictures of one's meal and post them; nowadays it's all a matter of cooking like a chef and sharing the entire process!

10.15 Conclusions

In conclusion, it seems clear that technology will increasingly come to change the way in which many of us interact with food and drink. Our prediction is that this will start at the tables of the cutting edge (including, but not restricted to, molecular gastronomy) restaurants, but that a number of the most successful technologies trialled there will, sooner or later, make their way onto the

---

[15] Presumably this may be another loss-leading branded food venture (cf. Steinberger 2010). What is even more peculiar is that the restaurant's exact location is meant to be a secret; how long it remains so is anyone's guess.

home dining table. It is however important to bear in mind here that there are at least two distinct routes by which technology will make its way onto the dining table: either supplied by the restaurant or else brought to the table (either in the restaurant or home dining setting) by the diners themselves. Indeed, given the explosion of handheld mobile technologies in recent years (see http://www.portioresearch.com/en/reports/current-portfolio/smartphone-futures-2012-2016.aspx), it would seem probable that the latter will be the primary route for mixing the two. Here, it should also be noted how technology companies such as Philips Research, Electrolux and Microsoft Research are all thinking about the ways in which technology may be introduced to the kitchens and restaurants of the future (and, what is more, investing research money in supporting the latest developments in this area). This might provide an additional route by which the latest in technology eventually appears at the dining table.

In conclusion, in this the penultimate chapter, we have reviewed a number of the ways in which digital technology may change the experience of food and drink. To recap: technology may be used to facilitate our interaction with (knowledge about) that which we are eating and drinking; it may be used to enhance the entertainment value of the meal; it may be used to change the taste/flavour of the food (a kind of 'digital seasoning', if you will) or it may be used to provide targeted multisensory interventions to diners; it may be used to provide a subtle nudge to those who may wish to eat more healthily; and it may also be used to provide company for those who, for whatever reason, eat alone (Anonymous 2013b). It may soon make all those old cookbooks you have on the shelf in the kitchen redundant. And, before too long, it may even be used to enable the busy international executive to be virtually present with his or her family when working away from home on business, say (see the next and final chapter).

On the flip side, the biggest danger here is perhaps that the technology becomes nothing more than a distraction for the diner (and possibly results in an unwanted, and likely unnoticed, increase in food intake). However, in the best case scenario we believe the benefits of bringing technology to the dinner table hold the potential to transform our dining experiences in a manner similar to the way that the introduction of new technology/techniques at the back of house (that is, in the kitchens) have helped to usher in the era of molecular gastronomy. It is our firm belief that if the technology is linked to the relevant behavioural science (this is where the gastrophysics comes in), then the benefits of introducing technology at the dining table are likely to outweigh the costs. Who knows, in the years to come the perfect meal may even come to require the presence of technology at the table (a thought that is no doubt horrifying to many). Having taken a look at the imminent arrival of technology at the table, we move on in the final chapter to see what the perfect meal might look like in the decades to come.

References

Anonymous (2007) Fat Duck dons an iPod. *Manchester Evening News*, 16 April.

Anonymous (2011) Futurism for foodies. *Artnet* August 12. Available at http://www.artnet.com/magazineus/news/artnetnews/the-futurist-cookbook.asp (accessed February 2014).

Anonymous (2013a) Food photos: How to do it properly. *BBC News Magazine*. Available at http://www.bbc.co.uk/news/magazine-21235195 (accessed February 2014).

Anonymous (2013b) Lone diners use their phones for company. *The Times*, 26 April, 31.

Anonymous (2013c) Musical spoons to go with your Heinz beans. *Advertising Age*, 28 March. Available at http://adage.com/article/creativity-pick-of-the-day/bompas-parr-design-musical-spoons-heinz-beans/240605/ (accessed February 2014).

Beigl, M., Gellersen, H.-W. and Schmidt, A. (1999) MediaCups: Experience with design and use of computer-augmented everyday artefacts. Available at http://docis.info/docis/lib/goti/rclis/dbl/connet/(2001)35%253A4%253C401%253AMEWDAU%253E/www.comp.lancs.ac.uk%252F~hwg%252Fpubl%252Fmediacups.pdf (accessed February 2014).

Bellisle, F. and Dalix, A. M. (2001) Cognitive restraint can be offset by distraction, leading to increased meal intake in women. *American Journal of Clinical Nutrition*, **74**, 197–200.

Berghaus, G. (2001) The futurist banquet: Nouvelle Cuisine or performance art? *New Theatre Quarterly*, **17 (1)**, 3–17.

Blass, E. M., Anderson, D. R., Kirkorian, H. L., Pempek, T. A., Price, I. *et al.* (2006) On the road to obesity: Television viewing increases intake of high-density foods. *Physiology and Behavior*, **88**, 597–604.

Blumenthal, H. (2008) *The Big Fat Duck Cookbook*. Bloomsbury, London.

Braude, L. and Stevenson, R. J. (2014) Watching television while eating increases energy intake. Examining the mechanisms in female participants. *Appetite*, **76**, 9–16.

Butz, A. and Schmitz, M. (2005) Design and application of a beer mat for pub interaction. Available at http://www.medien.ifi.lmu.de/pubdb/publications/pub/butz2005ubicomp/butz2005ubicomp.pdf (accessed February 2014).

Caballero, B. (2007) The global epidemic of obesity: An overview. *Epidemiologic Reviews*, **29**, 1–5.

Choi, J. H.-J., Foth, M. and Hearn, G. (eds) (in press). *Eat, Cook, Grow: Mixing Human–Computer Interactions with Human–Food Interactions*. MIT Press, Cambridge, MA.

Cowen, T. (2012) *An Economist Gets Lunch: New Rules for Everyday Foodies*. Plume, New York.

Crisinel, A.-S. and Spence, C. (2010a) A sweet sound? Exploring implicit associations between basic tastes and pitch. *Perception*, **39**, 417–425.

Crisinel, A.-S. and Spence, C. (2010b) As bitter as a trombone: Synesthetic correspondences in non-synesthetes between tastes and flavors and musical instruments and notes. *Attention, Perception and Psychophysics*, **72**, 1994–2002.

Crisinel, A.-S., Cosser, S., King, S., Jones, R., Petrie, J. and Spence, C. (2012) A bittersweet symphony: Systematically modulating the taste of food by changing the sonic properties of the soundtrack playing in the background. *Food Quality and Preference*, **24**, 201–204.

Crisinel, A.-S., Jacquier, C., Deroy, O. and Spence, C. (2013) Composing with cross-modal correspondences: Music and smells in concert. *Chemosensory Perception*, **6**, 45–52.

de Lange, C. (2012) Feast for the senses: Cook up a master dish. *New Scientist*, **2896** (18 December).

Fellett, M. (2011) Smart headset gives food a voice. *New Scientist*. Available at http://www.newscientist.com/blogs/nstv/2011/12/smart-headset-gives-food-a-voice.html (accessed February 2014).

Gernsback, H. (1911) *Ralph 124C 41+*. Fawcett Publications, Minnesota.

Gill, A. A. (2007) *Table Talk: Sweet and Sour, Salt and Bitter*. Weidenfeld & Nicolson, London.

Gonzalez, L. (2013) Ultraviolet by Chef Paul Pairet incorporates thematic video and perfumed air into his dining experience. Available at http://www.psfk.com/2013/05/multisensory-dining-sight-sound-smells.html (accessed February 2014).

Gopnik, A. (2011) Sweet revolution. *The New Yorker*, 3 January. Available at http://www.newyorker.com/reporting/2011/01/03/110103fa_fact_gopnik (accessed February 2014).

Gore, S. A., Foster, J. A., DiLillo, V. G., Kirk, K. and Smith West, D. (2003) Television viewing and snacking. *Eating Behaviors*, **4**, 399–405.

Grimes, A. and Harper, R. (2008) Celebratory technology: New directions for food research in HCI. In: *Proceedings of the Twenty-Sixth Annual SIGHCI Conference on Human Factors in Computing Systems (CHI'08)*, pp. 467–476. ACM, Florence.

Gyekye, L. (2011) Electronic labels light up beverage bottles. Available at http://www.packagingnews.co.uk/news/electronic-labels/ (accessed February 2014).

Hamrick, K., Andrews, M., Guthrie, J., Hopkins, D. and McClelland, K. (2011) How much time do Americans spend on food? *Economic Information Bulletin*, **EIB-86**, 1–64.

Hashimoto, Y., Nagaya, N., Kojima, M., Miyajima, S., Ohtaki, J., Yamamoto, A., Mitani, T. and Inami, M. (2007) Straw-like user interface: Virtual experience of the sensation of drinking using a straw. In: *Proceedings World Haptics 2007*, pp. 557–558. IEEE Computer Society, Los Alamitos, CA.

Hashimoto, Y., Inami, M. and Kajimoto, H. (2008) Straw-like user interface (II): A new method of presenting auditory sensations for a more natural experience. In: *Eurohaptics 2008, LNCS, 5024*. (ed M. Ferre), pp. 484–493. Springer-Verlag, Berlin.

Hirsch, J. (2011) A fine tune to match your entrée. Available at http://www.nytimes.com/2011/07/22/us/22bcculture.html?_r=0 (accessed February 2014).

Hoonhout, J., Gros, N., Geleijnse, G., Nachtigall, P. and van Halteren, A. (in press) What are we going to eat today? Food recommendations made easy, and healthy. In: *Eat, Cook, Grow: Mixing Human–Computer Interactions with Human–Food Interactions* (eds J. H.-J. Choi, M. Foth and G. Hearn). MIT Press, Cambridge, MA.

Howard, S., Adams, J. and White, M. (2012) Nutritional content of supermarket ready meals and recipes by television chefs in the United Kingdom: Cross sectional study. *BMJ*, **345**, e7607.

Jakubik, A. (2012) The workshop of Paco Roncero. *Trendland: Fashion Blog and Trend Magazine*. Available at http://trendland.com/the-workshop-of-paco-roncero/ (accessed February 2014).

Kadomura, A., Nakamori, R., Tsukada, K. and Siio, I. (2011) EaTheremin. In: *SIG-GRAPH Asia 2011 Emerging Technologies*, p. 7. ACM, New York.

Kessel, J. (2013) Ultra meal at Ultraviolet. *The New York Times*. Available at http://www.nytimes.com/video/dining/100000002498301/ultra-dining-at-ultraviolet .html (accessed February 2014).

Knöferle, K. M. and Spence, C. (2012) Crossmodal correspondences between sounds and tastes. *Psychonomic Bulletin and Review*, **19**, 992–1006.

Koizumi, N. Tanaka, H., Uema, Y. and Inami, N. (2011) Chewing jockey: Augmented food texture by using sound based on the cross-modal effect. In: *Proceedings of ACE'11, Proceedings of the 8th International Conference on Advances in Computer Entertainment Technology*, Article No. 21. ACM, New York.

Leahy, T. (2012) 'The perfect host' a new app by San Pellegrino. Available at http://www.foodepedia.co.uk/articles/2012/oct/san_pellegrino_app.htm (accessed February 2014).

López, C. (2013) MIT student invents LED ice cubes to track alcohol intake. Available at http://abcnews.go.com/blogs/technology/2013/01/mit-student-invents-led-ice-cubes-to-track-alcohol-intake/ (accessed February 2014).

Marks, P. (2013) Electrode recreates all four tastes on your tongue. *New Scientist*, **2944**, 22.

Marteau, T. M., Hollands, G. J. and Fletcher, P. C. (2012) Changing human behaviour to prevent disease: The importance of targeting automatic processes. *Science*, **337**, 1492–1495.

Masuda, M. and Okajima, K. (2011) Added mastication sound affects food texture and pleasantness. Poster presented at the *12th International Multisensory Research Forum* in Fukuoka, Japan, October 17–20.

Mesz, B., Trevisan, M. and Sigman, M. (2011) The taste of music. *Perception*, **40**, 209–219.

Mesz, B., Sigman, M. and Trevisan, M. A. (2012) A composition algorithm based on crossmodal taste-music correspondences. *Frontiers in Human Neuroscience*, **6 (71)**, 1–6.

Meyer, D. (2010) *Setting the Table: Lessons and Inspirations from one of the World's Leading Entrepreneurs*. Marshall Cavendish International, London.

Miller, G. (2012) The smartphone psychology manifesto. *Perspectives on Psychological Science*, **7**, 221–237.

Narumi, T., Kajinami, T., Tanikawa, T. and Hirose, M. (2010) Meta cookie. In: *ACM SIGGRAPH 2010 Emerging Technologies*, Article No. 18. ACM, New York.

Narumi, T., Nishizaka, S., Kajinami, T., Tanikawa, T. and Hirose, M. (2011) Augmented reality flavors: Gustatory display based on edible marker and cross-modal interaction. In: *Proceedings of the 2011 Annual Conference on Human Factors in Computing Systems (CHI'11)*, pp. 93–102. ACM, New York.

Narumi, T., Ban, Y., Kajinami, T., Tanikawa, T. and Hirose, M. (2012) Augmented perception of satiety: Controlling food consumption by changing apparent size of food with augmented reality. In: *Proceedings 2012 ACM Annual Conference Human Factors in Computing Systems; CHI 2012,* May 5–10, 2012. ACM, Austin, Texas.

Oberfeld, D., Hecht, H., Allendorf, U. and Wickelmaier, F. (2009) Ambient lighting modifies the flavor of wine. *Journal of Sensory Studies*, **24**, 797–832.

O'Hara, K., Helmes, J., Sellen, A., Harper, R., ten Bhömer, M. and van den Hoven, E. (2012) Food for talk: Phototalk in the context of sharing a meal. *Human-Computer Interaction*, **27 (1–2)**, 124–150.

Okajima, K. and Spence, C. (2011) Effects of visual food texture on taste perception. *i-Perception*, **2 (8)**, Available at http://i-perception.perceptionweb.com/journal/I/article/ic966 (accessed February 2014).

Pelaccio, Z. (2012) *Eat With Your Hands*. Ecco, New York.

Pelgrim, P. H., Hoonhout, H. C. M., Lashina, T. A., Engel, J., Ijsselsteijn, W. A. and de Kort, Y. A. W. (2006) *Creating Atmospheres: The Effects of Ambient Scent and Coloured Lighting on Environmental Assessment*. Paper presented at the Design and Emotion conference 2006, Gothenburg.

Piqueras-Fiszman, B., Alcaide, J., Roura, E. and Spence, C. (2012) Is it the plate or is it the food? Assessing the influence of the color (black or white) and shape of the plate on the perception of the food placed on it. *Food Quality and Preference*, **24**, 205–208.

Pollan, M. (2013) *Cooked: A Natural History of Transformation*. Penguin Books, London.

Poole, S. (2012) *You Aren't What You Eat: Fed Up With Gastroculture*. Union Books, London.

Ranasinghe, N., Cheok, A. D., Fernando, O. N. N., Nii, H. and Gopalakrishnakone, P. (2011) Digital taste: Electronic stimulation of taste sensations. *Ambient Intelligence: Lecture Notes in Computer Science*, **7040**, 345–349.

Sakai, N. (2011) Tasting with eyes. *i-Perception*, **2 (8)**, Available at http://i-perception.perceptionweb.com/journal/I/article/ic945 (accessed February 2014).

Schira, R. (2011) Daniel Facen, the scientific chef. Available at http://www.finedininglovers.com/stories/molecular-cuisine-science-kitchen/ (accessed February 2014).

Schöning, J., Rogers, Y. and Krüger, A. (2012) Digitally enhanced food. *Pervasive Computing*, **11 (3)**, 4–6.

Scourboutakos, M. J., Semnani-Azad, Z. and L'Abbe, M. R. (2013) Restaurant meals: Almost a full day's worth of calories, fats, and sodium. *JAMA Internal Medicine*, **173**, 1373–1374.

Sheridan, B. (2007) *Mind over platter. Stanford*, September/October, 65–69.

Spence, C. (2011a) Sound design: How understanding the brain of the consumer can enhance auditory and multisensory product/brand development. In: *Audio Branding Congress Proceedings 2010* (eds K. Bronner, R. Hirt and C. Ringe), pp. 35–49. Nomos Verlag, Baden-Baden.

Spence, C. (2011b) Wine and music. *The World of Fine Wine*, **31**, 96–104.

Spence, C. (2012a) Auditory contributions to flavour perception and feeding behaviour. *Physiology and Behaviour*, **107**, 505–515.

Spence, C. (2012b) Synaesthetic marketing: Cross sensory selling that exploits unusual neural cues is finally coming of age. *The Wired World in 2013*, November, 104–107.

Spence, C. (2013a) On crossmodal correspondences and the future of synaesthetic marketing: Matching music and soundscapes to tastes, flavours, and fragrance. In: *(((ABA))) Audio Branding Academy Yearbook 2012/2013* (ed K. Bronner, R. Hirt and C. Ringe), pp. 39–45. Nomos Verlag, Baden-Baden.

Spence, C. (2013b) Sound advice. *The Cocktail Lovers, Winter* (**6**), 18–19.

Spence, C., Shankar, M. U. and Blumenthal, H. (2011) 'Sound bites': Auditory contributions to the perception and consumption of food and drink. In: *Art and the Senses* (eds F. Bacci and D. Melcher), pp. 207–238. Oxford University Press, Oxford.

Spence, C., Harrar, V. and Piqueras-Fiszman, B. (2012) Assessing the impact of the tableware and other contextual variables on multisensory flavour perception. *Flavour*, **1**, 7.

Steinberger, M. (2010) *Au Revoir to All That: The Rise and Fall of French Cuisine.* Bloomsbury, London.

Stuckey, B. (2012) *Taste What You're Missing: The Passionate Eater's Guide to Why Good Food Tastes Good.* Free Press, London.

Tanaka, H., Koizuni, N., Uema, Y. and Inami, M. (2011) Chewing Jockey: Augmented food texture by using sound based on the cross-modal effect. Paper presented at *Siggraph Asia 2011*, Hong Kong, 12–15 December.

Thaler, R. H. and Sunstein, C. R. (2008) *Nudge: Improving Decisions About Health, Wealth and Happiness.* Penguin, London.

Toet, E., Meerbeek, B. and Hoonhout, J. (in press) Supporting mindful eating: InBalance chopping board. To appear in: *Eat, Cook, Grow: Mixing Human–Computer Interactions with Human–Food Interactions* (eds J. H.-J. Choi, M. Foth and G. Hearn). MIT Press, Cambridge, MA.

van der Wal, R. C. and van Dillen, L. F. (2013) Leaving a flat taste in your mouth: Task load reduces taste perception. *Psychological Science*, **24**, 1277–1284.

Ward, A. (2013) Mechanic masterchef: Robots cook dumplings, noodles and wait tables at restaurant in China. Available at http://www.dailymail.co.uk/news/article-2261767/Robot-Restaurant-Robots-cook-food-wait-tables-Harbin.html#ixzz2SWtkRc9S (accessed February 2014).

Wei, J. and Nakatsu, R. (2012) Leisure food: Derive social and cultural entertainment through physical interaction with food. *Entertainment Computing-ICEC 2012*, 256–269.

Winter, K. (2012) The fork that talks! New Japanese gadget makes bizarre sounds while you eat. Available at http://www.dailymail.co.uk/femail/article-2254192/The-fork-talks-New-Japanese-gadget-makes-bizarre-sounds-eat.html (accessed February 2014).

World Health Organization (1998) *Obesity: Preventing and Managing the Global Epidemic.* World Health Organization, Geneva.

Zampini, M. and Spence, C. (2005) Modifying the multisensory perception of a carbonated beverage using auditory cues. *Food Quality and Preference*, **16**, 632–641.

11
On the Future of the Perfect Meal

11.1 Introduction

We are finally coming to the end of the perfect meal. The last of the plates have been cleared from the table, and we are hopefully all feeling suitably sated. All that is left to do now is contemplate what the perfect meal might look like in the years to come! We could pick any year – 10 years hence, 25 years or more – but we are going to roll the clocks forward to the year 2084, for no reason other than its resonance with the title of George Orwell's famous novel, and think about what might be on the dining table (assuming, of course, that we are still eating from dining tables). Before we do that however, it's worth looking at some of the predictions that have been made over the last century or so regarding what, and how, we would be eating in the future. As we hope to show in the pages that follow, if you really want to know what we will be eating in 2084 perhaps the best place to start looking is in the fiction of some of the best-selling novelists, rather than what the scientists and chefs have been (mostly erroneously as it turns out) predicting.[1] In fact, it is rather surprising to see how few academic researchers seem to have wanted to speculate on the future of food previously (but see Belasco 2006 for an authoritative review of the limited exceptions to such a claim). Indeed, the response of the scientific community to Roosth's (2013) question of *"What will we eat tomorrow?"* has mostly been a disappointing silence or a rather unimaginative projection

---

[1] This is not because novelists necessarily make good food futurologists, but rather because so many of today's top chefs take their inspiration from the futuristic culinary ideas of the past. That said, some science fiction writers have turned out to have correctly predicted future developments in science more often than one might have expected (in particular, see Hugo Gernsback 1911).

The Perfect Meal: The Multisensory Science of Food and Dining, First Edition.
Charles Spence and Betina Piqueras-Fiszman.
© 2014 John Wiley & Sons, Ltd. Published 2014 by John Wiley & Sons, Ltd.

of today's immediate food concerns onto some seemingly arbitrarily chosen future date.

So, what might the future hold in terms of the constantly evolving experience of dining, at least at the top-end of contemporary modernist dining? Or, better said, what has *not* yet been tried, but which might be by one of those slightly unhinged chefs who has let the fame associated with appearing on our TV screens every week go to his or her head? At one level, it is already difficult to think of other ways in which the chefs could obtain louder "Wows!!" than some of them are already achieving.[2] Challenging though this task is, let's give it a go. The predictions here revolve around not only what the food and drink might consist of, and where (not to mention what) they may be sourced from, but also what the restaurant (or dining room) of the future will actually look like not to mention whether we will literally be seated around a table with our friends or rather merely virtually co-habiting the same dining space. As we saw in the last chapter, a variety of emerging technologies are already promising to transform the dining landscape at the front of house. But just where might we end up if the current trend towards mixing elements of theatre, science, technology, design and modernist cuisine continues to develop as it has over the last few years? Or, put a little differently, what might post-modernist cuisine look like?

As we saw back in Chapter 1, periods of great culinary change always (well, often at least) seem to be accompanied by equally vigorous movements in the opposite direction. Think only of the move towards 'back-to-basics' foods, whether slow food, which developed at around the same time as molecular cuisine really started to take off[3] or, going back a century, the rise of Regionalism (Beaugé 2012) which emerged in roughly the same period that the Italian Futurists were scribbling some of their own wonderfully crazy manifestos concerning cuisine (Apollinaire 1912–1913; Marinetti 1930).[4]

Any reading of historical predictions concerning the future (whether of food or anything else) is always going to provide something of a salutary lesson in terms of promises that went unfulfilled and new directions that were simply not seen or anticipated by those writing in the years gone by. It seems clear that many of the modernist chefs working around the globe today have been seeking their inspiration from the writings of the past, such as evidenced by their desire to bring to life some of the seemingly bizarre ideas committed to print by the novelists of yesteryear (particularly inspirational here have been Roald Dahl with his novel *Charlie and the Chocolate Factory* and Lewis Carroll's *Alice in Wonderland*).

---

[2] But not everyone is equally impressed, for example the famous French chef Alan Ducasse uses the term '*the wow effect*' "*to describe, disparagingly, cooking he thought offered more sizzle than substance*" (Steinberger 2010, p. 176).

[3] The target of the *slow* food movement's ire was, of course, the ubiquitous rise of *fast* food (see Petrini 2003).

[4] As well as the future of pretty much everything else (see Apollonio 1973 for a representative collection).

Whatever the perfect meal happens to looks like 70 years hence (that is, in the year 2084), it may well bear even less of a resemblance to the modernist restaurants of today than did the service *à la Russe* of the meals served by the likes of Carême and Escoffier.[5]

11.2 On the history of predicting the future of food

In the previous chapter, we saw how a variety of digital technologies currently have the potential to transform the culinary landscape over the next few years. But what other changes are we likely to see over the medium to longer term? Well, this is perhaps more the domain of the so-called 'food futurologists' (e.g. http://www.morgainegaye.com/ or http://www.lyndongee.co.uk/food-futurology) than of the chefs, restaurateurs, designers, sensory scientists or even psychologists. That said, looking back over the last century or so one soon starts to see how many of the predictions from the past regarding the future of food have ended up falling a rather long way short of the reality of the situation. Given the very large number of fascinating novels, TV shows and movies that have been set in the future, here we only have space to discuss a few of the most important (and/or entertaining) examples (see Belasco 2006 for an authoritative treatment of the topic). In this chapter, we focus primarily on those predictions that we consider most relevant in terms of food formats and eating behaviours and, more specifically, the future of the perfect meal.

11.2.1 A meal (or even a day's food) in a single dose

Going back to the end of the nineteenth century, one finds the scientists predicting that one day soon humankind would be eating meals that consisted of nothing more than a pill[6] (Berthelot 1894; see also Belasco 2006, p. 27, pp. 115–118). In her novel *The Republic of the Future* Anna Dodd (1887) wrote about a time (her work is set in New York c. 2050) when pneumatic tubes known as 'culinary conduits' would deliver bottles and pellets of food directly into kitchen-less apartments. Of course, as a good feminist, this would have been an ideal (utopian) solution to the kitchen drudgery forced upon the average housewife back then! As Dodd has one of her characters say *"When the last pie was made into the first pellet, woman's true freedom began"*

---

[5] Menu *à la Russe* (dishes served consecutively) as opposed to menu *à la Française* (where many dishes are served simultaneously).

[6] Nowadays, this is referred to as 'neutraceuticals', that is, products including drugs, dietary supplements or specific food ingredients usually taken as part of a diet (Belasco 2006, p. 10).

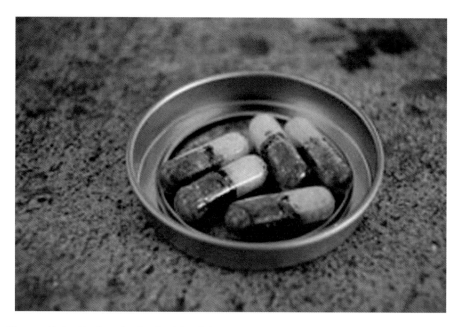

Figure 11.1 'Perfected pizza': this dish consists of dehydrated tomatoes, fresh basil, Parmigiano–Reggiano and sea salt in a vegetable capsule. It was served at The Tactile Dining Car, a participatory dining and performance installation held at the Flashpoint Gallery, Washington DC, 9–24 September 2011. *Source*: Reproduced with permission of banished? productions/Carmen C. Wong. *See colour plate section*

(Dodd 1887, p. 31). Although in Dodd's case this transition was more of a satirical prediction than a serious one, she wasn't alone among writers and artists in thinking that such a form of nutrition might well be commonplace in the opening years of the twenty-first century (see Belasco 2006; see Figure 11.1). Take the following from Arthur Bird's (1899) *Looking Forward: A Dream of the United States of the Americas in 1999* where one finds the following prediction:

> *"In order to save time, people [in 1999] often dined on a pill – a small pellet which contained highly nutritious food. They had little inclination to stretch their legs under a table for an hour at a time while masticating an eight-course dinner. The busy man of 1999 took a soup pill or a concentrated meat-pill for his noon day lunch. He dispatched these while working at his desk."*

Some of you may be old enough to remember the futuristic Jetson family who first appeared on the TV back in the early 1960s. The members of this family would sometimes eat food pills for dinner, helpfully dispensed by Rosie the household robot. Meanwhile, if you had switched channels during the same decade you might have seen the characters from the TV series 'Lost in Space' swallowing the protein pills that were meant to provide all of their daily nutrition needs.

A rather less-appealing future situation, one that we certainly don't want to imagine, is having to eat Soylent Green (a delicious green wafer advertised as containing high-energy plankton, actually secretly made of ... well ... human remains). This dystopian prediction comes from Richard Fleischer's movie of the same name set in the year 2022 (based on Harrison's 1966 novel *Make Room! Make Room!*). Although the idea behind this 1970s movie was partly to raise public awareness about where humankind was likely headed (namely massive food shortages, etc.), it can also be read as an attempt to highlight the possibility that algae might emerge as an increasingly important source of food (see also Vonnegut 1973). It seems rather unlikely that such a dystopian view (that is, people subsisting on processed human remains sold as plankton-based food) will be upon us in 70 years time. However, not too far from this idea of eating *real* plankton, other authors have been predicting that we will need to increase our consumption of (and dependence on) various single-celled proteins (algae, yeasts; Cohn 1956; Kahn 1976; Ford 1978; Belasco 2006).

The 1960s was also the time when Roald Dahl's[7] fictional character Willy Wonka first proudly presented his new chewing gum, the 'Three-Course Dinner', which was meant to sequentially reproduce the flavours of each of the courses in, well, a three-course dinner. Nowadays, one can find all kinds of concoctions that are supposedly going to help you to get your recommended daily dose of nutrients; that said, they certainly cannot substitute for one's daily meal. Regarding the actual development of these ideas, it is undoubtedly true to say that the food technologists can now encapsulate flavours in pill form more easily than ever before (see Madene *et al.* 2006 for a review of flavour encapsulation technologies). If you don't believe us, just take a look at the range of Harry Potter Bertie Bott's Every Flavour Jelly Beans (www.jellybelly.com)! Note however that you would need to eat a veritable handful of them if you wanted to taste all of the flavours that you would normally expect to find during the course of a meal. Of course, the one thing that is most definitely missing here when foodstuffs start to be delivered in pill form are all the oral-somatosensory textural attributes that we normally associate with food (see Spence *et al.* 2013 and Chapter 6). As yet, no one has managed to simulate such qualities in pill form. Ultimately, this may turn out to be all the more important when it is realized that many flavours are hard, if not impossible, to identify when presented in the absence of the appropriate textural cues (e.g. Stuckey 2012).

---

[7] During his childhood, Cadbury's would often send Dahl's school chocolate samples for the kids to taste. This is apparently where the inspiration came for him to include chocolate bars in many of his books (http://en.wikipedia.org/wiki/Roald_Dahl).

> *"'This gum,' Mr. Wonka went on, 'is my latest, my greatest, my most fascinating invention! It's a chewing gum meal! It's ... it's ... That tiny little strip of gum lying there is a whole three-course dinner all by itself!' ... 'This piece of gum I've just made happens to be tomato soup, roast beef, and blueberry pie, but you can have almost anything you want!'"* (Dahl 1964, p. 121)[8]

Perhaps, everything that we learnt about the use of sensory descriptive labelling (see Chapter 3) may turn out to be particularly important in this area. One shouldn't however forget that, when our stomach is empty and we bite into a food (or even just see it), our body starts to release various hormones that signal our brain to begin muscle contractions and the release of acids and other digestive fluids, thus preparing our stomachs for the meal that is about to arrive. If those expectations regarding the imminent arrival of food are not met, one might wonder whether in the long term the build-up of all those digestive juices could have a detrimental effect on our gut lining.

Although Dahl's *Charlie and the Chocolate Factory* wasn't set at any particular point in the future, a number of Wonka's most creative inventions may well make their way into the marketplace in the not-so-distant future. Note here that in the year 2010 a group of scientists from the Institute of Food Research in Norwich had already confidently announced that advances in nanotechnology would soon enable them to capture different flavours within microscopic capsules which could then be released as flavours in a precisely controlled sequence. As Professor Dave Hart, the lead researcher, explained to the press:

> *"Wonka's fantasy concoction has been nothing but a dream for millions of kids across the world. But science and technology is changing the future of food, and these nanoparticles may hold the answer to creating a three course gourmet gum ... Tiny nanostructures within the gum would contain each of the different flavours. These would be broken up and released upon contact with saliva or after a certain amount of chewing – providing a sequential taste explosion as you chew harder."* (see Gray 2010)[9]

More than three years later, the technology is still apparently 'under development'. What is certain though is that, as soon as this eccentric invention is successfully turned into an operational prototype (with a fabulous taste and without turning people into blueberries, as happened to that silly girl Violet Beauregarde in Dahl's classic novel), we may start to see it on the menu

---

[8] It is presumably no coincidence that Heston Blumenthal, who recently had the guests on one of his TV shows licking the (apparently tasty) wallpaper, is a good friend of Liccy Dahl, Roald Dahl's widow (see Whittle 2013). As Mrs Dahl said while dining at The Fat Duck *"How wonderful, if only my husband had been alive to see the real Willy Wonka"* (Whittle 2013).

[9] Note that many chewing gums already contain patented mixtures of sugars, specifically incorporated into the gum in order to dissolve at different rates on the tongue. The idea here (somewhat ironically given what is said above) is that by ensuring a longer-lasting and more continuous sensation of sweetness, the consumer's experience of one and the same minty (or whatever) flavour will actually be prolonged in the mouth (see Davidson *et al.* 1999).

of some of the world's more futuristic restaurants. Already moving us a step in this direction are the 'WikiPearls' developed by Professor David Edwards from Harvard University. This product allows chefs to do something very similar without having to get their fingers dirty. WikiPearls consist of spheres of ice-cream, yoghurt or cheese covered with a thin edible layer that was inspired by the skin of a grape (http://www.wikipearl.com/). This novel and edible packaging technology allows chefs to create concoctions that have a much longer shelf life than the espherified dishes one finds at many modernist restaurants.

One fundamental point to bear in mind here is that all of these ideas about future meals delivered in pill form (or as a stick of chewing gum) would seem to be based on an outdated notion of what constitutes flavour. Up until the 1980s (e.g. Schutz and Wahl 1981) researchers tended to think of flavours as nothing more than merely the combination of taste and smell (see Chapter 6). As we have seen throughout this book, it actually involves so much more. For instance, these examples clearly miss out on the array of different oral-somatosensory textures that we all experience during the course of eating a proper meal. Indeed, isn't the fact that good chocolate melts at body temperature a large part of what so many of us like about it? Even if, in some imagined future, the 'Three-course Dinner' chewing gum were to be developed, and should it happen to contain all of the appropriate time-locked retronasal flavours, our guess is that it would never be more than the shadow of the rewarding fully multisensory food experience that we have known and loved up until today.[10]

"*The snack food of the future could rely more on sensations in the mouth than flavour or texture. Food companies are experimenting with 'sensates' ... to make your mouth tingle, warm, cool, salivate, or tighten ... the next step is to manipulate the sensates to change the length of intensity of the sensation.*" (Fitzsimmons 2003)

11.2.2   On the mechanization of feeding

Over the decades, many writers have worried about a future in which the process of eating has become automated (see Giedion 1948). In 1911 for example, Hugo Gernsback (who has been described as the prophetic father of modern science fiction) talked about a "Scientific Restaurant" in his novel *Ralph 124C 41+* where, in the year 2660, it was anticipated that tubes would pump easily digestible liquid foods (or semi-liquids in the case of meat, etc.) directly into the diner's mouth:[11]

---

[10] Were the famous early psychologist Edward B. Titchener (1909) still alive, he would probably want to point out that the peach-flavoured chewing gum would miss out on all the luscious tactile sensations that are such a distinctive part of eating the ripe fruit. In his words, it would miss out on "*the softness and stringiness of the pulp, the pucker feel of the sour*".

[11] For those who ever had the pleasure, this sounds uncomfortably like being in one of Edmund Rolls' neuroimaging experiments trying to uncover the brain mechanisms underlying flavour perception!

> " *... a flexible tube hung down to which one fastened a silver mouthpiece, that one took out of a disinfecting solution, attached to the board. The bill of fare was engraved in the board and there was a pointer which one moved up and down the various food items and stopped in front of the one selected. The silver mouthpiece was then placed in the mouth and one pressed upon a red button ... If spices, salt or pepper were wanted, there was a button for each one which merely had to be pressed till the food was as palatable as wanted. Another button controlled the temperature of the food ... They did not have to use a knife and fork, as was the custom in former centuries. Eating had become a pleasure.*" (Gernsback 1911, pp. 43–44)

The idea (or better said, fear) here was also captured in Charlie Chaplin's 1936 film 'Modern Times', where at one point an assembly line is seen feeding the protagonist (see also Muecke 2004, pp. 124–126). As we'll discuss in the following sections, despite these new eating formats being seen by many as the future of feeding (who knows, perhaps they were simply over-awed by the new automated processes that were starting to develop around the time), we see future trends in dining going in a number of other directions. However, the notion of having tubes inserted into one's mouth and having food delivered in a liquefied or puréed form seems just as far from being realized as it ever was.[12]

11.2.3   Air 'food'

Anyone who remembers Mel Brook's film 'Spaceballs' will perhaps recollect the supply of canned air ('Perri-air') that President Skroob (who had wasted all of the air) had amassed in his office on Planet Spaceball (http://www.youtube.com/watch?v=SiabeNR_q0U). On this planet, clean air had become such a scarce commodity that companies were canning and then selling it at a hefty premium to the rich. OK, while this example isn't exactly a prediction concerning the future of food, back in 1889 Jules Verne (or his son Michel, it's uncertain quite which one of the two was the author) was suggesting that in the year 2889 scientists would be on the verge of the next major discovery: nutritious air that would enable us all to take our nourishment by doing nothing more than breathing (see Verne 1889). Meanwhile, in the year 2660 in Hugo Gernsback's Scienticafe, the guests would start their meals by inhaling "*several harmless gases for the purpose of giving you an appetite*".

> "*The other starring concept to make its debut this season is something Adrià calls 'air' ...* [the chef continues] *Foams are out – for us ... I have created something five times lighter than the foams. The new texture that I create is air. In the bathroom there is the bath foam. This is the same texture.*" (Lubow 2003)

---

[12] We admit that this reminds us of watching people ingesting some 750 kcal or so while sucking their Starbucks' mega Frappuccino (or similar) through a straw (see also Donawerth 1994).

The truth is that nowadays in some restaurants you can literally inhale a meal. Yes, that's right, simply inhale it! During Alinea's opening year back in 2005, chef Grant Achatz was already serving up the 'virtual Shrimp cocktail', a dish brought to the table with a black plastic atomizer. "Open your mouth and press the pump at least five times," the waiter would instruct the curious diner. When journalist Pete Wells tried this vaporized dish for the first time, he reported:

"Flavours bob into your consciousness – shrimp … horseradish … tomatoes – and crystallize. Shrimp cocktail. But better. The sensation that materializes in the back of your throat is so eerily precise that you wonder if artificial flavorings are involved. In fact, this effect is achieved by honest means. Achatz and company stew the shrimp together with the other ingredients, then run the slurry through a wine press for maximum concentration." (Wells 2005)

Why dine on vapourized food you might ask? Well, for one thing: it presumably delivers rather fewer calories than the real thing.

"He lay back for a little in his bed thinking about the smells of food, of the greasy horror of fried fish and the deeply moving smell that came from it; of the intoxicating breath of bakeries and the dullness of buns … He planned dinners, of enchanting aromatic foods that should be carried under the nose, snuffed and then thrown to the dogs … endless dinners, in which one could alternate flavour with flavour from sunset to dawn without satiety, while one breathed great draughts of the bouquet of old brandy." (Evelyn Waugh 1930)

More recently, American scientist David Edwards and French culinary designer Marc Bretillot invented a gadget that they christened 'Le Whaf', which can transform food into a cloud of flavour (http://labstoreparis.com/#& p=foodlab). How? First the food, say a lemon tart, is boiled down into liquid form. Next, the resulting mixture is strained and transferred to a carafe that turns the liquid to vapour by means of ultrasound. Finally, one simply pours a cloud of lemon tart into a glass and sips it with a specially designed straw (see the company's website http://www.aerodesigns.com/ for a number of other instantiations of this concept). Le Whaf is quite like a nebulizer – one of those devices that are used to deliver drugs direct to your lungs in the form of a vapour or mist – but one that, in this case, has been beautifully designed. Perhaps unsurprisingly, this concept has slowly started to make its way into a few high-end restaurants. This would seem entirely fitting when put together with the foams and liquid spheres that are now increasingly commonplace (e.g. Barba 2013)!

A number of chefs have already had the opportunity to play with Le Whaf at the *Laboratoire* (where it was created) during its launch party in 2012. Massimo Bottura from the Osteria Francescana (in Modena, Italy), Ben Shewry from Attica (in Melbourne, Australia) and chef Homaro Cantu from Moto (in Chicago, USA) all prepared their own inhalable recipes. Massimo prepared a *canard' à l'orange* and Ben mixed four clouds from rice, rice vinegar, soy sauce and ginger to create a surprising gaseous concoction. Meanwhile, Homaro used some of the miracle berries of which he is so fond, together with a handful of other ingredients including a mixture of hazelnut liquor and lemon vodka, in order to deliver what was supposed to be the first breathable chocolate cake. While it's definitely a (calorie-free) dining experience that's well worth a shot (or should that be, 'a spritz'?), we guess that, after sipping a whole 'glass' of cloud (which may consist of nothing more than just 40 micro-litres of the original liquid), one might well end up craving the oral-somatosensory textures associated with the consumption of the real stuff.[13] But, for anyone who is happy with a vaporized dinner, then what better way to finish off your meal than with an inhalable coffee. Once again, the means of delivery now exists (see Anonymous 2013a); you can even take it with a chocolate 'capsule' (as they call it to make it sound more futuristic) after having inhaled the coffee (http://virtualcoffeelab.com/).[14] Taking the topic to the extreme, there's even sprayable caffeine-based energy available now too (http://sprayable.co/) which one applies directly to the skin like a fragrance and which apparently delivers the kick of caffeine that you crave.

Although there are multiple ways in which to provide diners with a 'puff' of flavour, as we are seeing, Marinetti and his fellow Futurists opted to capture scented air in balloons. Guests would simply squeeze the balloon and sniff, in a dish that went by the name of 'Aerofood'. Nowadays, such dishes are a popular feature at the growing number of Futurists-inspired experimental dinners that one sees advertised (see Figure 11.2).

11.2.4   Artificial flavours

The 1970s saw the rise of a host of chemists excitedly predicting the dominance of the artificial over the natural in terms of the future of flavour (Toops 1998; Classen et al. 2005). Some of the more outspoken among them even went so far as to suggest that the children of tomorrow would believe that the artificial fruit flavours then being created for the first time in the labs of the big flavour houses were 'the genuine article' (e.g. Rosenbaum 1979; Shell 1986 for a hint of this optimism). And let's not forget George Orwell's dystopian novel *Nineteen Eighty-Four*, first published in 1949. Orwell envisioned a time when certain

---

[13] According to the Daily Mail, one woman apparently dropped 10 dress sizes simply by baking gooey rich cakes and just enjoying their aroma instead of tucking into them (as she couldn't help herself from doing until she realized that she was a size 22; Kirkova 2013).

[14] Note that while there may be sensory substitution devices for those who have lost their sight or hearing, there is currently nothing that can substitute effectively for a loss of taste, smell or touch, the three key flavour senses.

Figure 11.2 'Aerofood': a mylar balloon holding food scent, one of the dishes served at The Tactile Dining Car, a participatory dining and performance installation held at the Flashpoint Gallery, Washington DC, 9–24 September 2011. *Source*: Reproduced with permission of banished? productions/Carmen C. Wong. *See colour plate section*

products that we take for granted today (especially those that we most crave, such as chocolate and sugar) would have been censored or strictly rationed for consumption by the Outer Party. The speculation was that some people would end up not even remembering their real flavour/taste (and the delight that they had once elicited). While the former prediction might sound a little closer to the truth (who, after all, can always distinguish natural from artificial when it comes to flavour?) it fortunately hasn't quite come to pass. By the year 2084, though, who knows? That we have already moved some way in this direction can perhaps be inferred from Eric Schlosser's claim in his best-selling book *Fast Food Nation* a little over a decade ago that "*about 90% of the money spent by North Americans on food is used to buy processed food*" (2001, p. 121).

> "*In 20 years … I'll bet you that only 5 percent of the people will have tasted fresh strawberry, so whether we like it or not, we people in the flavour industry will really be defining what the next generation thinks is strawberry. And the same goes for a lot of other foods that will soon be out of the average consumer's reach.*"
> (Rosenbaum 1979, p. 92)

"I think it's the best blueberry flavor that's ever been made. And there's not a scrap of blueberry in it." (Shell 1986)

And why not stretch the concept of synthetic flavours to synthetic foods too? Frederick Edward Smith, the Earl of Birkenhead, is another of the line of authors who ventured to predict the food trends that would be popular a century hence. In his book *The World in 2030* he suggested:

> *"It will no longer be necessary to go to the extravagant length of rearing a bullock in order to eat its steak. From one 'parent' steak of choice tenderness, it will be possible to grow as large and juicy a steak as can be desired. So long as the 'parent' is supplied with the correct chemical nourishment, it will continue to grow indefinitely, and perhaps, internally."* (Smith 1930, pp. 18–20)

A related example here comes from Gernsback's *Ralph 124C 41+* (Gernsback 1911, pp. 70–71) where the novel's protagonists go to New York city in order to see where all the synthesized 'milk' comes from (milk, it turns out, that is made directly from grass, bypassing the pesky mammal in the middle).

> *"Only the microwave seemed remotely revolutionary in the 1980s, and even that excitement settled down as it became clear that while this handy appliance could heat frozen foods, it was inferior to the conventional toaster oven when it came to baking, broiling, and browning."* (Belasco 2006, p. 237)[15]

One might have realized already that most of the above-mentioned concepts, whether predictions concerning the future of food or else tales about those living on another planet or in an imaginary world, appear to be related to rapid consumption (that is, eating without the need for a plate on a table and, more appealingly, without the need for lengthy food preparation either). Most of the future forecasts here seem to revolve around the idea of just having a simple quick 'dose' of food (or else being force-fed by machine). It might undoubtedly seem more convenient in terms of the amount of time that would be saved but, while pre-prepared meals (requiring just a few of minutes in the microwave) have been a common feature on the supermarket shelves for decades now, we don't foresee this trend evolving into pills or other similar formats (and even less so in a restaurant context unless, of course, we are looking at something being served in an experimental restaurant 70 years hence; Haden 2005).

---

[15] In fact, quite what an advance in food that the last two decades has seen is made very clear if one looks at the sentiment being expressed around 1990 as far as the future of food was concerned (see Belasco 2006). Take, for example, Edward Dolnick's (1990) conclusion: *"What kind of a puny future is this? In a culture almost unable to resist the magical words 'new and improved', the foods we eat seem hardly to be changing at all"* or the observation of Rogers (1989) in an article regarding futurology, where the greatest enthusiasm in terms of the "marvel of the future" for food revolved around nothing more exciting that a genetically engineered tomato that might actually taste like something!

In addition, we shouldn't forget the pleasures of preparing a meal with others and enjoying that social time (Petrini 2007). Indeed, many writers have subscribed to a number of the ideas that underpin the 'slow food' movement. For example, in his latest book '*Cooked*', Michael Pollan (2013) describes how he bonded with his teenage children by reconnecting with the slow art of baking bread. Without having read the book, your author's brother has done the same with his own daughters. Certainly, many food companies still seem to want to approach their marketing from the 'home-made with care and passion, and natural ingredients' angle (what one might call the 'Grandma's style' approach; think only of Aunt Jemima products or Bonne Maman jams). Even if some of us living in the future would prefer to continue further reducing the amount of time we spend in matters related to food preparation and eating, we believe that most people will still prefer a 'Mr Jones' turkey and first-growth cranberry sandwich' than to take care of all their nutritional requirements by swallowing a pill dispensed from the vending machine.

11.3   From the past to the future of food

Having seen how at least a few of the predictions that have been made by novelists over the last century or so have come to pass, let's now take a look at some of the dramatic changes in the culinary landscape that have been brought about by the qualitatively different ways in which food is prepared nowadays. We then move on to making some predictions of our own concerning what will (or better said, might) and what probably won't be in the perfect meal *c*. 2084 (with apologies to George Orwell).

11.3.1   *Sous vide* as the twenty-first century microwave

Throughout the history of the development of modern cooking techniques (or appliances), many gadgets have been launched with the promise that they would revolutionize the way in which people prepare and cook food. At the present time, much of the hype revolves around the revolutionary potential of *sous vide* cooking. This French gastronomy term refers to the style of cooking in a vacuum, placing sealed meat or vegetables into a water bath at an exact (normally relatively low) temperature for much longer than one would normally cook food (up to 72 hours in some cases; Rayner 2004). The result of 'sous-viding' your meat is supposed to be a hassle-free (and moron-proof) perfect medium-rare steak. So just forget about the 10-min meal! This approach to cooking first made its appearance in Switzerland back in the 1960s as a way of preserving and sterilizing the food served in hospitals. The technique was further refined in 1967 by chef Pralus in his Roanne restaurant in France. There can be no doubt but that it sounds professional and, what is more, the very

Frenchness of the name likely gives the technique a touch of sophistication that the equivalent term in any other language simply couldn't hope to achieve (see Chapter 3). We believe that it may well come to be among the most common of cooking methods in the year 2084 (who knows, *sous vide* may become the new microwave in the kitchen). If the predictions are right, the perfect meal in the year 2084 may well be prepared, at least in part, using this technique. Although the idea behind *sous vide* has been around for almost half a century now (i.e. roughly as long as the microwave; Anonymous 1971, 1975), the former approach is nowadays being championed (often as part of a multi-stage cooking process) by many of the world's top chefs, including Thomas Keller and Heston Blumenthal, in a way that the microwave never was.[16]

Things are slowly starting to change, however. Significant here is likely to be the arrival of the SousVide Supreme (https://www.sousvidesupreme.com/). This machine started to appear on the shelves around 2009. For around $400 (plus another $100 or so for the vacuum packing device and a handy supply of bags), this machine allows the adventurous home chef to prepare perfectly cooked, highly controlled and flavour-preserving meals (at least, that's how it is being sold). Anyone who doesn't want to shell out for such a device just yet can perhaps get their feet (if hopefully nothing else) wet with the technique by trying it out in their dishwasher first (see Casali 2011; Rivalta 2012; Graf 2013). The highly controlled temperature offered by *sous vide* cooking is a feature that contrasts with the microwave (and, for that matter, with conventional) ovens, where the lack of precision can make it hard to cook meats, fish or egg-based dishes consistently, especially since they will quickly toughen if overcooked (McGee 2008).[17] Imagine the results should the enthusiastic home-cook have microwaved a 5 lb Christmas turkey for 35 minutes following the instructions given back in the 1970s!

> "*Dr. This might advise you to wrap your turkey in a large plastic bag and then roast it in the dishwasher. But for those having the in-laws around for lunch, cooking the Christmas turkey on the Electrolux's economy cycle might be seen more as a sign of too much sherry than of a dazzling affinity with the principles of molecular gastronomy.*" (see Bartley 2009, p. 12)

---

[16] This reminds one of your authors of the good old boil-in-the-bag Findus 'Cod in butter sauce' that he would eat after school as a kid. Presumably, this is not the high-end association that many of this technique's proponents would like us to have.

[17] This kind of cooking obviously appeals to the international chef who has aspirations of becoming a global brand, for it offers the promise of precision cooking in all of your restaurants even if you happen to be off filming your next TV series! There is also a link here to Grant Achatz's Chicago restaurant, Alinea. The name of the restaurant literally means 'off the line'. The restaurant's symbol, known as the pilcrow, indicates the beginning of a new train of thought or, literally, a new paragraph. However, there's a double meaning here: on the one hand, Alinea represents a new way of thinking about food but, being a restaurant, everything still has to come 'off the line'. (See http://gizmodo.com/5344393/thought-for-food-alineas-reinvention-of-cooking-and-eating.)

11.3.2   3D printed food: an astronomical idea

These days one hears a lot about 3D printers being used to create customized and precisely engineered foods. The technology itself has been around for a few of years now (see Sandhana 2010; Anonymous 2012a). At Cornell University in Ithaca, NY for example, Jeffrey Lipton developed a 3D food printer capable of laying down liquid (or *puréed*) versions of foods, dot by dot and layer by layer, to build up edible dishes (http://creativemachines.cornell.edu/node/194). *"So far we have printed everything from chocolate, cheese and hummus to scallops, turkey and celery"* Lipton said (Purvis 2012). Meanwhile, Amit Zoran and Marcelo Coelho, MIT design engineers, announced a rather similar concept going by the name of the Digital Fabricator (see Zoran and Coelho 2011). All well and good, but if the price of your 3D printed meal is anything like the cost of printing a 'mini-me' – that's a miniature replica of oneself (currently *c.* £200) – then these devices will need to become much cheaper before it becomes practical to utilize such technology in the setting of a commercially viable restaurant or home kitchen. (More likely, the big food companies may end up utilizing such technologies first.) We can certainly foresee the use of 3D printing to print foods that would perfectly match the shape of a diner's tongue. This could potentially offer a novel means of delivering a more uniform flavour hit. Just imagine, say, a chocolate melting in your mouth with no sharp edges. All you experience would be the pure flavour hit as it melted uniformly all over the surface of your tongue (see Spence 2013)!

As we have already seen several times throughout this book, fine dining chefs (especially those of the modernist or molecular persuasion) are continually developing innovative new techniques and seeking the latest in technology that will allow them to push the boundaries of their culinary art (not to mention develop their unique selling point; see Rayner 2004). Solid freeform fabrication (SFF) promises to be one of the most important of these enabling technologies. Wielding such a device, the chefs of tomorrow will be able to create unimaginably complex geometries that no mould could ever get right. The technology even opens up the possibility of writing hidden messages inside a sponge cake should one so desire. In the meantime, we have already seen the innovative chef Homaro Cantu printing his menus and sushi on edible paper with food-based inks using a standard inkjet printer since 2005 (see Chapter 2). Through his company Cantu Designs, this commercially minded chef already has a number of patents pending covering diverse fields that include edible surfaces.

He hopes that his projects could help to change the way in which people are fed in times of crisis, not to mention helping to distribute medicines and even sustain astronauts embarking on long space missions. In fact, Cantu has worked together with NASA in order to develop a food 'replicator' (remember Star Trek?), a converted ink-jet printer that is already producing tasty prints.[18] A sample space menu showing what the astronauts would typically have been eating during the Apollo missions to the moon (taken from Bourland and Vogt 2010, p. 13) is shown below.[19]

DAY 1

Meal A

Peaches (R), Bacon Squares (IMB), Cinnamon Bread Toast Cubes (DB), Breakfast Drink (R)

Meal B

Corn Chowder (R), Chicken Sandwiches (DB), Coconut Cubes (DB), Sugar Cookie Cubes (DB), Cocoa (R)

Meal C

Beef and Gravy (R), Brownies (IMB), Chocolate Pudding (R), Pineapple-Grapefruit Drink (R)

Abbreviation Key

R= Rehydratable

DB= Dry bite

IMB= Intermediate Moisture Bite

Source: Bourland, C. T. & Vogt, G. L. (2010) The astronaut's cookbook: Tales, recipes, and more. Springer, New York. Reproduced with permission of Springer Science+Business Media

Having already started to talk about space food, it is relevant here to note that in many ways humanity's hopes and fears about the future of food have been projected onto our beliefs about the future development of space food (cf. Horwitz 2004). That is, many of us seem to naturally associate the concept of future food with space exploration and astronauts (in fact, there are already some space-themed restaurants out there; Marriott 1999). In their intriguing book *The Astronaut's Cookbook*, Bourland and Vogt (2010) include a chapter dedicated to this most esoteric of topics. They describe NASA planners contemplating the practicalities associated with a two-year round trip manned mission to Mars. As they make all too clear, the acceptability of the food that is served (or, better said, consumed) in space becomes more not less important as the length of the space mission increases: "*After all, when you are 250 million miles from Earth, you can't just send out for a pizza. If you don't have*

---

[18] That said, we are still a very long way from realizing Cantu's crazy suggestion that we would be able to feed the world's starving by simply dropping 3D food printers over Africa.

[19] In her book *The Sex Life of Food* the wonderfully named Bunny Crumpacker describes how the strawberry cubes were so unpopular with the astronauts on the early Apollo missions that they were eventually dipped in Lucite and sold off as souvenirs to space tourists in Houston (Crumpacker 2006, pp. 114–115). As Space.com put it: "*Along with the hazards of space travel, early astronauts proved their bravery again during meal times*" (Belasco 2006).

a pizza maker on board, it is best not to think about pizzas" (Bourland and Vogt 2010, p. 181). The very idea of an astronaut eating pizza in space would presumably have sounded like nothing more than a pipe dream even just a few years ago. Given what is shown on the menu card above, a synthetically flavoured pizza cube might have been the closest that an astronaut would have come to this delicacy.[20] That said, recent developments have apparently turned the idea of space pizza into a reality (well, hypothetically at least).

In May 2013, NASA officials confirmed that they had awarded a small $125,000 contract to the Systems and Materials Research Consultancy (SMRC) company in order to study how to make nutritious space food using a 3D printer while on long space missions (Jayakumar 2013; Klotz 2013). As of today, the concept doesn't actually involve printing a full pizza with pepperoni slices and cheese, as perhaps you might have been imagining, but instead the technique would be used to allow the operator to print out layers of carbohydrates, proteins, and macro- and micro-nutrients which could then all be fed into the machine (initially in powder form). Reassuringly, or perhaps not, these powders have a shelf-life of more than 15 years.

11.3.3 On the future of cultured meat

One of the other hot topics these days in terms of advancing food technology relates to cultured meat. In fact it turns out that Frederick Smith's prediction (Smith 1930), made 80 years ago, about growing meat for us to eat is surprisingly close to the current situation. Concern about the consequences of humanity continuing to eat large quantities of meat has been around for a long time (Lappé 1971; Belasco 2006).

Unlike the 3D pizza and pills described above, cultured meat looks like today's food, has the texture of meat and even tastes (or 'flavs', as some of our colleagues would have us say) of meat. Just as it should, since it really is meat; it is just that it has never been anywhere near a living animal. Instead, artificial, *in vitro* or cultured meat is grown from stem cells in fetal calf serum in giant vats (see Vogelzang 2010). Ok, so maybe it doesn't sound all that appealing, but scientists argue that the search for meat substitutes is becoming increasingly critical because Western eating habits are now spreading to China and other emerging economies. In addition, the beef industry has a carbon footprint of 10.7–22.6 kg of carbon dioxide (equivalent to almost ten litres of petrol) per kilo of beef produced while, for instance, the carbon footprint when producing chicken is less than 5 kg per kilo.[21]

Much of the research into artificial meat products is taking place in Europe, with scientists in Holland and Britain already developing edible muscle tissue grown from stem cells in their laboratories. It would seem that 20,000

---

[20] Given the decline of the American space program in recent years, one wonders here what the Chinese and Russian cosmonauts crave. Pizza? Surely not.

[21] For those who are interested, cows release the greatest amount of methane during the calf-feeding stage (Osborne 2013).

strips of that muscle have been brought together in a hamburger currently retailing at \$325,000 (see Butler 2013). Before the end of 2013, if Professor Post (one of the leaders in this field of research at Maastricht University) gets his way, who knows whether we'll be seeing Heston Blumenthal flame-grilling the world's first 'test-tube burgers' and serving them to celebrity guests while the nation watches the live video feed sitting entranced in their living rooms. After conducting some informal tastings, Post reported that the tissue tastes 'reasonably good' (Butler 2013). Obviously, he will need to seek out the services of an energetic marketing team if this is really to be the new meat product that will fly off the supermarket shelves. Post launched the lab-grown burger in August 2013 during the world's first public tasting event of this new product in London. Chef Richard McGeown cooked one such burger in front of an invited audience and gave it to two food experts. They agreed that it tasted pretty "close to meat" (The Associated Press 2013). While chefs will have to make their own minds up about whether to put their patties in the pan or in the SousVide, we guess that it will take longer for diners to embrace this new meat. Come 2084, genetically modified (GM) or non-GM may be the least of a diner's worries!

11.3.4   Note-by-note cuisine

Now that we've donned our imaginary lab coats, let's move on to discuss the possibility of reconstructing our food and drink from scratch. Imagine a bottle of 1976 Chateau Margaux from Bordeaux, say. There are very few bottles of this legendary vintage left. As one can well imagine, they are nowadays extremely expensive (and when tasted, the wine has in most cases already started its graceful decline). One obvious thought here is to wonder: wouldn't it be great if we could just create (i.e. synthesize) some new wine artificially with exactly the same flavour profile as one of the classic vintages simply by throwing together the key flavour compounds found in the real wine (when it was showing at its best, that is)? As incredible as this may seem, the idea of synthesizing foods has already been passing through some people's minds for nearly a century (see Williams 1907; Slosson 1919; Belasco 2006). Nowadays, thanks to the work of Hervé This, the idea is once again very much back on the menu. Thankfully, however, not in the way that Belgian scientist Jean Effront suggested back in 1913 when he said "*It would be a hundred times better if foods were without odor or savor. For then we should eat exactly what we needed and would feel a good deal better*" (cited in Belasco 2006, p. 181). It is more likely that everyone would simply become depressed, as happens when people actually lose their sense of smell.

Hervé This, one of the fathers of 'molecular gastronomy', is the maverick scientist who wants to lead us all much further in this direction, not just preparing meals inspired by the latest chemical and physical insights but rather "*to make food directly from the basic constituent chemicals themselves*" (Ashley 2013). What is more, the latest research by scientists at Oxford University estimated that cultured meat uses far less energy than most other forms (Connor 2013). This futuristic culinary trend – creating and cooking from scratch – is what the Frenchman would have us call 'note-by-note cuisine' (This 2012a–c, 2013). The fundamental question here is why we have to limit ourselves to the combinations of odours, tastes, colours and mouthfeels that nature has given us, instead of creating our own from scratch? "*Using chemical compounds opens up billions of new possibilities,*" he suggests. "*It's like a painter using a palette of primary colors or a musician composing note by note*" (Ashley 2013). In fact, the recipes sound more like notes taken in a chemistry lecture than instructions from a cookbook. Take, for example, the following instructions for creating a wineless wine sauce:

> "*Melt 100 g of glucose and 20 g of tartaric acid (one of the main acids in wine) in 200 ml of water. Add 2 g of polyphenol, which has been extracted from grape juice using reverse osmosis filtering. Boil and add sodium chloride (table salt) and piperine – the pungent agent of black pepper. Bind the sauce with amylose, a polysaccharide, one of the two components of starch. Remove the preparation from off the heat and stir in 50 g of triacylglycerol (a triglyceride or fat and oil component). Serve as a sauce.*" (This 2012b)

Even if note-by-note cuisine is technically feasible, however, the results are still mixed (at least if your author's experience of the results at one London event attended by Hervé This was anything to go by). Perhaps crucially, these synthetic dishes will surely lack one key element – authenticity – in the minds of the diners or consumers (who, This admits, are still rather conservative in this regard). The danger is that one just ends up with something like Aldous Huxley's (1932) 'Chemical wine', which makes an appearance in his dystopian novel *Brave New World*. Remember also that we already encounter many such synthetically flavoured products in the supermarket, products that actually only have a very small percentage (if any) of the real ingredient that are supposed to be delivering the primary flavours that are promised (think of those prawn-cocktail-flavoured crisps that contain absolutely no prawns). In this case, consumers really do not seem to care too much since the crisps taste exactly as we expect a synthetic prawn cocktail to taste (i.e. nothing like the real thing, but still identifiably like the synthesized version of that distinctive flavour that so many of us have come to know and love).

In the case of the note-by-note dishes being championed by Hervé This however, the stress is very much on the 'chemical manipulation', making the diner who is sitting in front of the dish more aware of the absence of any of those traditional ingredients (what some would call 'the real stuff'). Will we care about this in the future? Certainly, when it comes to the wine expert's willingness to pay astronomical sums for a classic vintage, this would seem to have relatively little to do with the sensory properties of the wine itself (as long as it is still drinkable) and rather more to do with the connection to the past that drinking the wine seems to promise us (Wallace 2009; Spence 2010). Obviously, simply synthesizing the wine loses something crucial about the link to the past and the way and place in which the wine was made (cf. Goode 2005). Hence, even though such an artificial wine might come in at a fraction of the cost of the real thing (this isn't always the case, however; see Wallace 2009), we wonder just how many aficionados such a product will really attract.

Wrapping up these two last sections, what are the implications for the perfect meal? Might a majority of the flavours, textures and possibly even foodstuffs that we eat end up being synthetic by the year 2084? Certainly the rapid development of technology means we'll have better control over the delivery of flavours than ever before, but the craving for authenticity and naturalness (whatever natural means; see Chapter 3) remains strong (and will, we believe, still be with us in 2084). So while the synthetic meal may be a very real possibility 70 years from now, we think that people will crave authenticity. While synthetic foods will undoubtedly be increasingly available, and likely very tasty too, the perfect meal will still be made up of a majority of natural food ingredients (just perhaps not the same natural ingredients that we are familiar with today).

11.3.5   Eating insects for pleasure: bug burger with insect paste, anyone?

Much more interesting, at least to our way of thinking, is the growing fascination with insects as a rich source of protein. Grasshoppers, spiders, wasps, worms, ants and beetles may not be on many Western menus at the present time, but it turns out that there are at least 1400 species of insect eaten across parts of Africa, Latin America and Asia.[22] In fact, there are thought to be around 1900 potentially edible varieties worldwide (Ceurstemont 2013). Fried witchetty grubs, for example, are a favoured delicacy among the Aboriginals of Australia. "*With grasshoppers you can roast them or fry them, or you can go wild with spice blends*" (Ben Reade, Head of Culinary R&D at Nordic food Lab, quoted in Eriksen 2013, p. 32). They apparently have the taste, if not the texture, of almonds. With rising food prices and worldwide land shortages, it is surely just a matter of time before insect farms start to appear in many other parts of the world. Ceurstemont (2013, p. 37) however notes that "*Discerning palates will tell you that farmed insects simply do not taste as good as those grown in the wild.*" As the caterpillar living in Roald Dahl's giant peach told us:

_____

[22] The first author's wife, a Colombian lady, is particularly partial to the crispy flying ants from the Santander region of the country, colloquially known as 'big-assed ants' that are in season for just a few months each year. (Her husband remains to be convinced of the culinary benefits of this particular proteinacious, not to mention currently sustainable, entomophagic treat.)

"I've eaten fresh mudburgers by the greatest cooks there are,
And scrambled dregs and stinkbugs' eggs and hornets stewed in tar,
And pails of snails and lizards' tails,
And beetles by the jar.
(A beetle is improved by just a splash of vinegar.)" (Dahl 1961, p. 18)
Source: Dahl, R. (1961) James and the Giant Peach, with permission of Penguin Books.

Of course, while the subject of eating bugs might be particularly topical at the present time, what with all the press coverage of the Nordic Food Lab's mission to get us all subsisting on a diet that is a lot richer in bugs before too long, it is worth remembering that some people in the West have actually been promoting the consumption of insects for well over a century now (see Holt 1885; Bodenheimer 1951 for early examples). As Holt put it: *"I can never understand the intense disgust with which the appearance at the dinner-table of a well-boiled caterpillar, served with cabbage, is always greeted"* (cited in MacClancey 1992, p. 39).

Could the insect-based perfect meal of the future look a little like the early insect-based menu from the Victorian naturalist Vincent Holt (1885; cited in MacClancy 1992, p. 39)?

Menu
Snail Soup
Fried Soles with Woodlouse Sauce
Curried Cockchafers
Fricassee of Chicken with Chrysalids
Boiled Neck of Mutton with Wireworm Sauce
Ducklings with Green Peas
Cauliflowers with Caterpillars
Moths on Toast

An UN report (http://www.fao.org/docrep/018/i3253e/i3253e.pdf) published in 2013 emphasized the case for diets that are much richer in insects given the widely anticipated future shortages of food. The UN's Food and Agriculture Organization (FAO) noted that more than 2 billion people already eat insects as part of their diet. Indeed, insects are highly nutritious and packed with protein, vitamins, fibre and minerals, while being low in fat. According to the report (p. 2), crickets need 12 times less feed than cattle in order to produce the same amount of protein, making them a far more sustainable food source. So far, it would seem like a good idea. What is more, insect farms don't need much space and are environmentally friendlier than, say, cattle. The biggest obstacle that a growing insect-based diet faces is overcoming the aversion, if not revulsion, that many people currently feel towards the idea of eating bugs.[23] Indeed, there can't be many other foodstuffs that we make so many jokes about. Hasn't everyone heard the one about a diner calling the waiter over to say that there's a fly in his/her soup? "Waiter! Waiter! There's a dead fly in my soup." "Yes, I know! It never learned to swim." With this potential new eating trend slowly emerging, we might start to hear this conversation instead (not as a joke): "Waiter, there are three flies in my soup!"; "Then you're lucky. According to the recipe, there should only be one." Joking apart, the truth is that the EU recently offered its member states US$ 3 million to promote the use of insects in cooking (Vidal 2013) and researchers have been asked to see whether they can find ways to extract the protein that many bugs contain instead of forcing people to eat the whole creature.

The culinary team at the Nordic Food Labs is already thinking hard about how to achieve this, and you certainly don't need to pay £10 an ant as did the guests when the Noma chefs 'popped up' at Claridge's hotel in London while the UK was hosting the 2012 Summer Olympics. The two-night culinary

Figure 11.3 Chimp stick dish. A carved down liquorice stick coated in juniper-wood-infused honey, with two species of ant (Lasius fuliginosus and Formica rufa) stuck to it with shiso, buckwheat, flax and freeze-dried raspberries. *Source:* Photograph by Chris Tonnesen. Reproduced with permission of Nordic Food Lab. *See colour plate section*

---

[23] Although nobody seems to care too much about the fact that many of their favourite naturally red-coloured foods have the hue they do thanks to the carminic acid that some scaly insects produce, for example red Smarties used to be coloured with cochineal before the manufacturers switched to red cabbage instead. The name itself, cochineal, indicates the insect from which this red pigment is made (see Wilson 2009). Furthermore, few of us have any qualms about eating honey either which, when you think about it, happen to be another insect product (see MacClancy 1992, p. 39).

event 'Pestival' was also organized by Noma and the Wellcome Collection in London in 2013. At the latter event, the audience/guests were served delights such as a French-style mousseline that contained butter-roasted crickets and pureed wax moth larvae, and a tangy ant cocktail (Ceurstemont 2013; see the full menu below). This approach to cuisine could certainly give a whole new meaning to Fergus Henderson's notion of nose-to-tail eating (see also Animal Vegetable Mineral; www.avmcuriosities.com).

Pestival Menu (reproduced with permission of Ben Reade, http://nordicfoodlab .org/blog/2013/5/pestival)

Anty-Gin and Tonic
Bespoke gin of wood ants (*Formica rufa*) and botanicals with carmine and tonic *with The Cambridge Distillery*

Chimp Stick (see Figure 11.3)
Liquorice root with seeds, fruits, herbs, and ants (*Formica rufa* and *Lasius fuliginosus*)

Moth Mousse
Wax moth larvae mousseline (*Galleria mellonella*) and morels

Cricket Broth
House cricket broth (*Acheta domestica*) with grasshopper garum (*Locusta migratoria*)

Wormhole
Oatmealworm stout (*Tenebrio molitor*) *with Siren Craft Brewery*

Roasted Locust
Butter-roasted desert locust (*Schistocerca gregaria*) with wild garlic and ant emulsion (*Formica rufa*)

Bee Brood
Honeybee drone comb (*Apis mellifera*)

The Whole Hive
Beeswax ice cream, honey kombucha sauce, 'bee bread', propolis tincture, and crisp honey

Mead
Lindisfarne Mead from St Aidan's Winery

Giving bugs a good name is an aim that is being pursued by the culinary and intellectual might of the adventurous modernist chefs (and who better than the star chefs and research scientists from the Noma kitchen to spearhead this; see also Garric 2013). Who knows how rapidly attitudes may be changed in the decades to come. Let's hope they change fast, given the predicted food shortages that so many of us will soon be facing. So what is it going to be, cultured meat or bug burger? At present, the choice is still ours to make. Everyone seems agreed that before too long something has clearly got to give. What we saw in Chapter 3 about encouraging the populace to switch to 'variety meats' back in the 1940s may, before too long, be with us again; it's just that the source of protein may be rather different from anything that most of us are used to.

11.3.6   The new algal cuisine

Algae lie at the very bottom of the food chain and are eaten by everything from the tiniest shrimp to the largest of great blue whales. Those living in countries such as Japan and China already eat lots of algae, primarily in the form of seaweed. Importantly, algae provide many of the vitamins that we need including vitamins A, B1, B2, B6, C and niacin. They are also rich in iodine, potassium, iron, magnesium and calcium. In the food industry, common food additives such as agar-agar and carrageena, both polysaccharide compounds, are extracted, it just so happens, from seaweed. In North America, and in some parts of Europe, there is a growing interest in using raw and processed seaweed as a foodstuff and/or as a food supplement (Arthur 1997). Indeed, a growing number of companies are now producing and marketing various forms of dried and salted seaweed products for human consumption (Mouritsen 2012; see also This 2012a). Many western countries are starting to consider the possibility of switching to commercial algae farms. Algae has the advantage that it can grow very rapidly in the sea and in conditions where other food crops just wouldn't survive (see Collis 2013). *"They range from giant seaweeds and kelps to microscopic slimes, they are capable of fixing CO_2 in the atmosphere and providing fats, oils and sugars"* (Vidal 2013).

   Algae are already widely used in the field of molecular gastronomy. After all, who hasn't heard of the famous spherification technique, popularized by Ferran Adrià? Sodium alginate, which is the gelling agent that is dissolved in a flavourful liquid and jellifies when the ensuing mixture is submerged in a calcium solution, is extracted from algae. As Mouritsen (2012) puts it: *"To explore the full gastronomical potential of this resource, there is a need for fundamental research into the gastrophysics of seaweeds"*. So, if algae or their extracts are increasingly cementing their prominent position as essential ingredients in modernist cuisine, we may all be eating more of this than any of us realized anyway. Compared to the other foods mentioned above, it is certainly easy to imagine that we may all soon end up eating far more algal food extracts. They could, for example, be used to give foods wholly new structures/textures, etc. and will not necessarily be presented in the recognizable form of, well, the algae that we all think of floating in the sea.

Before we get too excited about algae however, it is important to remember that we have been there before. In his fascinating book *Meals to Come*, Warren Belasco (2006, pp. 201–213) does a great job of highlighting the phenomenal interest that grew up around chlorella algae in the 1950s. However, all the promise of the early research on the commercial farming of this predicted future food, and its successor spirulina, failed to materialize. It turned out that estimates of the ease and cost of production were wildly inaccurate. There is also the small matter of how to disguise the nasty taste/flavour, not to mention the vivid green colour, of what some have been tempted to label 'essence of plankton' (Wilson, cited in Belasco 2006, p. 130)!

Putting all these futuristic topics together, in the year 2084 we will perhaps be sitting in a restaurant polishing off a plate of tasty witchetty grubs *sous vide*, or a lab-grown steak with a handful of 3D printed veggies or a side of free-range organic algae. Hard to imagine? Well, at this pace, who among us can really profess to know what the future holds?

11.4 Anyone for a spot of neo-Futurist cuisine?

Throughout this book we have seen numerous examples of how many of the culinary ideas that were first suggested by the Italian Futurists have been redis-covered/reinvented by modernist chefs. Everything from playing soundtracks in the background to enhance the dining experience (whether croaking frogs or the sound of the sea; Spence *et al.* 2011)[24] to the use of perfume squirted over the meal, first tried by the Italian Futurists (Berghaus 2001; Weiss 2002) before being recreated in dishes such as the lavender-scented oyster served at The Fat Duck or the chilled citrus soup served at Cantu's Moto in Chicago. For the latter, the waiter finishes the dish off at the table by spraying a little togarashi mist over the bowl. "*This is my favorite part of the meal*" says Cantu; "*I get to pepper-spray our guests*" (Vettel 2013).

You may be asking yourself: what exactly should one expect when next you are invited to a Futurist dinner party? The following extract from *The Futurist Cookbook* (Marinetti 1930) should give you some idea (see also Hoyle 2008; Anonymous 2011):

> "*Aerofood: A signature Futurist dish, with a strong tactile element. Pieces of olive, fennel, and kumquat are eaten with the right hand while the left hand caresses various swatches of sandpaper, velvet, and silk. At the same time, the diner is blasted with a giant fan (preferably an airplane propeller) and nimble waiters spray him with the scent of carnation, all to the strains of a Wagner opera.*" (Marinetti 1930)

---

[24] Marinetti and his colleagues worked on a number of dishes that were specially designed to be eaten while listening to particular sounds and, on occasion, while smelling certain artificial aromas. Take, for example, the Total Rice dish which consisted of rice and beans, garnished with frogs' legs and salami. This dish was to be "*served to the sound of croaking frogs*" (Berghaus 2001, p. 10).

While isolated elements of Futurist-inspired cuisine have been presented at various top restaurants (sometimes with the source acknowledged, other times not; see Poole 2012), we would argue that one can now see a more wholesale recognition of the Futurists and their crazy, brilliant ideas.[25] We hazard a guess here that high-end diners are going to see a lot more of the neo-Futurist dinner party (or at least neo-Futurist-inspired dishes) in the years to come.[26] It's time to bring out the kitchen foil (as Visionaire magazine did for the launch party of their 63rd issue: http://blog.bureaubetak.com/page/4) and prepare to be amazed.[27] In fact, it may turn out that the Futurists will have the last laugh here. Once written off by the Italian press as nothing more than *"a fart from the kitchen"* (Berghaus 2001, p. 15), there certainly can't be many other schools/movements that can have boasted so many predictions around the future of food that are now, finally, being brought to life by the world's greatest chefs.

11.4.1   Food theatre: food as entertainment

For those who have been lucky enough to enjoy the mind-blowing experience of one of the Punch Drunk theatre company's events (http://punchdrunk.com/), just imagine what the Punch Drunk of high-end cuisine would be like. Our prediction is that we are going to see a continued blurring of boundaries between theatrical and culinary experiences. The two will become ever more tightly intertwined, with a hint of magic possibly thrown into the mix as well (Goldberg 1979; Rayner 2006; Achatz 2009; Wang 2013). This is in some way inextricably linked to earlier ideas around the experience economy (Pine and Gilmore 1998, 1999), whether the Theatre of the Senses, a dining-in-the-dark experience (Chapter 8; see also www.ediblecinema.co.uk), an opera dinner such as that organized by the founders of El Celler de Can Roca (http://www.elsomni.cat/en/el---somni /dinner/) or something like The Banquet for Ultra Bankruptcy, developed for Art Laboratory Berlin in 2013. During one of the five performances (each for six guests) of the six-course meal served at the latter event, selected foods were combined with images, sounds and scents. Each course had been

---

[25] One shouldn't forget that the movement itself had some rather unsavoury links to Fascism. What is more, a number of the early players in the Futurist movement were male chauvinists (see Halligan 1990). That said, the essence of their ideas and their creativity, if not their politics or philosophy, undoubtedly still have relevance and provide inspiration today (and will likely do so for some decades to come).

[26] But note that Heston Blumenthal's attempt to apply the same technique (encouraging customers to spray 'the smell of the chippy' – actually the atomized juice of pickled onions – over the fish and chips) at Britain's Little Chef Motorway restaurant chain couldn't prevent the company's steady decline into administration (see Anonymous 2013b). That this particular intervention should have failed comes as little surprise to your first author, who spent his formative years working at Sunnybank Fisheries in Horsforth, Leeds. He remembers all too well the near impossible task of trying to get the smell of the chippie off his clothes when he finished his shift.

[27] Note that aluminium foil was a very futuristic material at the time that the Futurists were writing.

designed as an aesthetic experience, allowing the audience to participate in a host of simultaneous multisensory experiences (see also Wang 2013).

The merging of technology and food, and the possibility of multisensory atmospherics responding to whatever the diner happens to be eating and drinking, will certainly contribute to these experiences (Peralta 2012; Houge 2013; Irwin 2013). Here, we are also thinking of the performance of the staff or by the diner (as instructed by the staff) during the course of the meal experience, or the mists and special effects that make us believe we're sitting in a theatre seat rather than at a dining table (if, that is, we *are* sitting). As the metaphor of dinner as theatre or magic show suggests, one can see the consumption of food not just as a social event (Martin 2007) but increasingly as a participatory one as well. In the future, we foresee a growing number of eating experiences in which the diners themselves become actors in their very own show or performance. As we are beginning to see, if we start to pull the various strands that have been brought out in this chapter together, restaurant, theatre and food laboratory will soon be fused into one. Key to the success of this approach will be the need to provide some form of coherent narrative structure (or story) to proceedings as well as a seamless multisensorial experience for the diner (Pine and Gilmore 1998, 1999; Goldstein 2005; Mielby and Bom Frøst 2012; Wells 2012; Wang 2013).

> "*The salad will be made again for several hundred spectators ... Beginning the event, a Mozart duo for violin and cello is followed by production of the salad by the artist and eating of the salad by the audience. The salad is always different as Mozart remains the same.*" (Alison Knowles 1962 *Make a Salad*, from Jones 2006, p. 19)

In the year 2084, could it even be that you go out for entertainment and it is unclear whether you are being fed, experimented upon or having a magic trick played on you? Restaurants such as Madeleine's Madteater in Copenhagen have been described as a free-form experimental food theatre where various performances are 'served up', all with food as the centre of the diner's (or audience's) attention (Sekules 2010). And there are many more culinary performance artists (for want of a better term) out there delivering similar experiential meals (some of them fairly ephemeral, as we'll see in the following section).[28]

11.4.2 Plating art

On the subject of the future of plating, our prediction is that there will continue to be an increasing emphasis on the artistic presentation of food (Tefler 2002; Deroy *et al.* in press; Spence *et al.* in press). We expect that this trend will go all

---

[28] For example, Carolee Schneeman's *Meat Joy*, group performance using raw fish, chickens, sausage, etc. (Jones 2006; cf. Kirshenblatt-Gimblett 1999).

the way down to everyday eateries (fast food included). In the past, food was most definitely not considered a form of art. Anyone who dared to suggest that it was, or even that it ever could be, would be rudely shot down (e.g. Carey 2005). Things are now starting to change however (see Spence *et al.* 2011) and, as we saw in Chapter 4, chefs are increasingly becoming inspired by the great visual artists, creating what we could call 'art'; some of the best in visual illusions are being created not simply on the computer screen, but with the ingredients on the plate[29]. Before too long, as the act of eating is increasingly removed from its origins as merely providing a source of nutrition, as meals become vaporized and as dining becomes more entertainment than merely a matter of sustenance, then some culinary creations may start to lay a stronger claim to being considered as a genuine art form. One can see the trend to move from considering food as merely nutrition to food as but one component of a multisensory entertainment experience (that may just so happen to be nutritional, but then again, maybe it won't).

In a related vein, a growing number of artists such as Gayle Chong Kwan in the UK (http://gaylechongkwan.com/) are now working with food as a medium (see also McCowan 2013; Quay 2013). Indeed, people are increasingly talking about food as art – or at least that it has the potential to be considered as such (Brooks 2008; http://www.somersethouse.org.uk/visual-arts /elbulli-ferran-adria-and-the-art-of-food).[30]

> *"Consider a world-class meal at a restaurant like Cyrus. Chef Keane has been through professional training and practice, and he clearly shows creativity and talent at the stove. A dish at Cyrus is a work of art. Extremely perishable art, but art nonetheless."* (Stuckey 2012, p. 56)

11.5   Interim summary

Up to this point, we have seen a number of foods and ways of cooking that we expect to see by the year 2084. Summarizing them is rather complex because some food concepts are not exactly 'food' and some cooking methods (or ways of preparing a meal) are not exactly 'cooking' either; rather, they are more like 'creating'. While *"amongst historians it is almost axiomatic that predictions of the future reflect contemporary problems"* (Belasco 2006, p. 21), our predictions concerning the food and cooking of the future of the perfect meal are also similarly directed. We see this counterbalanced with the eating process however, which will likely be much more experiential in nature (while at the same time not forgetting about the quality of the ingredients themselves).

---

[29] See Curated dinner: Culinary trickery with Clare Anne O'Keefe (https://dublin.sciencegallery.com /events/2013/08/curated-dinner-culinary-trickery-clareanne-okeefe) and Dr Andrew Szydlo (http: //www.flickr.com/photos/guerillascience/4812027704/in/set-72157624423865637) who convinces people that his violin playing can change the colour of a drink.
[30] *"I have never eaten a Velázquez. Or a Picasso, come to that. But I have eaten an Adrià. And it was pretty tasty"* (Pollard 2007, p. 6).

11.6 Acknowledging our differences

Having reviewed some of our own ideas about the future of food, which may sound more or less appetizing depending on your point of view (not to mention on whether you happen to be a neophobe or a neophile), one mustn't forget the fact that certain people have up to 16 times more taste buds than others (see Miller and Reedy 1990). As Linda Bartoshuk (1980, p. 49) *put it: "When it comes to food tastes, we all speak in different tongues ... People inhabit separate taste worlds."* Add to that the fact that we have a rapidly growing aging population (US Senate Special Committee on Aging, 1985–1986), with many more people now starting to live with a severely impaired sense of taste and smell and all the negative health consequences that implies (Spence 2002, 2012). In the future, we imagine that food delivery will become increasingly personalized rather than 'one-size-fits-all' as is mostly the case currently. Given that we already buy shoes to fit the size of our feet, why not order a meal that matches our palate as well? After all, these days you may well pay about the same for both. It would make sense, for example, to think of the food matching our taster status.

So why it is that everyone is given the same food when they dine out at a restaurant? In the years to come, there is going to be an increasing realization by top chefs (and global food manufacturers alike) that we consumers/diners really do live in very different taste worlds! No matter whether it is differences in sensitivity between young and old diners, or between supertasters and non-tasters, we predict that before too long we will start to see foods (and in the context of the present book, dishes, targeted at different 'sensotypes', to borrow Wober's 1966, 1991 intriguing term). Perhaps when we look back on 2014 some years from now it will seem altogether bizarre that, despite the fact that we have always lived in different taste worlds, we all used to eat exactly the same food.

11.7 The meal as catalyst for social exchange

We saw in Chapter 2 how the social elements of dining exert a profound effect on our enjoyment of a meal. Will this change by the year 2084? For many centuries now, meals have constituted the focus of human social interaction (see Jones 2008): it's the main (and perhaps only) daily activity that makes us gather around a table and socialize. In the context of a restaurant, we're used to gathering only with those people we know (friends, family, colleagues, etc.). Basically, we sit around a table, we order, we eat, we chat, we pay and then we leave. That's pretty much how it goes. However, in the restaurant of the future we see a much wider possibility of interaction with the waiting staff (who will still be human in most cases) and with the chefs. We have already seen an evolution in this regard with the chefs/cooks from being in a kitchen totally separate and far away from the dining area to actually having only a glass partition between the stoves and our table (in some cases, as in many Asian restaurants, the chef is even cooking just in front of us or finishing the dish *at* your table). We see this trend extending into more open spaces without borders between the kitchen and the dining area. If you stop and think about it, why does the diner of the future have to be prevented from experiencing all of those wonderful sounds and smells emanating from the ovens and pans? After all, hasn't the kitchen been the centrepiece of the home for a long time now?

Regarding our interaction with those at the table, what about those cases in which the people with whom we're used to gathering around the table are simply not there? Indeed, one problem facing a growing number of people is that they increasingly find themselves working in a different city (or even country) from the rest of their family. The traditional Sunday roast with several generations of the family coming together has become but a distant memory for many of us (see Figure 11.4 as a meal moment in which diners are literally connected). As a result, these individuals often miss out on shared family time,

Figure 11.4 'Coffee seeks its own level', by Allan Wexler in 1990. Here a coffee break turns into a communal experience in which none of the drinkers can raise their cup before the others, without the remaining cups overflowing. *"Drinking is no longer unconscious but precise, poetic and precious"* (Schwartzman 2011, p. 56). This is certainly one low-tech way of making diners much more connected. *Source:* Ronald Feldman Gallery. Photograph by Dennis Cowley. Reproduced with permission of Ronald Feldman Fine Arts.

time that is typically centred around the dinner table. Barden *et al.* (2012; see also Comber *et al.* in press) have recently investigated the use of technology to allow those who find themselves far apart from one another to share meaningful virtual mealtime/dining experiences. That said, further research is most definitely needed on what goes by the name of the 'telematic dinner party' before any workable solution emerges.

In the year 2084 will we be seeing more of these technologies at the table enabling us to continue to consider dining as a fundamentally social activity (even when the other diners are not technically sitting *at* our table)? We certainly hope that that is the case, rather than technology being used as it is at the moment to take people's mind away from where they are currently dining. Given that experts in social (and more specifically consumer) behaviour are aware of the current problem that we are facing (namely technology distracting diners), we hope and predict that 'Skype-ing' will not only be used to discuss with a group during office/work meetings but to bring people far away close to our table.

11.8   Is it a restaurant or is it a science laboratory?

Let's put on our imaginary lab coats one last time to discuss the concept of the restaurant of the future. We already see signs that the increasing appearance of technology at the front of house may, in some cases at least, start to blur the boundaries between the restaurant dining table and the science laboratory[31] (see Figure 11.5). Indeed, given our particular backgrounds, this is obviously something that your authors have a particular affinity for. Some of the world's top culinary research institutes already boast of having dining spaces that have been wired up in such a way that those who are eating/drinking can be more-or-less unobtrusively observed by means of cameras, directional microphones, hidden weighing scales to measure the amount of food that has been served (not to mention consumed) and even facial recognition software to detect their diners' expressions[32] (e.g. the Restaurant of the Future in Wageningen, Holland, http://www.restaurantvandetoekomst.wur.nl/UK/; the experimental restaurant at the Institut Paul Bocuse in Lyon, France, http://www.institutpaulbocuse.com/us/food-hospitality/; Anonymous 2013c). But don't panic, we don't see this coming to real restaurants with the purpose of listening in on our conversations as happens in Big Brother societies

---

[31] This doesn't sound too different from Gernsback's (1911) novel, mentioned earlier, in which the characters have their meals delivered via tubes in a restaurant that goes by the name of the Scienticafé!
[32] Achieving through technology what the great restaurateur may be able to do intuitively (see Meyer 210, p. 83).

Figure 11.5 The workshop of Paco Roncero: restaurant or experimental laboratory? The introduction of technology is increasingly starting to blur this boundary. *Source*: Photograph by Gerald Kiernan. Reproduced with permission of Estudio Baselga and Gerald Kieran. *See colour plate section*

(and some reality TV shows), nor is it likely that it will extend to restaurants that are not directly affiliated with a research institution.

> "*At Grant Achatz's one-of-a-kind restaurant, the chefs are scientists and the kitchen is a laboratory*" (taken from a description of Grant Achatz's Alinea restaurant in Chicago, http://www.youtube.com/watch?v=P7t0EPGKpdM).

Elsewhere, it is now becoming harder and harder to distinguish between high-end dining spaces and the cutting-edge (albeit exceedingly well-funded) science lab focused on the study of food and its perception under more or less ecologically valid testing conditions (e.g. Schira 2011; Jakubik 2012). One can perhaps see this blurring of the boundaries as the natural extension of the increasingly scientific research laboratories that have, over the last few years, started to spring up in support of some of the world's top restaurants (e.g. Ulla 2012; René Redzepi's Noma Lab in Copenhagen, http://uk.phaidon.com/agenda/design/picture-galleries/2012/march/28/first-look-at-nomas-food-lab/).

"For example, in "FoodLab," a supper club in the basement of an art-science atelier near the Place des Victoires, the otherwise spare dining space featured circulating water baths, test tubes, and lab-grade glassware, giving it the air of a laboratory, or at least a set designer's imagination of a laboratory." (Roosth 2013, p. 9)

11.9 Pop-up dining, story telling and the joys of situated eating

Regarding *where* (the physical location) we will be eating in 2084, we also predict more ephemeral restaurants (or dining spaces) such as itinerant circuses or fairs that move from one town/city to the next, all with the purpose of bringing delight to diners. For instance, Electrolux (who appear to have something of a penchant for trying to develop innovative ideas around future cooking and dining) launched The Cube in 2012, a $140\,m^2$ crystal cube designed by the studio of architecture Park Associati. The Cube was situated in locations that offered spectacular (and unusual) views of cities such as Brussels, London, Stockholm and Milan. It was placed above the most important buildings and monuments, over the sea or right on the water itself, and a series of special lunches and dinners were hosted for up to 18 guests at a sitting. In addition to enjoying the great views the diners also had the chance to see how their food was being prepared by local Michelin-starred chefs (and even have a chat with them), all for the princely sum of £175–215 (see http://www.electrolux.co.uk/Cube/). Meanwhile, for the last few years David Ghysels has been organizing pop-up meals in even more unusual locations, for example 50 m up in the air suspended from a crane (http://www.eventsinthesky.asia/dits_dinner/about.php).

Apart from these gastronomic events that for some are certainly astronomically expensive, one increasingly finds pop-up restaurants and dining venues that, in the same way as art exhibitions in galleries, come and go within a few weeks or months (Anonymous 2012b). We can also see the slow but steady growth of situated dining in fields and in art galleries; who knows where next (e.g. Brooks 2008; www.stirringwithknives.com)? If the diner is not brought to the fields the fields can be brought closer to the diner, so that one can still enjoy a picnic at the table as happens currently in one of the courses at Eleven Madison Park. Everything is valid in communicating a nice story (Fleming 2013). In the same way as for art exhibitions, the aim is to surprise and engage the visitor; one can therefore expect, as mentioned above, a concept of dining moving away from its traditional base and shifting towards the adoption of more theatrical culinary practices.

To conclude, where will we be consuming our food in 2084? Will we end up having to invent a new term for the eclectic artistic, theatrical, experiential, experimental and most definitely multisensory approach to dining, being fed and entertained at one and the same time? What we have seen is that there are currently two contrasting trends: one involving taking the diner to particular (likely spectacular) venues to eat and the other equipping the futuristic restaurants in a way in which they can recreate any atmosphere in the dining room itself, no matter how dingy the location. Think of the synthesized infamous wet British weather as the perfect backdrop to accompany the experience of having some good old fish and chips at Ultraviolet in Shanghai (Bergman 2012).

11.10 Conclusions

With a certain degree of trepidation, our top predictions for what the perfect meal might look like in the year 2084 are as follows:

1. Increasingly individualized cuisine, that is, cuisine that matches an individual diner's unique sensory world.
2. Technologically facilitated social dining given the growth of distributed living (and the technology available for that purpose).
3. *Sous vide* cooking and insects – entomophagia will be 'in' in a big way as will the new algal cuisine and the serving up of other sustainable bottom feeders and proteinacious unicellular organisms.
4. Lab-grown meat will be widely available and increasingly accepted, although the authentic sort will still be preferred (at least for special occasions).
5. An increasing number of the foods being served on our plates will have emerged from 3D food printers. Printing will be used to create innovative shapes with real ingredients (such as a spiral of purée, say) for garnishes, etc. rather than necessarily being used to substitute a slice of pizza or steak.
6. More and more we'll be using digital artefacts while eating, most likely without even giving a thought to the electronics that will undoubtedly have been embedded within the flatware and tableware (see Chapter 10).

What else should we be looking out for in the year 2084? Will the cutlery and plateware in the perfect meal be the same as that currently in use? After having seen their rapid evolution over the last decade or so (see Chapters 4 and 5), we would be surprised if this doesn't change still further over the coming years (at least in the modernist or neo-modernist restaurants, anyway). But the main point here is not that the fork and spoon won't look at all like they do now, rather that there will be other utensils on the table as well: utensils that will have been adapted to the changing eating behaviours of diners. For example, perhaps there'll be different sizes of spoons at the table, say. Not dictated by the type of food or drink being consumed (soup, dessert or coffee) that is, but more for the pace at which a single dish has to be eaten or the individual food components delivered to our mouths.[33]

Dining and entertainment will become increasingly synonymous. It will very definitely be multisensory and, increasingly, it will be a participatory activity. It is hard to say whether the use of sensory incongruity as a means of surprise will still be as popular as it is today, or whether instead diners will no longer be quite sure what is authentic and what is not (i.e. what is synthetic; see Chapter 7). Perhaps we can count on augmented and virtual reality cuisine to give the pensioners in 2084 the nostalgic feel of what it used to be like to eat a good steak. (Have you noticed how there are already joints advertising 'Hamburgers like in the good old days!' whatever that is supposed to mean.)

Finally, one can ask how the spaces in which we dine will look, and where they will be in the year 2084. As we've seen throughout this chapter, it is currently very difficult to predict what the future will bring in terms of dining; technology is changing this landscape in so many ways, both at the eating table and hidden from the eyes of the expectant diners in the kitchen. Even the chefs themselves are currently striving to keep surprising and delighting their diners in ever new ways but, whatever it is that they present, a more sustainable approach will need to be taken some day soon.

Another point to mention before closing is to acknowledge the fact that throughout this book we have described examples of what top chefs and restaurants are delivering to their diners. We are aware that these restaurants are currently affordable only to those with the fattest of wallets. But, bearing in mind that the most successful trends and techniques will likely end up percolating down to (or being adapted by) what are currently <$50 restaurants, we hope to have described what we will be seeing in any of the future restaurants that provide a meal at an affordable price.

So having reached this point, we hoped you have enjoyed the experience of imagining the future of the perfect meal. Perhaps the next time you sit down for dinner in a restaurant or at home you will share with friends some of the stories presented here, or perhaps use them as an icebreaker at a cocktail party. "Did you know that you might be a supertaster?", "Did you know that ice cubes can have a taste?" or "Apparently, in the future, you might be smelling rather than eating your dinner?" We hope this book is read by someone in 2084 who might find that at least a few of the predictions outlined here have indeed come to pass, as well as many others that no one has yet had the imagination or insight to foresee.

---

[33] Bear in mind that all design evolutions tend to be slow; inertia is a powerful inhibitor. We know that there are other configurations of the keyboard than the QWERTY that would enable us to type that much faster (not to mention allow your authors to have finished this book in less time), yet nothing has changed (see Norman 1998). Perhaps the same inertia will prevent too much change in cutlery and plateware.

References

Achatz, G. (2009) Food tasting or art installation? *The Atlantic Monthly*, 23 April. Available at http://food.theatlantic.com/back-of-the-house/food-tasting-or-art-installation.php (accessed February 2014).

Anonymous (1971) Introducing the microwave oven. Available at http://click americana.com/topics/food-drink/introducing-the-microwave-oven-1971 (accessed February 2014).

Anonymous (1975) *The Amana Guide to Great Cooking with a Microwave Oven*. Popular Library, New York.

Anonymous (2011) Futurism for foodies. *Artnet*, 12 August. Available at http://www.artnet.com/magazineus/news/artnetnews/the-futurist-cookbook.asp (accessed February 2014).

Anonymous (2012a) 3D printed pasta carbonara please. Available at http://3dprinting industry.com/2012/08/20/3d-printed-pasta-carbonara-please/ (accessed February 2014).

Anonymous (2012b) Grand Palais: Soup/No soup. Available at http://www.latriennale.org/en/grand-palais/ (accessed February 2014).

Anonymous (2013a) Le café monte en gamme. *Le Parisien*, 29 June, 35.

Anonymous (2013b) Big chef takes on Little Chef. Available at http://www.channel4.com/programmes/big-chef-takes-on-little-chef (accessed February 2014).

Anonymous (2013c) Así sera el restaurant del futuro. *Las Provincias*. Available at http://www.lasprovincias.es/20130125/mas-actualidad/sociedad/restaurante-futuro-interactivo-201301250211.html (accessed February 2014).

Apollinaire, G. (1912–1913) Le gastro-astronomisme ou la Cuisine Nouvelle. In *Le Poète assassin (The assassinated poet)*. Gallinard, Paris.

Apollonio, U. (ed) (1973) *Futurist Manifestos*. Viking, New York.

Arthur, C. (1997) Food and drink: Futurology – an alien at my table. *The Independent*, 15 November. Available at http://www.independent.co.uk/life-style/food--drink-futurology--an-alien-at-my-table-1294217.html (accessed February 2014).

Ashley, S. (2013) Synthetic food: Better cooking through chemistry. Available at http://www.pbs.org/wgbh/nova/next/physics/synthetic-food-better-cooking-through-chemistry/ (accessed February 2014).

Barba, E. D. (2013) My cuisine is tradition in evolution. Available at http://www.swide.com/food-travel/chef-interview/michelin-starred-chef-an-interview-with-massimo-bottura/2013/4/23 (accessed February 2014).

Barden, P., Comber, R., Green, D., Jackson, D., Ladha, C., Bartindale, T., Bryan-Kinns, N., Stockman, T. and Olivier, P. (2012) Telematic dinner party: Designing for togetherness through play and performance. In: *Proceedings of the ACM Conference on Designing Interactive Systems 2012 (DIS2012)*, pp. 38–47. ACM, New York.

Bartley, E. (2009) How to uncook an egg. *The Times*, 3 December, 12.

Bartoshuk, L. (1980) Separate worlds of taste. *Psychology Today*, **14**, 48–49, 51, 54–56, 63.

Beaugé, B. (2012) On the idea of novelty in cuisine: A brief historical insight. *International Journal of Gastronomy and Food Science*, **1**, 5–14.

Belasco, W. J. (2006) *Meals to Come: A History of the Future of Food*. University of California Press, Berkeley, CA.

Berghaus, G. (2001) The futurist banquet: Nouvelle Cuisine or performance art? *New Theatre Quarterly*, **17 (1)**, 3–17.

Bergman, J. (2012) Restaurant report: Ultraviolet in Shanghai. *The New York Times*, October 7. Available at http://www.nytimes.com/2012/10/07/travel/restaurant-report-ultraviolet-in-shanghai.html (accessed February 2014).

Berthelot, M. (1894) '*En l'an 2000*'. Paper given at the banquet for the Chambre Syndicale des Produits Chimiques. Available at http://trove.nla.gov.au/ndp/del/article/77542390 (accessed February 2014).

Bird, A. (1899/1971) *Looking Forward: A Dream of the United States of the Americas in 1999*. Arno, New York.

Bodenheimer, F. S. (1951) *Insects as Human Food: A Chapter in the Ecology of Man*. Junk, The Hague.

Bourland, C. T. and Vogt, G. L. (2010) *The Astronaut's Cookbook: Tales, Recipes, and More*. Springer, New York.

Brooks, R. (2008) Tate tosses up super-salad as art. The Sunday Times, 16 March, 9.

Butler, K. (2013) In-vitro meat: $325,000 lab-grown hamburger 'tastes reasonably good'. Available at http://www.upi.com/blog/2013/05/13/In-Vitro-meat-325000-lab-grown-hamburger-tastes-reasonably-good/1881368462399/ (accessed February 2014).

Carey, J. (2005). *What Good are the Arts?* Faber & Faber, London.

Casali, L. (2011) *Cucinare in lavastoviglie*. Gusto, sostenibilità e risparmio con un metodo rivoluzionario (Brossura) (Cooking with the dishwasher: Taste, sustainability, and saving with a revolutionary method). Gribaudo, Milan.

Ceurstemont, S. (2013) Grub's up. *New Scientist*, 6 July, 35–37.

Classen, C., Howes, D. and Synnott, A. (2005) Artificial flavours. In *The Taste Culture Reader: Experiencing Food and Drink* (ed. C. Korsmeyer), pp. 337–342. Berg, Oxford.

Cohn, V. (1956) *1999 - Our Hopeful Future*. Bobbs-Merrill, Indianapolis.

Collis, H. (2013) Waiter, waiter, there's gloop in my soup! Chinese chef makes use of an abundance of algae by serving it up to his customers. *DailyMail Online*, 13 July. Available at http://www.dailymail.co.uk/news/article-2362524/Waiter-waiter-s-gloop-soup-Chinese-chef-makes-use-abundance-algae-serving-customers.html #ixzz2YxRgdFip (accessed February 2014).

Colquhoun, K. (2007) *Taste: The Story of Britain through its Cooking*. Bloomsbury, London.

Comber, R., Barden, P., Bryan-Kinns, N. and Olivier, P. (in press) Not sharing sushi: Exploring social presence and connectedness at the telematic dinner party. In: *Eat, Cook, Grow: Mixing Human–Computer Interactions with Human–Food Interactions* (eds J. H.-J. Choi, M. Foth and G. Hearn). MIT Press, Cambridge, MA.

Connor, S. (2013) Special report: 'In vitro' beef – it's the meat of the future. The *Independent on Sunday*, 28 July. Available at http://www.independent.co.uk/news/science/special-report-in-vitro-beef--its-the-meat-of-the-future-8735104.html (accessed February 2014).

Crumpacker, B. (2006) *The Sex Life of Food: When Body and Soul Meet to Eat.* Thomas Dunne Books, New York.

Dahl, R. (1961) *James and the Giant Peach.* Alfred A. Knopf, USA.

Dahl, R. (1964) *Charlie and the Chocolate Factory.* Alfred A. Knopf, USA.

Davidson, J. M., Linforth, R. S. T., Hollowood, T. A. and Taylor, A. J. (1999) Effect of sucrose on the perceived flavor intensity of chewing gum. *Journal of Agriculture and Food Chemistry,* **47**, 4336–4340.

Deroy, O., Michel, C., Piqueras-Fiszman, B. and Spence, C. (in press) Plating manifesto (I): From decoration to creation. *Flavour.*

Dodd, A. B. (1887/2012) *The Republic of the Future.* Cassell & Co., New York (republished by Forgotten Books).

Dolnick E. (1990) Future schlock. *In Health,* September–October, 22.

Donawerth, J. L. (1994) Science fiction by women in the early pulps, 1926–1930. In *Utopian and Science Fiction by Women* (eds J. L. Donawerth and C. A. Kolmerten). Syracuse University Press, Syracuse.

Eriksen, L. (2013) Insect breeder: The latest gadget for grow-your-own cooks. *The Guardian,* 29 June, 32.

Fitzsimmons, C. (2003) Snacks to be a real sensation. *The Australian,* 20 August.

Fleming, A. (2013). Food with a story to tell. Available at http://www.theguardian.com/lifeandstyle/2013/sep/11/food-with-story-to-tell (accessed February 2014).

Ford, B. (1978) *Future Foods: Alternative Protein for the Year 2000.* William Morrow, New York.

Garric, A. (2013) Les petites bêtes qui épicent le menu du Festin nu (The small beasts that spice up the menu of Le Festin nu). *Le Monde,* 12 October. Available at http://www.lemonde.fr/planete/article/2013/10/12/les-petites-betes-qui-epicent-le-menu-du-festin-nu_3494724_3244.html (accessed February 2014).

Gernsback, H. (1911) *Ralph 124C 41+.* Fawcett Publications, Minnesota.

Giedion, S. (1948) *Mechanization Takes Command: A Contribution to Anonymous History.* W. W. Norton, New York (republished 1975 and 2013).

Goldberg, R.-L. (1979) *Performance Art: From Futurism to the Present.* Thames & Hudson, London.

Goldstein, D. (2005) The play's the thing: Dining out in the new Russia. In *The Taste Culture Reader: Experiencing Food and Drink* (ed C. Korsmeyer), pp. 359–371. Berg, Oxford.

Goode, J. (2005) *Wine Science.* Mitchell Beazley, London.

Graf, V. (2013) Hungry to save cash? Cook dinner in the dishwasher! The latest money-saving technique to cut down on fuel bills. *Daily Mail Online,* 9 September. Available at http://www.dailymail.co.uk/femail/food/article-2415722/Cooking-dinner-dishwasher-Latest-money-saving-technique-cut-fuel-bills.html (accessed February 2014).

Gray, R. (2010) Willy Wonka chewing gum could become reality. Available at http://www.telegraph.co.uk/news/newstopics/howaboutthat/8053260/Willy-Wonka-chewing-gum-could-become-reality.html (accessed February 2014).

Haden, R. (2005) Taste in an age of convenience. In *The Taste Culture Reader: Experiencing Food and Drink* (ed C. Korsmeyer), pp. 344–358. Berg, Oxford.

Halligan, M. (1990) *Eat My Words.* Angus & Robertson, London.

Harrison, H. (1966/1967) *Make Room! Make Room!* Berkeley Medallion, New York.

Holt, V. M. (1885) *Why Not Eat Insects?* Field and Tuer, London.

Horwitz, J. (2004) Eating space. In *Eating Architecture* (eds J. Horwitz and P. Singley), pp. 259–275. MIT Press, Cambridge, MA.

Houge, B. (2013) Food opera: Merging taste and sound in real time. Available at http://www.newmusicbox.org/articles/food-opera-merging-taste-and-sound-in-real-time/?utm_source=rss&utm_medium=rss&utm_campaign=food-opera-merging-taste-and-sound-in-real-time (accessed February 2014).

Hoyle, B. (2008) Recipe for revolution takes diners back to the Futurists. *The Times*, 28 January, 23.

Huxley, A. (1932) *Brave New World*. Harper & Row, New York.

Irwin, J. (2013) Listen up for Food Opera – a live, sonic interactive dinner. Available at http://killscreendaily.com/articles/articles/food-opera-your-new-favorite-interactive-eating-experience/ (accessed February 2014).

Jakubik, A. (2012) The workshop of Paco Roncero. *Trendland: Fashion Blog and Trend Magazine*. Available at http://trendland.com/the-workshop-of-paco-roncero/ (accessed February 2014).

Jayakumar, A. (2013) Home-baked idea? NASA mulls 3D printers for food replication. Available at http://www.guardian.co.uk/technology/2013/jun/04/nasa-3d-printer-space-food (accessed February 2014).

Jones, C. A. (ed) (2006) *Sensorium: Embodied Experience, Technology, and Contemporary Art*. MIT Press, Cambridge, MA.

Jones, M. (2008) *Feast: Why Humans Share Food*. Oxford University Press, Oxford.

Kahn, H. (1976) *The Next 200 years: A Scenario for America and the World*. William Morrow, New York.

Kirkova, D. (2013) 'I sniffed my way thin!' Baker loses 7 STONE while still enjoying cakes (but now she just smells them instead). *Dailymail*, 26 August. Available at http://www.dailymail.co.uk/femail/article-2402165/Baker-Lynne-Gadd-South-Wales-lost-7-STONE-sniffing-cakes-instead-eating-them.html (accessed February 2014).

Kirshenblatt-Gimblett, B. (1999) Playing to the senses: Food as a performance medium. *Performance Research*, **4 (1)**, 1–30.

Klotz, I. (2013) NASA investing in 3-D food printer for astronauts. Available at http://uk.reuters.com/article/2013/05/22/us-space-food-idUKBRE94L1B420130522 (accessed February 2014).

Lappé, F. M. (1971) *Diet for a Small Planet*. Ballantine Books, New York.

Lubow, A. (2003) A laboratory of taste. *The New York Times*, August 10. Available at http://www.nytimes.com/2003/08/10/magazine/a-laboratory-of-taste.html?pagewanted=all&src=pm (accessed February 2014).

MacClancy, J. (1992) *Consuming Culture: Why You Eat What You Eat*. Henry Holt, New York.

Madene, A., Jacquot, M., Scher, J. and Desobry, S. (2006) Flavour encapsulation and controlled release – a review. *International Journal of Food Science and Technology*, **41(1)**, 1–21.

Marinetti, F. T. (1930) *The Futurist Cookbook* (translated by S. Brill). Bedford Arts, San Francisco (republished 1989).

Marriot, M. (1999) Mars 2112: A space odyssey. *The New York Times*, 18 February. Available at http://www.nytimes.com/1999/02/18/technology/mars-2112-a-space-odyssey.html (accessed February 2014).

Martin, D. (2007) *Evolution*. Editions Favre, Lausanne.

McCowan, D. (2013) It's Alive! Designing Living Food. *Chicago Foodies*, 6 August. Available at http://www.chicagofoodies.com/2013/08/its-alive-designing-living-food.html (accessed February 2014).

McGee, H. (2008) On food and zapping. *New York Times*, 2 April. Available at http://www.nytimes.com/2008/04/02/dining/02curious.html?ref=microwaveovens&_r=1& (accessed February 2014).

Meyer, D. (2010) *Setting the Table: Lessons and Inspirations from one of the World's Leading Entrepreneurs*. Marshall Cavendish International, London.

Mielby, L. H. and Bom Frøst, M. (2012) Eating is believing. In *The Kitchen as Laboratory: Reflections on the Science of Food and Cooking* (eds C. Vega, J. Ubbink and E. van der Linden), pp. 233–241. Columbia University Press, New York.

Miller, I. J. and Reedy, D. P. (1990) Variations in human taste bud density and taste intensity perception. *Physiology and Behavior*, **47**, 1213–1219.

Mouritsen, O. G. (2012) The emerging science of gastrophysics and its application to the algal cuisine. *Flavour*, **1**, 6.

Muecke, M. (2004) Food to go: The industrialization of the picnic. In *Eating Architecture* (eds J. Horwitz and P. Singley), pp. 228–246. MIT Press, Cambridge, MA.

Norman, D. (1998) *The Design of Everyday Things*. MIT Press, London.

Orwell, G. (1949) *Nineteen Eighty-four*. Secker and Warburg, London.

Osborne, H. (2013) Beef industry greenhouse gas emissions: Cows produce most methane while feeding calves. Available at http://www.ibtimes.co.uk/articles/430077/20130131/beef-industry-greenhouse-gasses-cows-methane-production.htm (accessed February 2014).

Peralta, E. (2012). The sounds of asparagus, as explored through opera. *The Salt*. Available at http://www.npr.org/blogs/thesalt/2012/05/29/153950254/the-sounds-of-asparagus-as-explored-through-opera (accessed February 2014).

Petrini, C. (2007) *Slow Food: The Case for Taste* (translated by W. McCuaig). Columbia University Press, New York.

Pine II,, B. J. and Gilmore, J. H. (1998) Welcome to the experience economy. *Harvard Business Review*, **76(4)**, 97–105.

Pine, II,, B. J. and Gilmore, J. H. (1999) *The Experience Economy: Work is Theatre and Every Business is a Stage*. Boston, MA: Harvard Business Review Press.

Pollan, M. (2013) *Cooked: A Natural History of Transformation*. Penguin, London.

Pollard, S. (2007) Food as art, or pie in the sky? *The Times*, 17 May (times2), 6.

Poole, S. (2012) *You Aren't What You Eat: Fed Up With Gastroculture*. Union Books, London.

Purvis, A. (2012) Will 3D printers make food sustainable? Available at http://www.guardian.co.uk/environment/2012/may/18/3d-printers-food-sustainable (accessed February 2014).

Quay, A. (2013) In Photo series, Bauhaus-influenced food designs. Available at http://designtaxi.com/news/357758/In-Photo-Series-Bauhaus-Influenced-Food-Designs/ (accessed February 2014).

Rayner, J. (2004) The man who mistook his kitchen for a lab. *The Observer*, 15 February. Available at http://www.guardian.co.uk/lifeandstyle/2004/feb/15/food anddrink.restaurants (accessed February 2014).

Rayner, J. (2006) 'Molecular gastronomy is dead.' Heston speaks out. *The Observer*, Food Monthly, 17 December. Available at http://observer.guardian.co.uk /foodmonthly/futureoffood/story/0,,1969722,00.html (accessed February 2014).

Rivalta, F. (2012) Dishwasher haute cuisine: Earning time, but not the taste. *Fine Dining Lovers*. Available at http://www.finedininglovers.com/stories/cooking -techniques-dishwasher-low-temperatures/ (accessed February 2014).

Rogers, M. (1989) Marvels of the future. *Newsweek*, 25 December, 77–78.

Roosth, S. (2013) Of foams and formalisms: Scientific expertise and craft practice in molecular gastronomy. *American Anthropologist*, **115**, 4–16.

Rosenbaum, R. (1979) Today the strawberry, tomorrow … In *Culture, Curers and Contagion* (ed. N. Klein), pp. 80–93. Chandler and Sharp, Novato, CA.

Sandhana, L. (2010) The printed future of Christmas dinner. Available at http://www .bbc.co.uk/news/technology-12069495 (accessed February 2014).

Schira, R. (2011) Daniel Facen, the scientific chef. Available at http://www.finedining lovers.com/stories/molecular-cuisine-science-kitchen/ (accessed February 2014).

Schlosser, E. (2001) *Fast Food Nation: The Dark Side of the All-American Meal*. Houghton Mifflin, New York.

Schutz, H. G. and Wahl, O. L. (1981) Consumer perception of the relative importance of appearance, flavour, and texture to food acceptance. In *Criteria of Food Acceptance: How Man Chooses What He Eats* (eds J. Solms and R. L. Hall), pp. 97–116. Forster, Zurich.

Schwartzman, M. (2011) *See Yourself Sensing: Redefining Human Perception*. Black Dog Publishing, London.

Sekules, K. (2010) Food for thought. Copenhagen's coolest dinner theatre. *New York Times*, 19 January. Available at http://tmagazine.blogs.nytimes.com/2010/01/19 /food-for-thought-copenhagens-coolest-dinner-theater/ (accessed February 2014).

Shell, E. R. (1986) Chemists whip up a tasty mess of artificial flavors. *Smithsonion*, **17(1)**.

Slosson, E. E. (1919) *Creative Chemistry*. Century Co., New York.

Smith, F. E. (1930) *The World in 2030 AD*. Hodder & Stoughton, London.

Spence, C. (2002). *The ICI Report on the Secret of the Senses*. The Communication Group, London.

Spence, C. (2010) The price of everything – the value of nothing? *The World of Fine Wine*, **30**, 114–120.

Spence, C. (2012) The development and decline of multisensory flavour perception. In: *Multisensory Development* (eds A. J. Bremner, D. Lewkowicz and C. Spence), pp. 63–87. Oxford University Press, Oxford.

Spence, C. (2013) Unravelling the mystery of the rounder, sweeter chocolate bar. *Flavour*, **2**, 28.

Spence, C., Shankar, M. U. and Blumenthal, H. (2011). 'Sound bites': Auditory contributions to the perception and consumption of food and drink. In *Art and the Senses* (F. Bacci and D. Melcher, eds), pp. 207–238. Oxford University Press, Oxford.

Spence, C., Hobkinson, C., Gallace, A. and Piqueras-Fiszman, B. (2013) A touch of gastronomy. *Flavour*, **2**, 14.

Spence, C., Piqueras-Fiszman, B., Michel, C. and Deroy, O. (in press) Plating manifesto (II): the art and science of plating. *Flavour*.

Steinberger, M. (2010) *Au Revoir to All That: The Rise and Fall of French Cuisine*. Bloomsbury, London.

Stuckey, B. (2012) *Taste What You're Missing: The Passionate Eater's Guide to why Good Food Tastes Good*. Free Press, London.

Tefler, E. (2002) Food as art. In: *Arguing About Art: Contemporary Philosophical Debates*, 2nd Edition (eds A. Neill and A. Riley), pp. 11–29. Routledge, New York.

The Associated Press (2013) World's 1st lab-grown burger cooked and eaten. Available at http://www.cbc.ca/news/technology/story/2013/08/05/technology-lab-grown -burger.html (accessed February 2014).

This, H. (2012a) *Molecular Gastronomy: Exploring the Science of Flavor*. Columbia University Press, New York.

This, H. (2012b) *La cuisine note à note en 12 questions souriantes* (Note-by-note cuisine in 12 smiling questions). Belin, Paris.

This, H. (2012c) Molecular gastronomy is a scientific activity. In *The Kitchen as Laboratory: Reflections on the Science of Food and Cooking* (eds C. Vega, J. Ubbink and E. van der Linden), pp. 242–253. Columbia University Press, New York.

This, H. (2013) Molecular gastronomy is a scientific discipline, and note by note cuisine is the next culinary trend *Flavour*, **2**, 1.

Titchener, E. B. (1909) *A Textbook of Psychology*. Macmillan, New York.

Toops, D. (1998) Forecasts for the millennium: Advertising gurus predict meals on wheels, pizza-flavoured corn. *Food Processing*, November, 59.

Ulla, G. (2012) The future of food: Ten cutting-edge restaurant test kitchens around the world. Available at http://eater.com/archives/2012/07/11/ten-restaurant-test -kitchens.php (accessed February 2014).

US Senate Special Committee on Aging (1985–1986): *Aging America, Trends and Projections*. US Senate Special Committee on Aging in association with the American Association of Retired Persons, the Federal Council on the Aging, and the Administration on Aging, pp. 8–28.

Verne, M. (1889) The day of an American journalist in 2889. Available at http://www .eastoftheweb.com/short-stories/UBooks/DayAmer.shtml (accessed February 2014).

Vettel, P. (2013) *Good Eating's Fine Dining in Chicago*. Agate Digital, Chicago.

Vidal, J. (2013) Breed insects to improve human food security: UN report. Available at http://www.theguardian.com/environment/2013/may/13/breed-insects-improve -human-food-security-un (accessed February 2014).

Vogelzang, M. (2010) TEDxMunich - Marije Vogelzang - Food Love. Available at http://www.youtube.com/watch?v=1WZv459bOCQ (accessed February 2014).

Vonnegut, K. (1973) *Breakfast of Champions*. Random House, New York.

Wallace, B. (2009) *Billionaire's Vinegar: The Mystery of the World's Most Expensive Bottle of Wine*. Broadway Paperbacks, New York.

Wang, Q.J. (2013) Music, mind, and mouth: Exploring the interaction between music and flavour perception. MSc thesis, Massachusetts Institute of Technology, Cambridge, MA.

Waugh, E. (1930) *Vile Bodies*. Chapman & Hall, London.

Weiss, A. (2002) *Feast and Folly: Cuisine, Intoxication and the Poetics of the Sublime*. State University of New York Press, Albany, NY.

Wells, P. (2005) Brain food | Grant Achatz. Available at http://www.foodandwine.com/articles/brain-food-grant-achatz (accessed February 2014).

Wells, P. (2012) Talking all around the food: At the reinvented Eleven Madison Park the words fail the dishes. *The New York Times*, 17 September. Available at http://www.nytimes.com/2012/09/19/dining/at-the-reinvented-eleven-madison-park-the-words-fail-the-dishes.html?pagewanted=all (accessed February 2014).

Whittle, N. (2013) Chef talk: Simon Hopkinson. *FT Magazine*, 25/26 May, 45.

Williams, H. S. (1907) The miracle-workers: Modern science in the industrial world. *Everybody's Magazine*, October, 497–498.

Wilson, B. (2009) *Swindled: From Poison Sweets to Counterfeit Coffee – The Dark History of the Food Cheats*. John Murray, London.

Wober, M. (1966) Sensotypes. *Journal of Social Psychology*, **70**, 181–189.

Wober, M. (1991) The sensotype hypothesis. In: *The Varieties of Sensory Experience: A Sourcebook in the Anthropology of the Senses* (ed D. Howes), pp. 31–42. University of Toronto Press, Toronto.

Zoran, A. and Coelho, M. (2011) Cornucopia: The concept of digital gastronomy. *Journal of the International Society for the Arts, Sciences and Technology*, **44**, 425–431.

Index

Note: Page numbers with suffix 'f' refer to Figures, those with suffix 't' refer to tables and boxes and those with suffix 'n' refer to footnotes

"A Darker Sunset," 254
à la carte menu, 58
à la Russe, 341, 341n
Abas, S., 119
Achal, 347
Achatz, G., 4, 93, 113, 129–31, 139n, 170, 237, 237f, 256, 257f, 277, 352n
Adagio for Strings, 40n
ADNY, 159
Adrià, F., 14, 113, 116, 172, 216, 218, 233–4, 238f, 331, 362
Aduriz, A.L., 21, 152
"Aerofood," 348, 349f
affluent
    modernist cuisine for, 22–5
Against Nature, 256
air "food," 346–8
airlines
    cutlery on, 160n
    food served on, 296–7
à la Française, 341n
algae cuisine, 362–3
Alice in Wonderland, 222n
Alícia Foundation, 116
Alija, J., 168

Alinea, 4, 93, 123n, 129f, 130, 132, 170, 171f, 200, 257f, 277, 289, 347, 352n
    improvised plateware at, 134
    menu of, 61–2, 61f
    Osetra "dish" at, 167, 167f
    oyster leaf at, 237f
Allen, J.S., 13–14, 13n, 14n, 255n
Alleyne, R., 261–2
amuse bouche, 25–7
amygdala
    activation in considering appetitive incentive value of foods, 83
An Anthropologist on Mars, 258, 258f
An Economist Gets Lunch, 24n, 118n, 274n
An Officer and a Gentleman, 40–1
"Angels on horseback," 76
Apician Dish, 127
Apicius, 127, 215
Apostle spoons, 154
appraisal of meal
    atmosphere effects on, 280
AR food. *see* augmented reality (AR) food

The Perfect Meal: The Multisensory Science of Food and Dining, First Edition.
Charles Spence and Betina Piqueras-Fiszman.
© 2014 John Wiley & Sons, Ltd. Published 2014 by John Wiley & Sons, Ltd.

Arana, F.S., 83
Argyra, M., 155
Ariely, D., 45, 79
Aronson, E., 221
Arrop, 44
Art Laboratory Berlin, 364
artificial flavours, 348–51
Artisanal, 289n
Arzak, J.M., 330
Asiana Airlines
    cutlery on, 160n
atmosphere, 271–309
    appraisal of meal effects of, 280
    context and, 286–7
    ethnicity of meal and, 280–2
    expectation and, 286–7
    experience economy and, 275–7
    feel of restaurant, 291–4, 292f, 294f
    importance of, 271–309
    introduction, 271–5
    lighting, 287–8
    money and time spent at
            restaurant, 282–6
    multisensory, 298–9, 299f
    music, 282–6. see also music
    olfactory, 288–93
    Provencal Rose paradox, 278–9
    in taste and flavour perception, 294–7
    technology at dining table in
            controlling, 323–4
attentional capture
    as diner's response to sensory
            incongruity at dinner
            table, 232
Attica, 348
augmented reality (AR) food
    technology related to, 317–19, 318f,
            319f
Australasia restaurant, 62
Avianca
    food served on, 296

Bakewell tart, 190
balance
    on plate, 136–41, 137f, 141f
Balthazar, 49
Barber, S., 40n
Barden, P., 369

Barnett-Cowan, M., 175, 193
Barnham, P., 8–9, 8n, 168
Barrós-Loscertales, A., 82
Bartoshuk, L., 327n, 367
Basinger, K., 251n
Baumann, Z., 8
Beaugé, B., 8
"Beetroot and Orange Jelly," 222, 223,
            223f, 236
behaviour
    menu price effects on, 55–9
Belasco, W., 363
belief(s)
    food-centered, 215
Bell, R., 85, 280–1, 281n
Berlyne, D.E., 138
Bethaz, L., 164
Beverly Hills Hotel, 257
Bilson, G., 94
Bird, A., 342
Blackout talks and music events, 254
Blanch & Shock, 131, 131f
Blind Cow restaurant, 249
blind persons
    dining in the dark by, 264
blindness
    colour, 259
Blinkekuh (Blind Cow) restaurant, 249
Bloor, L., 172
Blue Curaçao, 227, 239
blue raspberry drinks, 226
Blumenthal, H., 5, 7, 10, 21, 23–4, 43, 73,
            74f, 80, 134, 135f, 163, 175–6,
            216, 216n, 233, 274n, 296, 315,
            315n, 316f, 344n, 352, 364n
Bocado de Luz (serving plate), 330
Bocuse, P., 3, 3n, 113
Bolhios, D.P., 165
Bom Frøst, M., 92, 266
Bompas & Parr, 125
Bompas, S., 327
Bonaparte, N., 111, 158
Boring, E.G., 186n
"Boston blues," 75
Bottura, M., 347–8
Bourdain, A., 142n
Bourland, C.T., 77n, 354
Bourn, D., 87

Bournemouth University, 47
  Grill Room at, 280–1
  training restaurant in, 42–3
Bowen, J.T., 50, 51
brain on flavour, 12–16
  neurogastronomists and great-tasting
      food, 14–16
  neurogastronomy, 12–14
Bras, M., 136, 137f
Brasserie Zédel, 51n
Brave New World, 357
Brearley, H., 158
Bretillot, M., 347
Brillat-Savarin, J.A., 5n
British Airways
  food served on, 296
Broackes, J., 259
Brook, M., 346
Bruni, F., 15
Bryant, J., 175
Bryden, R., 96n
Brydon, S., 96n
"Bubble and Squeak," 98
bug burger with insect paste, 358–62,
      360f
Burton, S., 54
Byram, J., 190

Cadbury's, 343n
Cage Ke'ilu, 276n
calories
  on menu, 54–5
Camarena, R., 44, 232
Cambridge University, 45
camouflage plateware, 132–3, 133f
"Can't Say No" Sundae, 57–8
Cantu Designs, 353–4
Cantu, H., 4, 60, 60f, 60n, 170, 171f,
      229–30, 348, 353–4,
      354n, 363
Caparoso, R., 56
Caraulani, J., 328
Cardello, A.V., 130, 221
Carême, M.-A., 75–6, 86, 93, 111–14,
      112f, 112n, 129, 136,
      341
Carlsmith, J.M., 221
Carlson, A.J., 263n

Carnevino
  wine at, 56
Carter, E., 176
casserole
  stew vs., 86
Castel, A., 16n
Cavanagh, K.V., 89
celebrity chef, 7
  rise of, 7
Center for Hospitality Research
  at Cornell University, 56
Ceurstemont, S., 358
Cezzar, J., 58n
Chandon, P., 90
Chapel, A., 3
Chaplin, C., 346
Charlie and the Chocolate Factory, 344
Charlie Brown's restaurant chain, 43
"Cheers," 326
"Cheese Cake Vesuvius," 57
chef(s)
  celebrity, 7
  tips from, 331
"Chewing Jockey," 318, 318f
Chicago School of Restaurants, 4
Chicago's O'Hare International
      Airport, 53
Chicken à la King and Rice dish, 280
"Chicken on a brick," 133n
Chilean sea bass, 75
chimp stick dish, 360f
chopsticks, 164–5
Chossat, V., 279
Chung, J., 56
CIA. see Culinary Institute of America
      (CIA)
Cilione, 119
cinema
  popcorn in, 262–3, 262n
Claridge's Hotel, 360
Clear Tab cola, 222, 226
Clore, G.L., 41
cochlea, 153
Cockburn, A., 3
Coelho, M., 353
cognitive neuroscience
  of multisensory flavour
      perception, 200–2

Cohen, J., 96
cola
    Clear Tab, 222, 226
Colbert, 51*n*
"collaborative properties," 138, 138*n*
*Collectors' Handbook for Grape
        Nuts,* 157
Collins English Dictionary, 3
colour
    of cutlery, 168–9, 169*f*
    flavour perception effects of, 196–200
    of food, 255–60, 257*f*, 258*f*, 260*f*
    of plateware, 115–19
colour blindness, 259
Colour Society of Australia, 256*n*
colour-flavour incongruity, 226–7
Committee on Food Habits, 221
company
    in social aspects of dining, 44–7
Concorde
    cutlery on, 160*n*
Condiment Junkie, 315, 324, 327, 328
"*Contraires,*" 229, 229*f*
Coogan, S., 96*n*
Cooked, 351
*Cookery and Dining in Imperial
        Rome,* 127
cooking
    in the dark, 265
Cooking with Fernet Branca, 72–3
"Cooking without Looking," 265
"Cool Blue raspberry" fruit
        drink, 226–7
Coppola, F., 281
Coques Saint-Jacques, 130*n*
Corbin, C., 51, 51*n*
Cordon Bleu, 20
Coren, G., 39*n*, 44*n*, 90*n*, 92*n*, 93, 94
Cornell University, 353
    Center for Hospitality Research at, 56
Coughlan, M.R., 41
Courvoisier, 325
coutellerie, 151
Covent Garden, 51
Cowen, T., 24*n*, 58, 118*n*, 274*n*, 323*n*
Cradock, F., 225
Crawford, I., 292
Crisinel, A.-S., 296, 324, 325*n*

crispiness, 195–6
croquembouche, 112*f*
cross-cultural differences
    in multisensory flavour
        perception, 190–1
Crossmodal Research Laboratory
    of Oxford University, 140, 168, 172,
        217, 315*n*, 324, 325, 327, 328
Crucial Detail, 129, 129*f*, 170, 171*f*
Crumpacker, B., 118, 156*n*, 157*n*, 168,
        354*n*
crunchiness, 195–6
Csergo, J., 137*n*
Culinary Institute of America (CIA), 56
culinary movements
    celebrity chef, 7
    history of, 2–7
    modernist cuisine, 5–7
    molecular gastronomy, 3–5
    nouvelle cuisine, 2–3
culinary techniques
    labelling of, 92–4, 94*f*, 95*f*
cultured meat
    future of, 355–6
cutlery, 151–81
    on Asiana Airlines, 160*n*
    chopsticks, 164–5
    colourful, 168–9, 169*f*
    on Concorde, 160*n*
    defined, 151
    designers of, 170–3, 171*f*, 173*f*
    future of, 170
    history of, 153–9, 158*f*
    introduction, 151–3
    material qualities of, 159–65, 163*f*
    at Per Se, 168
    quality of, 160–1
    silver, 156
    silver-plating, 157
    size of, 165–6
    stainless steel, 158–60
    story of, 153–9, 158*f*
    tasting, 161–5, 163*f*
    texture/feel of, 166–8, 167*f*
    that is not cutlery, 169–73, 171*f*, 173*f*

Dahl, L., 344*n*
Dahl, R., 343, 343*n*, 344, 344*n*, 358–9

Daily Mail, 348*n*
Dalton, P., 188–90, 189*f*
Damrosch, P., 277
Dand, D., 326
Dans Le Noir, 249*n*, 261
dark
  dining in, 249–60. *see also* dining in
    the dark
Dark Dining, 254
darkness
  cooking in, 265
David Bellamy cocktail
  at The House of Wolf, 186
David, E., 97*n*, 290
Davidson, A., 76*n*
de Broglie, E., 249*n*
de Castro, J.M., 46
De Essientes, 256
de la Reynière, G., 256*n*
de Medici, C.
  cutlery and, 155–6
Debrett's, 153, 175
Delboeuf illusion, 121, 122*f*
Delfina restaurant, 285
Delwiche, J., 194
Desbuissons, F., 112
deuteranopia, 259, 260*f*
"Devils on horseback," 76
Diagnostic Kitchen, 321
"Dialog im Dunkeln" exhibition
  at Zurich Museum of Design, 249
dichromacy
  types of, 259
diet
  menu items for people on, 54–5
Digital Fabricator, 353
Dimensions of the Meal, 136
diner response(s)
  to sensory incongruity at dinner
    table, 232–3, 239–40
  to visual presentation of food, 141–2
dining
  social aspects of, 42–7. *see also* social
    aspects of dining
dining in the dark, 249–60
  as how the blind experience food, 264
  introduction, 249–50
  popcorn, 262–3, 262*n*

  seeing or not seeing the food, 255–64,
    257*f*, 258*f*, 260*f*
  senses' effects on, 260–4
  social aspects of, 251
dining in the dark restaurants
  popularity of, 252–5
dining table
  technology at, 311–37. *see also*
    technology at dining table
  technology on, 312–14, 314*f*
Dinner restaurant
  in Mandarin Oriental Hotel in
    London, 7, 43, 74*f*, 274*n*
disconfirmation of expectations, 221–2
disconfirmed expectations, 221–2
Discover Your Inner Economist, 58–9
distraction
  technology at dining table and, 322–3
Dodd, A., 341–2
Doerfler, W., 49
dog food study, 17*n*
Dolnick, M., 350*n*
Dornenburg, A., 142*n*
Dover, P.A., 79
Downs, J.S., 54–5
Drake, 194–5
drink(s)
  blue raspberry, 226
  "Cool Blue raspberry" fruit, 226–7
  mouthfeel of, 191, 191*n*
DuBose, C.N., 198
Ducasse, A., 159, 340*n*
Duffy, E., 176
Duke University, 79
Duncker, K., 220
Dunkin' Donuts, 86
Dunlop, F., 162–3

Earl of Birkenhead, 350
Eat With Your Hands, 175*n*
eating
  without hands, 175–6
eating insects for pleasure, 358–62, 360*f*
Ebbinghaus–Titchener size-contrast
  illusion, 121, 122*f*
economy
  atmospherics and, 275–7
Edwards, D., 345, 347

Edwards, J.S.A., 42–3, 47, 280
Effront, J., 356
"Egg in the Hole," 98
El Celler de Can Roca restaurant, 313,
        314f, 364
    "Messi's Goal" dessert from, 316
elBulli, 4, 93, 94f, 114, 234, 331
    appetisers at, 230, 230f
    cutlery at, 170
    FACES collection of, 172
    planning timeline for 2008 menu
        from, 110
Electrolux, 332, 371
Eleven Madison Park, 58n, 61–2, 127,
        127f, 131, 131f, 136, 137f,
        138, 371
Eliot, T.S., 154n
Escoffier, A., 8, 112–13, 113n, 341
EsTheremine talking fork, 321
ethnicity of food
    labelling's influence on, 85–6
ethnicity of meal, 280–2
Evans, D., 288
EverCrisp App, 318
"everything else"
    food vs. perception of, 16–17
expectation(s)
    atmosphere and, 286–7
    disconfirmed, 221–2
    flavour, 219, 219f, 219n
    food-centered, 215
    food description–related, 96–8
    taste of, 10–11
experience(s)
    flavour, 219, 219f, 219n
Experimental Food Society, 325n
Eye of the Beholder (platter),
        330

Fabian, A., 170
Facen, D., 314
FACES collection
    of elBulli, 172
Fairmont Hotel, 275n
Fama (long plate), 330
FAO. see Food and Agriculture
        Organization (FAO)
Fast Food Nation, 349

feeding
    mechanization of, 345–6
feel
    of cutlery, 166–8, 167f
    of plateware, 123–8, 125f–7f
    of restaurant, 291–4, 292f, 294f
finger food, 174–5
flatware, 151–81. see also cutlery
flavour(s). see also multisensory flavour
        perception
    artificial, 348–51
    brain on, 12–16. see also brain on
        flavour
    colour effects on, 196–200
    defined, 183–4
    introduction, 183
    ISO on, 183–4
    labelling in enhancing, 77–81
    multisensory perception of, 183–213.
        see also multisensory flavour
        perception
    perceiving of, 183–6, 185f
    supertaster, 187
    taste and, 186–7
    visual, 80, 196–200
flavour expectations
    flavour experiences vs., 219, 219f,
        219n
flavour perception
    atmospheric contributions to, 294–7
Fleischer, R., 343
fMRI. see functional magnetic resonance
        imaging (fMRI)
food(s). see also specific types
    "air," 346–8
    AR, 317–19, 318f, 319f
    colour of, 255–60, 257f, 258f, 260f
    as entertainment, 364–6
    finger, 174–5
    history of predicting future
        of, 341–51, 342f, 349f
    labelling in enhancing taste and/or
        flavour of, 77–81
    labelling's influence on ethnicity
        of, 85–6
    matching sound to, 324–6
    mood, 38–41, 39f
    mouthfeel of, 191, 191n

naming of, 81–4
  from past to future of, 351–63, 360*f*
  perception of "everything else"
    and, 16–17
  QR codes in changing interactions
    with, 319–20
  seeing or not seeing, 255–64, 257*f*,
    258*f*, 260*f*
  served on airlines, 296–7
  sound of, 194–6
  as theatre, 11–12
  3D printed, 353–5
Food and Agriculture Organization
    (FAO)
  of UN, 359
food-centered beliefs, 215
food-centered expectations, 215
food description, 71–107
  art and science of, 71–107
  ethnic, 72–3
  exotic, 72
  expectations, 96–8
  food labelling, 91–2
  food naming, 99*t*
  health/ingredient labels, 88–90
  introduction, 71–3
  labelling as influence on perceived
    ethnicity of dish, 85–6
  labelling culinary techniques, 92–4,
    94*f*, 95*f*
  labelling enhancing taste and/or
    flavour of food, 77–81
  local labels, 90–1
  naming names in, 84–5
  natural and organic labels, 87–8
  neuroscience of naming food, 81–4
  reactions, 96–8
  "Snail Porridge," 73–7, 74*f*
  surprise in, 95–6
FOOD design probe, 330*n*
Food in History, 77
food items
  PET in processing incentive value
    of, 83
food labelling, 91–2
food labels. *see also* label(s); labelling
  descriptive, 82
Food Neophobia Scale, 240

food perception
  multisensory nature of, 21–2
"food porn," 142*n*
"food porn" images, 320
food presentation, 109–49
  contributions of visual appearance of
    dish, 135–41, 137*f*, 141*f*
  individual diner responses to
    visual, 141–2
  introduction, 109–11
  multisensory, 109–49
  potted history of, 111–14, 112*f*, 114*f*
food theatre, 364–6
Forestell, C.A., 89
fork. *see also* cutlery
  history of, 153–9
format–flavour incongruity, 227–30,
    228*f*–30*f*
Fourier, C., 18*n*
Fragrant Supper Club, 172
Frank, R.A., 190
"French Paradox," 122
Frewer, L.J., 87*n*
functional magnetic resonance imaging
    (fMRI)
  in multisensory flavour perception
    study, 201
future of food
  history of predicting, 341–51, 342*f*,
    349*f*
  from past to, 351–63, 360*f*
future of perfect meal, 339–81
  acknowledging our differences, 367
  air "food," 346–8
  algae cuisine, 362–3
  artificial flavours, 348–51
  cultured meat, 355–6
  eating insects for pleasure, 358–62,
    360*f*
  food theatre, 364–6
  history of predicting, 341–51, 342*f*,
    349*f*
  introduction, 339–41
  joys of situated eating, 371–2
  meal as catalyst for social
    exchange, 367–9, 368*f*
  meal in single dose, 341–5, 342*f*
  mechanization of feeding, 345–6

future of perfect meal (*continued*)
  neo-Futurist cuisine, 363–6
  note-by-note cuisine, 356–8
  from past to, 351–63, 360*f*
  plating art, 365–6
  pop-up dining, 371–2
  restaurant *vs.* science
              laboratory, 369–71, 370*f*
  Sous vide as twenty-first century
              microwave, 351–2
  story telling, 371–2
  3D printed foods, 353–5

Gal, D., 119
Gallace, A., 124
Gallup Organisation
  on diner's scanpath, 50
Garber, L.L., Jr., 239
gastronomy
  defined, 5, 5*n*
  molecular, 3–5. *see also* molecular
              gastronomy
"gastrophy," 18*n*
gastrophysics, 2, 18–21
gastroporn, 3
Gault, H., 3
Gault-Millau restaurant guide, 279
Gergaud, O., 279
Gernsback, H., 312*n*, 346, 350,
              369*n*
Ghysels, D., 371
Giacometti, A., 257
Gilbert, D., 13*n*
Gill, A.A., 9, 235*n*
Gilmore, J.H., 252, 272, 276, 276*n*
"Gin & Sonic," 328–9
"Glazed nun farts," 75
"Golden Opulence Sundae," 57
Good Food Guide, 176
Goss, E., 120
Gourmand syndrome, 14, 203
Grabenhorst, F., 79*n*, 83
grande, 86
Greenaway, P., 256
Greene, G., 159
Grill Room
  at Bournemouth University, 280–1
Guangxin, B., 165*n*

Guéguen, N., 53, 289
"gustemology," 18*n*

Haiko Cornelissen Architecten, 292*f*
"Halibut, Black pepper, coffee,
              lemon," 256, 257*f*
Halligan, M., 9, 72*n*, 86
"halo effect," 88*n*
Hamilton, L., 23
Hamilton-Patterson, J., 72
hand(s)
  eating without, 175–6
Hand-Fed Dark Dining Parties, 254
Hanig, D.P., 186*n*
Hannahan, M., 87
HAPIfork, 321
HAPILabs, 321
Hard Rock Café, 272
Hardee's
  atmosphere in, 298
Harding, A., 95*n*
Hargreaves, D.J., 283
harmony
  on plate, 136–41, 137*f*, 141*f*
Harrar, V., 116, 118
Harry Potter Bertie Bott's Every
              Flavour Jelly Beans, 343,
              343*n*, 344, 344*n*
Hart, D., 344
Hashimoto, Y., 317
Hayward, L., 257
HCI. *see* human–computer interaction
              (HCI)
HCI/Ubiquitous Computing
              arena, 319
health/ingredient labels, 88–90
healthy eating
  through technology at dining
              table, 320–2
Heineken, 326
Heinz Beanz Flavour Experience, 125,
              126, 126*f*
Heinz green/blue/purple ketchup, 226
Heldke, A., 84
Henderson, F., 93, 361
"Hendrick's gin infused cucumber
              Granita," 167, 167*f*
Henry II, 155

Henry VIII, 174
hidden incongruity, 222–3, 223f
"Historic Heston Blumenthal," 216n
Hitchcock, A., 224, 226
Hobkinson, C., 129, 129f, 167, 167f, 170, 176, 252
Holt, V.M., 359
Hoonhout, J., 321
Hopkinson, S., 97n
"Hot and Cold Pea Soup," 218
"Hot and Iced Tea" dish, 218, 222
House of Blues, 272
Huber, J., 58n
Huber, L., 170
Hultén, B., 119
human–computer interaction (HCI)
    technology at dining table
        and, 317–19, 318f, 319f
Hummel, T., 40
Humphrey Slocombe company, 229
Huron, D., 40n
Huxley, A., 357
Huysmans, J.K., 256
hyperdirectional loudspeakers, 327

Icebar, 293, 293n, 294f
Ignite bottle, 326
Ijzerman, H., 293
Il Cucchiaio d'Argento, 154n
image(s)
    on menu, 52–4
Inamo restaurant, 312
incongruity
    colour–flavour, 226–7
    format–flavour, 227–30, 228f–30f
    hidden, 222–3, 223f
    sensory, 215–47. see also sensory incongruity
    smell–flavour, 230–1
    visible, 222–3, 223f
information source
    mood as, 41
insect(s)
    eating, 358–62, 360f
Institut Paul Bocuse, 20, 117, 369
Institute of Food Research, 344

International Standards Organization (ISO)
    on flavour, 183–4
Irmak, C., 89
ISO. see also International Standards Organization (ISO)
Italian Futurists, 124, 126, 340, 363

Jelly Belly Bean Boozled collection, 239
Jeon, J., 125, 167, 167f
John Salt
    improvised plateware at, 133
Jones, C., 258
Jones, M., 44–5

Kacherdsky, L., 57n
Kandinsky, W., 140–1, 141f
Kaneko, R., 135f
Kayac Inc., 318
Keller, T., 23, 124, 125f, 157n, 352
Kim, H.M., 57n
King, J., 51, 51n
Kings Place, 254
knives. see also cutlery
    history of, 153–9
Koppert Cress BV, 237f
Kotler, P., 252, 275
Koza, B.J., 197–8
Krishna, A., 86
Kubrick, S., 78n
Kurti, N., 3–4
Kwan, G.C., 366

La Cocina de los Sentidos, 15
La Cucina Futuristica, 290
Labbe, D., 190
label(s)
    descriptive food, 82
    health/ingredient, 88–90
    local, 90–1
    natural, 87–8
    organic, 87–8
labelling
    of culinary techniques, 92–4, 94f, 95f
    in enhancing taste and/or flavour of food, 77–81
    food, 91–2

labelling (*continued*)
  influence on perceived ethnicity of
    dish, 85–6
Laboratoire, 347–8
"Lacquered mullet *facon*
    Mondrian*,*" 137
Lammers, H.B., 283
Laughlin, Z., 161
Lavazza, 172
Lawson, N., 331
Le Hollandais restaurant, 256
Le Patissier royan Parisien, 112*f*
"Le Whaf," 347
Lee, L., 79
Legal Sea Foods
  wine at, 56
L'Esguard de Sant Andreu de
    Llavaneres, 15
Levay, J., 45
Life in Fragments, 8
lighting
  in restaurants, 287–8, 298
Lincoln Park restaurant, 274*n*
Linder, L.W., 44*n*
Lipton, J., 353
Little Chef Motorway restaurant chain,
    24, 364*n*
local labels, 90–1
London Savoy, 8
Long, C., 251
*Looking Forward: A Dream of the
    United States of the Americas
    in 1999,* 342
Lorenz, O.A., 87
loudspeakers
  hyperdirectional, 327
Louis XIV
  cutlery and, 156
Ludden, G.D.S., 232
Lufthansa
  food served on, 297
Lunar Eclipse (bowl), 330
Lyman, B., 115–16

MacClancy, J., 87*n*, 240*n*
Macht, M., 40
Mad Hatter Tea Party, 134–5, 135*f*
Madeleine's Madteater, 176, 365

Maga, J.A., 196–7
Maillard, L-C, 4*n*
Maillard reaction, 4
Mandarin Oriental Hotel, 73*n*
  Dinner restaurant in, 7, 43, 74*f*, 274*n*
Marinetti, F.T., 290, 291, 348, 363, 363*n*
marketing
  sensory, 217
Martin, D., 38–40, 39*f*, 118, 163–4, 173,
    229, 229*f*, 328
Master Chefs of France, 279
McCabe, D., 16*n*
McCormick and Schmick's
  wine at, 56
McDonalds, 25*n*, 90, 291–2
McElrea, H., 284
McGee, H., 4
McGeown, R., 356
McNally, K., 49
Mead, M., 76
meal(s)
  ethnicity of, 280–2
  perfect. *see* perfect meal
  in single dose, 341–5, 342*f*
Meals to Come, 363
meat
  cultured, 355–6
"Meat Fruit," 73–5, 74*f*
Meat Joy, 365*n*
Mediacup, 326*n*
Meiselman, H.L., 42–3, 85, 136, 280
memorability
  as diner's response to sensory
    incongruity at dinner
    table, 232–3
mental imagery
  food-related, 84
menu(s). *see also specific types*
  à la carte, 58
  of Alinea, 61–2
  calories on, 54–5
  design of, 47–62
  diet and, 54–5
  format of, 59–62, 60*f*, 61*f*
  function of, 47
  goal of, 47
  images on, 52–4
  introduction, 47–8

nutritional information on, 54–5
optimum number of items per section
on, 47
price and behaviour effects of, 55–9
scanning of, 48–50, 49*f*
sensory, 252–3
"this dessert is literally calling
me," 50–2
time spent looking at, 48
merluza chilena, 75
merluza negra, 75
Messi, L., 316–17
"Messi's Goal" dessert
from El Celler de Can Roca, 316
Metcalfe, J., 254
Meyer, D., 44, 287, 313
Michel, C., 140–1, 141*f*, 172
Michelin Guide, 271–2, 278–9
Michelin star, 279, 314, 328, 330
Microsoft Research, 332
microwave
twenty-first century, 351–2
Mielby, L.H., 92, 140, 252, 266
Millau, C., 3
Milliman, R.E., 284
Mindless Eating, 166
Miodownik, M., 159*n*, 163
Mishra, A., 165, 166*n*
Miss Kō, 312
"MIT Brew," 79–80
"Modern Times," 346
modernist cuisine
for affluent only, 22–5
molecular gastronomy *vs.,* 5–7
Modernist Cuisine, 5
Moir, H.C., 196, 220, 224, 226
Molecular Cooking at Home, 142
molecular gastronomy
defined, 4–6
modernist cuisine *vs.,* 5–7
rise of, 3–5
surprise and, 233–6, 234*f*
molecular gastronomy/modernist
cuisine
rise of, 216–17
money spent at restaurant
atmosphere and, 282–6
Monjo, A., 163*f*, 164

monosodium glutamate (MSG), 189,
190, 201
Monroe, M., 257*n*
mood
as source of information, 41
mood food, 38–41, 39*f*
Morris, A.J., 50, 51
Moss and Cep, 114, 114*f*
Moto, 4, 60, 60*f*, 170, 171*f*, 348, 363
Mouritsen, O.G., 362
mouthfeel
of food and drink, 191, 191*n*
M&S marketing principle, 92*n*
MSG. *see* monosodium glutamate
(MSG)
Mueller, J., 40
Mugaritz restaurant, 152
multisensory atmospherics, 298–9, 299*f*
technology at dining table in
controlling, 323–4
multisensory design
rise of, 217
multisensory flavour
perception, 183–213. *see
also* flavour(s)
cognitive neuroscience of, 200–2
colour influencing, 196–200
cross-cultural differences in, 190–1
diagram of, 184–5, 185*f*
fMRI of, 201
introduction, 183
OFC in, 201–3
olfactory–gustatory
interactions, 188–91, 189*f*
oral-somatosensory contributions
to, 191–3
supertaster, 187
taste and, 186–7
multisensory nature of food
perception, 21–2
multisensory perception of
flavour, 183–213. *see also*
multisensory flavour
perception
multisensory presentation of
food, 109–49. *see also* food
presentation
Murphy, C., 188

music, 282–6, 298
  absence of, 286
  in restaurants, 298
  style and volume of, 282–4
  tempo of, 284–5
  too much sound, 285–6
Myhrvold, N., 5, 97n

name(s)
  food-related, 84–5
Narumi, T., 320, 322
NASA, 353–5
natural labels, 87–8
natural pine berries, 238f
"naturalness"
  sensory incongruity and, 236–9, 237f, 238f
need for uniqueness, 45
neo-Futurist cuisine, 363–6
Nerua, 168
neurogastronomist(s)
  as creators of great-tasting food, 14–16
neurogastronomy, 12–14
neuromania, 16
neuroscience
  of matching sound to food, 324–6
  of multisensory flavour perception, 200–2
  of naming food, 81–4
"neutraceuticals," 341n
New Scientist, 18
New York Review of Books, 3, 3n
9½ Weeks, 251n
Nineteen Eighty-Four, 348
1999 Dom Pérignon Champagne, 56
Nitschke, J.B., 78
Noma, 92, 111, 114, 114f, 292, 360–2
Nordic Food Lab, 359, 360
Norman, R., 56, 94
North, A.C., 281–3
note-by-note cuisine, 356–8
nouvelle cuisine, 2–3
novelty
  sensory incongruity and, 216
novelty and surprise
  food as theatre, 11–12
  search for, 8–12

taste of expectation, 10–11
nutritional information
  on menu, 54–5

Oberfeld, D., 295
"Object," 123
OFC. see orbitofrontal cortex (OFC)
Offline Glass, 323
Okamoto, M., 80n
O'Keefe, C.A., 366n
olfactory atmosphere, 288–93
olfactory–gustatory interactions, 188–91, 189f
olive oil jelly candy, 227, 228f
Olson, J.C., 79
Oppenheim, M., 123, 124
oral-somatosensory contributions
  to multisensory flavour perception, 191–3
orbitofrontal cortex (OFC)
  activation in considering appetitive incentive value of foods, 83, 83n
  described, 79
  in flavour perception, 201–3
organic labels, 87–8
Orwell, G., 339, 348
Osetra "dish"
  at Alinea, 167, 167f
Oshinsky, N.S., 77–8
Osteria Francescana, 347
Oxford Companion to Food, 76n
Oxford English Dictionary, 87n, 151
Oxford University, 356
  Crossmodal Research Laboratory of, 140, 168, 172, 217, 315n, 324, 325, 327, 328
oyster leaf
  at Alinea, 237f
"Oysters and Pearls," 124, 125f

Page, K., 142n
"Painting number 201," 140–1, 141f
Pairet, P., 330
Palmer-Watts, A., 74f
"Pan con tomaté," 229, 229f
"Panel for Edwin R. Campbell No. 4," 140–1, 141f

"panini"
  "stuffed melt" *vs.,* 86
Paris Ritz, 8
Park Associati, 371
Parr, H., 327
Patagonian toothfish, 75
Pelaccio, Z., 175*n*
Per Se, 42*n*, 124, 157*n*, 277, 289
  cutlery at, 168
  wine at, 56
perception
  of flavour, 183–213. *see also*
      multisensory flavour
      perception
perfect meal
  amuse bouche, 25–7
  brain on flavour, 12–16
  as catalyst for social exchange, 367–9,
      368*f*
  culinary movements, 2–7
  food and perception of everything
      else, 16–17
  food perception as fundamentally
      multisensory, 21–2
  future of, 339–81. *see also* future of
      perfect meal
  gastrophysics, 18–21
  introduction, 1–35
  modernist cuisine for affluent
      only, 22–5
  search for novelty and surprise,
      8–12
  start of, 37–69. *see also* start of
      perfect meal
"perfected pizza," 342*f*
Perry, N., 296
"Pestival," 361
PET. *see* positron emission tomography
      (PET)
Petr, C., 289
Peynaud, E., 289*n*
Pfeiffer, J.C., 189
phenylthiocarbamide (PTC)
  sensitivity to, 187
Philips Design, 330
Philips Research, 321, 323, 332
Pickering, G.J., 12*n*
Pier Four restaurant, 275

Pine, B.J., II, 252, 272, 276, 276*n*
Piqueras-Fiszman, B., 116, 117,
      117*n*, 119–20, 123, 124, 160,
      161
Planet Hollywood, 272
plate(s), 115–28. *see also* plateware
  balance on, 136–41, 137*f*, 141*f*
  camouflage, 132–3, 133*f*
  colour of, 115–19
  as essential element of everyday
      meal, 115
  harmony on, 136–41, 137*f*, 141*f*
  improvised, 133–4
  purpose-made, 134–5, 135*f*
  reaching new heights, 129, 129*f*
  shape of, 119–20
  size of, 121–2, 122*f*
  smell and sound of, 130–2, 131*f*
  that is not a plate, 128–35, 129*f*, 131*f*,
      133*f*, 135*f*
  weight of, 122–3
plateware, 109–49. *see also* food
      presentation; plate(s)
  balance on, 136–41, 137*f*, 141*f*
  camouflage, 132–3, 133*f*
  feel of, 123–8, 125*f*–7*f*
  haptic aspects of, 122–8, 125*f*–7*f*
  harmony on, 136–41, 137*f*, 141*f*
  improvised, 133–4
  introduction, 110–11
  purpose-made, 134–5, 135*f*
  selection of, 110–11
  smell and sound of, 130–2, 131*f*
  tablet as, 329–31, 329*f*
  that is not a plate, 128–35, 129*f*, 131*f*,
      133*f*, 135*f*
  weight of, 122–3
plating, 109–49. *see also* food
      presentation
plating art, 365–6
Platte, P., 40
Point, F., 113
Pollan, M., 351
Polpo restaurants, 56, 94
pop-up dining, 371–2
popcorn
  in cinema, 262–3, 262*n*
Porcini Mushroom Lasagna, 227–8

positron emission tomography (PET)
  in processing incentive value of food
    items, 83
Post, 355–6
Pralus, 351
Prescott, J., 87
price(s)
  menu, 55–9
Pringles potato chips, 195, 195n
PROP. see 6-n-propylthiouracil (PROP)
6-n-propylthiouracil (PROP)
  sensitivity to, 186
protanopia, 259
Proust phenomenon, 291
Provencal Rose paradox, 21n, 39,
    278–9, 299
Provencher, V., 89
PTC. see phenylthiocarbamide (PTC)
Punch Drunk theatre company, 364
Puto, C., 58n

QR code(s)
  interaction with food changes related
    to, 319–20
QR Code Cookies, 319–20
Qantas
  food served on, 296
Quilon, 162–3, 163f

Rainforest Café, 276
Ralph 124C 41+, 345, 350
Rausch Brothers, 296
Rayner, J., 25
reaction(s)
  food description–related, 96–8
regionalism, 340
Reidinger, P., 285
Reisfelt, H.H., 139
restaurant(s). see also specific names
  atmosphere related to money and
    time spent at, 282–6
  dining in the dark, 252–5. see also
    dining in the dark
  feel of, 291–4, 292f, 294f
  lighting of, 287–8, 298
  music in, 282–6, 298
  science laboratory vs., 369–71, 370f
Restaurant Story, 9

Revel, J.F., 234
Reynolds, D., 52
Richelieu, Cardinal, 154
"Rien," 163–4
"River and Sea" menu, 95
Robot Restaurant, 313
Rogers, E., 350n
Rolls, B.J., 121–2
Rolls, E., 345n
Roncero, P., 293, 329, 370f
Roosth, S., 4, 339–40
Roudnitzky, N., 192
Rourke, M., 251n
Royal Danish Academy of Sciences and
    Letters, 18
Rozin, P., 231
Rudolph, A., 249
rule of primacy and recency, 49
Ruscalleda, C., 271, 272, 301

Sacks, O., 258, 258f
Sagartoki, 227–8
Sainsbury's, 75
Saint Laurent, Y., 257
Saison, 292
Salisbury, D., 254
Salisbury District Hospital, 118–19
Salve Jorge, 323
Samper, G., 72–3
San Roman, M.J., 225n
Sánchez Romera, M., 14–15
Sant Pau restaurant, 271
Satie, E., 257
scanpath
  defined, 49n
  factors determining, 49–50
  Gallup Organisation on, 50
Schaeeman, C., 365n
Schlosser, E., 349
Schuldt, J.P., 87, 88
Schutz, H.G., 87
Schwartz, N., 41, 88
science laboratory
  restaurant vs., 369–71, 370f
Scienticafé, 346, 369n
Scientific American, 4
"Scientific Restaurant," 345
Semin, G.R., 293

Senderens, A., 271–2, 312
sense(s)
   dark effects on, 260–4
sensory dominance, 219–20
sensory incongruity
   colour–flavour, 226–7
   defined, 218–19, 219f
   diner's responses to, 232–3
   excitement of, 216–18
   format–flavour, 227–30, 228f–30f
   globalization of, 217–18
   history of, 224–6
   individual differences in diner's
      responses to, 239–40
   in meal, 215–47
   "naturalness" and, 236–9, 237f, 238f
   noticing, 219–23
   novelty and, 216
   popularity of, 216–18
   smell–flavour, 230–1
sensory marketing
   rise of, 217
sensory menu, 252–3
Seo, H.S., 40
Seremetakis, C.N., 72n, 133
Serendipity3, 57
SFF. see solid freeform fabrication
   (SFF)
Shankar, M.U., 198n, 226
shape
   of plateware, 119–20
Shepherd, G., 12
Shewry, B., 347–8
Show Cooking Anteprima, 314
sight course, 252, 253
silver cutlery, 156
silver-plating cutlery, 157
Simner, J., 120
Singapore airlines
   food served on, 297
situated eating
   joys of, 371–2
69 Colbrooke Row, 327
size
   of cutlery, 165–6
   of plateware, 121–2, 122f
Skenes, J., 292
Small, D.M., 15, 201–2

SmartPlate, 328
smell
   of plateware, 130–2, 131f
smell–flavour incongruity, 230–1
Smith, B., 278
Smith, F.E., 350
SMRC company. see Systems and
      Materials Research
      Consultancy (SMRC)
      company
snack brands
   healthy vs. unhealthy, 89
"Snail Porridge," 73–7, 74f
social aspects of dining, 42–7
   company, 44–7
   waiting staff, 42–4
social exchange
   meal as catalyst for, 367–9, 368f
solid freeform fabrication (SFF), 353
Sonic Cake Pop dish, 325
sound
   of food, 194–6
   matching to food, 324–6
   of plateware, 130–2, 131f
Sound course, 252n, 253–4
Sous vide
   as twenty-first century
      microwave, 351–2
Sous Vide Supreme, 352
Soylent Green, 343
"Spaceballs," 346
Space.com, 354n
Spalding, B., 133n
Spence, C., 123, 124, 160, 195, 195n, 324,
      325n
"Spherical egg of white asparagus with
      false truffle," 93, 94f
"split spoons," 156
spoon(s). see also cutlery
   Apostle, 154
   history of, 153–9, 158f
   "split," 156
"Spotted Dick," 77, 77n
"Spotted Richard," 77, 77n
Spurlock, M., 25n
"Squab," 134
St. John, 93
St. Peter Damian, 155

stainless steel cutlery, 158–60
Standing, L., 284
Starbucks coffee chain, 276, 325n, 346n
Starck, P., 312
start of perfect meal, 37–69
    introduction, 37–41, 39f
    menu design, 47–62. see also menu(s)
    mood food, 38–41, 39f
    social aspects of dining, 42–7
Steinberger, M., 159
Steingarten, J., 14
sterling cutlery
    Victorian period, 157, 158f
stew
    casserole vs., 86
Stewart, P.C., 120
Still Life with Old Shoe, 134
"Stinking Bishop" cheese, 79
story telling
    dining-related, 371–2
Stuckey, B., 186–7, 291
Studio William
    spoons designed by, 173, 173f
"stuffed melt"
    "panini" vs., 86
Subway, 90
Sunnybank Fisheries, 364n
Super Size Me, 25n
supertaster
    described, 187
surprise. see also sensory incongruity
    as diner's response to sensory
        incongruity at dinner
        table, 232
    in food description, 95–6
    in meal, 215–47
    molecular gastronomy and, 233–6,
        234f
    search for, 8–12. see also novelty and
        surprise
Swami, V., 46n
Symposium of Australian
        Gastronomy, 94–5
synaesthesia, 188, 188n
Systems and Materials Research
        Consultancy (SMRC)
        company, 355
Szydlo, A., 366n

Tabla restaurant, 173
tablet
    twenty-first century
        plateware, 329–31, 329f
tagesteller, 51
Tannahill, R., 77, 112
Tartufo al cerdo Ibérico y aceituna
        verde, 233, 234f
taste, 186–7
    of cutlery, 161–5, 163f
    labelling in enhancing, 77–81
taste perception
    atmospheric contributions to, 294–7
taste-related words, 82
Taste What You're Missing, 186–7,
        291
taster(s)
    thermal, 193
"Tasty Formula," 167, 167f
technology at dining table, 311–37
    AR food, 317–19, 318f, 319f
    companies involved in, 332
    in controlling multisensory
        atmosphere, 323–4
    distraction and, 322–3
    fostering healthy eating through
        incorporation of, 320–2
    future of, 326–7
    "Gin & Sonic," 328–9
    HCI and, 317–19, 318f, 319f
    introduction, 311–12
    matching sound to food, 324–6
    QR codes, 319–20
    SmartPlate, 328
    tablet as twenty-first century
        plateware, 329–31, 329f
    tips from chef at your fingers, 331
    in transforming dining
        experience, 315–17, 316f
    VR applications, 318–19, 319f
technology on dining table, 312–14,
        314f
texture
    of cutlery, 166–8, 167f
The Astronaut's Cookbook, 77n, 354
The Banquet for Ultra Bankruptcy,
        364
"The bittersweet symphony" dish, 325

The Champ, 40
"The Cook, the Thief, his Wife, and her
    Lover," 256
The Cube, 371
The Daily Telegraph, 261–2
The Delaunay, 51
The Edinburgh Science Festival, 176
"The Emerging Science of
    Gastrophysics," 18
The Experience Economy, 276
The Fat Duck, 4, 24, 25*n*, 44, 59*n*, 74*f*,
    134–5, 135*f*, 175–6, 222, 223,
    223*f*, 236, 277, 279, 290–1,
    315, 316*f*, 324, 327*n*,
    344*n*, 363
The French Laundry, 44
The Futurist Cookbook, 363
The Hi Lo Jamaican Eating House,
    59*n*
The House of Wolf, 166–7, 167*f*,
    252, 325
    David Bellamy cocktail at, 186
    Wolf's Liar at, 293*n*
The Icebar, 293, 293*n*, 294*f*
The Independent, 5
The New York Times, 15
The Omnivorous Ape, 274
The Omnivorous Mind, 255*n*
"The Perfect Host," 331
The Physiology of Taste, 5*n*
The Republic of the Future, 341
"the scientific study of deliciousness," 4
The Sex Life of Food, 354*n*
The Tactile Dining Car, 342*f*, 349*f*
The Theory of the Leisure Class, 158
The Trip, 96*n*
The Wolseley, 51*n*
The World in 2030, 350
"the wow effect," 340*n*
theatre
    food as, 11–12
Theatre of the Senses, 364
thermal taster
    described, 193
This, H., 4, 6*n*, 93, 356, 357
"Three-Course Dinner," 343, 343*n*, 344,
    344*n*, 345
3D printed foods, 353–5

Tiffany's
    cutlery at, 158
Tihany, A., 289*n*
time spent at restaurant
    atmosphere and, 282–6
Toet, E., 320–1
Tonga Room & Hurricane Bar, 275*n*
Total Rice dish, 363*n*
Tovée, M.J., 46*n*
tritanopia, 259
Troisgros, J., 3
Troisgros, M., 137
Troisgros, P., 3
Trotter, C., 274*n*
Tsuji, S., 3, 113
Tsunami restaurant
    in West Palm Beach, Florida, 119
twenty-first century microwave
    Sous vide as, 351–2
2001: A Space Odyssey, 78*n*

Ultraviolet restaurant, 330, 372
UN. *see* United Nations (UN)
Union Square Hospitality Group, 287,
    313
uniqueness
    need for, 45
United Nations (UN)
    FAO of, 359
Unsicht-Bar, 95, 249–50
Urban picNYC table, 291–2, 292*f*

van der Velde, H., 159
Van Ittersum, K., 298
Van Kleef, E., 121
VAST. *see* visual aesthetic sensitivity test
    (VAST)
Veblen, T., 57, 158
Velasco, C., 298–9, 299*f*
venti, 86
Verma, R., 53
Verne, J., 346
Vevey restaurant, 38–40, 39*f*
Vickers, Z., 195
Victor Hugo soup, 76
Victorian period sterling cutlery, 157,
    158*f*

Vilgis, T., 73
virtual reality (VR) applications
    technology at dining table
            and, 318–19, 319*f*
"virtual Shrimp cocktail," 347
visible incongruity, 222–3, 223*f*
Visser, M., 6*n*, 160*n*
visual aesthetic sensitivity test
            (VAST), 142
visual flavour, 80, 196–200
visual-flavour responses, 198
Vogelzang, M., 256, 257*f*
Vogt, G.L., 77*n*, 354
von Restorff, 51
von Restorff effect, 50–1
VR applications. *see* virtual reality (VR)
            applications

waiting staff, 42–4
Wan, X., 227
Wansink, B., 44*n*, 88–92, 99, 121, 122,
            124, 130, 166, 166*n*, 262, 298
Waters, A., 45
Watson, L., 274, 274*n*
weight
    of plateware, 122–3
Weinstein, S., 123
WEIRD (western, educated, intelligent,
            rich, and democratic), 18,
            18*n*
Wellcome Collection, 361
Wells, P., 347
Welsh, W., 173
Werle, C.O.C., 89
Wexler, A., 368*f*
Whale Inside Dark restaurant, 251
Wheatley, J., 235*n*
Wheeler, E., 132*n*
When Harry Met Sally, 40

"White funeral" meals, 256–7, 257*f*
"Wicked Blinding Mental Burger with
            Knobs On," 90*n*
"WikiPearls," 345
Willing, E., 173, 173*f*
Willy Wonka's Chocolate factory, 231
Wilson, B., 76*n*, 155*n*, 156*n*, 164*n*–5*n*
Wilson, S., 283
wine
    prices of, 56
Wine List Consulting Unlimited, 56
Wittgenstein, 240*n*
Wolf's Liar
    at The House of Wolf restaurant,
            293*n*
Wolfson, J., 77–8
Wonka, W., 343–4
Woods, A.T., 295
Wurgaft, B.A., 44

Yale School of Medicine, 12
Yang, J., 120*n*
Yang, S.S., 50, 52*n*
Yeoh, J.P.S., 281
Yeomans, M., 41, 80–1
Yo Sushi-type restaurants, 53*n*
Yo Sushi chain, 313
Young, C., 5, 97*n*
Youssef, J., 6, 93, 142, 293*n*

Zampini, M., 195, 195*n*, 199–200
Zampollo, F., 141–2
Zellner, D., 139–40
Zhao, X., 87
Zoran, A., 353
Zurich Museum of Design
    "Dialog im Dunkeln" exhibition
            at, 249
Zwart, P., 174*n*–5*n*